Astronomy and Astrophysics in the New Millennium

Panel Reports

Astronomy and Astrophysics Survey Committee
Board on Physics and Astronomy–Space Studies Board
Division on Engineering and Physical Sciences
National Research Council

NATIONAL ACADEMY PRESS
Washington, D.C.

NOTICE: The project that is the subject of this report was approved by the Governing Board of the National Research Council, whose members are drawn from the councils of the National Academy of Sciences, the National Academy of Engineering, and the Institute of Medicine. The members of the panels responsible for the report were chosen for their special competences and with regard for appropriate balance.

This project was supported by the National Aeronautics and Space Administration under Grant No. NAG5-6916, the National Science Foundation under Grant No. AST-9800149, and the Keck Foundation.

Cover: The montage on the front cover consists of one image from each of the seven panel reports in this volume. *Top left:* x-ray image showing loops of million-degree plasma in the solar corona (page 240). *Top middle:* optical/infrared image of the stars at the very center of our galaxy in orbit around the putative black hole located there (page 115). *Top right:* radio map of the giant elliptical galaxy Messier 87 showing the structure of the huge cloud of relativistic plasma encompassing the galaxy and powered by the central black hole (page 187). *Middle:* x-ray image of the nearby supernova remnant E0102-72.3 showing the expanding shell of hot gas formed by the supernova explosion (page 37). *Middle right:* simulated interferometric image reconstruction of how an Earth-like planet around a nearby star would appear to the Terrestrial Planet Finder (page 348). *Bottom right:* artist's conception of the Laser Inteferometer Space Antenna orbiting Earth and superimposed on the ripples in space-time produced by the gravitational power of merging supermassive black holes (page 144). *Bottom left:* computer visualization of the hot gas in a theoretical chunk of the universe from a cosmological simulation (page 292).

Library of Congress Control Number: 2001093504

International Standard Book Numbers: 0-309-07037-6

Additional copies of this report are available from:

National Academy Press, 2101 Constitution Avenue, N.W., Lockbox 285, Washington, DC 20055; (800) 624-6242 or (202) 334-3313 (in the Washington metropolitan area); Internet <http://www.nap.edu>

Board on Physics and Astronomy, National Research Council, HA 562, 2101 Constitution Avenue, N.W., Washington, DC 20418 Internet <http://www.national-academies.org/bpa>

THE NATIONAL ACADEMIES

National Academy of Sciences
National Academy of Engineering
Institute of Medicine
National Research Council

The **National Academy of Sciences** is a private, nonprofit, self-perpetuating society of distinguished scholars engaged in scientific and engineering research, dedicated to the furtherance of science and technology and to their use for the general welfare. Upon the authority of the charter granted to it by the Congress in 1863, the Academy has a mandate that requires it to advise the federal government on scientific and technical matters. Dr. Bruce M. Alberts is president of the National Academy of Sciences.

The **National Academy of Engineering** was established in 1964, under the charter of the National Academy of Sciences, as a parallel organization of outstanding engineers. It is autonomous in its administration and in the selection of its members, sharing with the National Academy of Sciences the responsibility for advising the federal government. The National Academy of Engineering also sponsors engineering programs aimed at meeting national needs, encourages education and research, and recognizes the superior achievements of engineers. Dr. Wm. A. Wulf is president of the National Academy of Engineering.

The **Institute of Medicine** was established in 1970 by the National Academy of Sciences to secure the services of eminent members of appropriate professions in the examination of policy matters pertaining to the health of the public. The Institute acts under the responsibility given to the National Academy of Sciences by its congressional charter to be an adviser to the federal government and, upon its own initiative, to identify issues of medical care, research, and education. Dr. Kenneth I. Shine is president of the Institute of Medicine.

The **National Research Council** was organized by the National Academy of Sciences in 1916 to associate the broad community of science and technology with the Academy's purposes of furthering knowledge and advising the federal government. Functioning in accordance with general policies determined by the Academy, the Council has become the principal operating agency of both the National Academy of Sciences and the National Academy of Engineering in providing services to the government, the public, and the scientific and engineering communities. The Council is administered jointly by both Academies and the Institute of Medicine. Dr. Bruce M. Alberts and Dr. Wm. A. Wulf are chairman and vice chairman, respectively, of the National Research Council.

ASTRONOMY AND ASTROPHYSICS SURVEY COMMITTEE

CHRISTOPHER F. McKEE, University of California, Berkeley, *Co-chair*
JOSEPH H. TAYLOR, JR., Princeton University, *Co-chair*
DAVID J. HOLLENBACH, NASA Ames Research Center, *Executive Officer*
TODD BOROSON, National Optical Astronomy Observatories
WENDY FREEDMAN, Carnegie Observatories
DAVID C. JEWITT, University of Hawaii
STEVEN M. KAHN, Columbia University
JAMES M. MORAN, JR., Harvard-Smithsonian Center for Astrophysics
JERRY E. NELSON, University of California Observatories
R. BRUCE PARTRIDGE, Haverford College
MARCIA RIEKE, University of Arizona
ANNEILA I. SARGENT, California Institute of Technology
ALAN TITLE, Lockheed Martin Space Technology Center
SCOTT TREMAINE, Princeton University
MICHAEL S. TURNER, University of Chicago

NATIONAL RESEARCH COUNCIL STAFF

DONALD C. SHAPERO, Board on Physics and Astronomy, Director
JOSEPH K. ALEXANDER, Space Studies Board, Director
ROBERT L. RIEMER, Senior Program Officer
JOEL R. PARRIOTT, Program Officer
GRACE WANG, Administrative Associate (1998-1999)
SÄRAH A. CHOUDHURY, Project Associate (1999-2000)
MICHAEL LU, Project Assistant (1998-2000)
NELSON QUIÑONES, Project Assistant (2000)

PANEL ON ASTRONOMY EDUCATION AND POLICY

ANDREA K. DUPREE, Harvard-Smithsonian Center for Astrophysics, *Chair*
R. BRUCE PARTRIDGE, Haverford College, *Vice Chair* (education)
ANNEILA I. SARGENT, California Institute of Technology, *Vice Chair* (policy)
FRANK BASH, McDonald Observatory, University of Texas
GREGORY BOTHUN, University of Oregon
SUZAN EDWARDS, Smith College
RICCARDO GIACCONI, Associated Universities, Inc.
PETER A. GILMAN, National Center for Atmospheric Research
MICHAEL HAUSER, Space Telescope Science Institute
BLAIR SAVAGE, University of Wisconsin
IRWIN SHAPIRO, Harvard-Smithsonian Center for Astrophysics
FRANK SHU, University of California, Berkeley
NEIL DE GRASSE TYSON, American Museum of Natural History

PANEL ON BENEFITS TO THE NATION FROM ASTRONOMY AND ASTROPHYSICS

STEPHEN E. STROM, National Optical Astronomy Observatories, *Chair*
DAVID J. HOLLENBACH, NASA Ames Research Center, *Vice Chair*

CONTRIBUTORS TO THE PANEL

ROGER ANGEL, Steward Observatory, University of Arizona
DOUGLAS DUNCAN, American Astronomical Society; University of Chicago
ANDREW FRAKNOI, Foothills College
PAUL GOLDSMITH, National Astronomy and Ionosphere Center, Cornell University
NEAL KATZ, University of Massachusetts, Amherst
EUGENE LEVY, University of Arizona
STEPHEN MARAN, NASA Goddard Space Flight Center
DAVID MORRISON, NASA Ames Research Center
LEIF ROBINSON, Sky Publishing Corporation

CHARLES LADA, Harvard-Smithsonian Center for Astrophysics
JAMES W. LIEBERT, University of Arizona
CHARLES C. STEIDEL, California Institute of Technology
CHRISTOPHER STUBBS, University of Washington
DAVID C. JEWITT, University of Hawaii, *Ex Officio*

PANEL ON PARTICLE, NUCLEAR, AND GRAVITATIONAL-WAVE ASTROPHYSICS

THOMAS K. GAISSER, University of Delaware, *Chair*
MICHAEL S. TURNER, University of Chicago, *Vice Chair*
BARRY BARISH, California Institute of Technology
STEVEN WILLIAM BARWICK, University of California, Irvine
EUGENE BEIER, University of Pennsylvania
JOSHUA FRIEMAN, Fermi National Accelerator Laboratory
ALICE KUST HARDING, NASA Goddard Space Flight Center
RICHARD ALWIN MEWALDT, California Institute of Technology
RENE ASHWIN ONG, University of Chicago
BOHDAN PACZYNSKI, Princeton University Observatory
BERNARD SADOULET, University of California, Berkeley
PIERRE SOKOLSKY, University of Utah
RAINER WEISS, Massachusetts Institute of Technology

PANEL ON RADIO AND SUBMILLIMETER-WAVE ASTRONOMY

MARTHA P. HAYNES, Cornell University, *Chair*
JAMES M. MORAN, JR., Harvard-Smithsonian Center for Astrophysics,
    Vice Chair
GEOFFREY A. BLAKE, California Institute of Technology
DONALD B. CAMPBELL, Cornell University
JOHN E. CARLSTROM, University of Chicago
NEAL J. EVANS, University of Texas at Austin
JACQUELINE N. HEWITT, Massachusetts Institute of Technology
KENNETH I. KELLERMANN, National Radio Astronomy Observatory
ALAN P. MARSCHER, Boston University
STEVEN T. MYERS, University of Pennsylvania

MARK J. REID, Harvard-Smithsonian Center for Astrophysics
WILLIAM J. WELCH, University of California, Berkeley
DONALD BACKER, University of California, Berkeley, *Consultant*

PANEL ON SOLAR ASTRONOMY

MICHAEL KNOELKER, University Corporation for Atmospheric
    Research, *Chair*
ALAN TITLE, Lockheed Martin Space Technology Center, *Vice Chair*
DALE EVERETT GARY, New Jersey Institute of Technology
PHILIP R. GOODE, New Jersey Institute of Technology
JOSEPH B. GURMAN, NASA Goddard Space Flight Center
SHADIA RIFAI HABBAL, Harvard-Smithsonian Center for Astrophysics
DANA WARFIELD LONGCOPE, Montana State University
RONALD LEE MOORE, NASA Marshall Space Flight Center
THOMAS RIMMELE, National Solar Observatory
JOHN H. THOMAS, University of Rochester
ELLEN GOULD ZWEIBEL, University of Colorado, Boulder

PANEL ON THEORY, COMPUTATION, AND DATA EXPLORATION

WILLIAM H. PRESS, Los Alamos National Laboratory, *Chair*
SCOTT TREMAINE, Princeton University, *Vice Chair*
CHARLES ALCOCK, Lawrence Livermore National Laboratory/
    University of Pennsylvania
LARS BILDSTEN, University of California, Berkeley/Santa Barbara
ADAM BURROWS, University of Arizona
LARS HERNQUIST, Harvard-Smithsonian Center for Astrophysics
CRAIG JAMES HOGAN, University of Washington
MARC PAUL KAMIONKOWSKI, Columbia University
MICHAEL NORMAN, University of Illinois at Urbana-Champaign
EVE OSTRIKER, University of Maryland
THOMAS A. PRINCE, California Institute of Technology
ALEX SANDOR SZALAY, Johns Hopkins University
ROBERT F. STEIN, Michigan State University, *Consultant*

PANEL ON ULTRAVIOLET, OPTICAL, AND INFRARED ASTRONOMY FROM SPACE

AD HOC CROSS-PANEL WORKING GROUPS

BOARD ON PHYSICS AND ASTRONOMY

Preface

In 1997, the Board on Physics and Astronomy asked BPA member Anthony Readhead and director Don Shapero to convene a small group of leading astronomers to consider the need for a new decadal survey of astronomy and astrophysics. The group concluded that the time was ripe for a new decadal survey in the 50-year series of such studies. It recommended the establishment of a new Astronomy and Astrophysics Survey Committee to carry out a broad scientific assessment of the field and to recommend new ground- and space-based programs for the decade 2000 to 2010. It also considered the framework for the survey, which ultimately led to the following detailed charge to the committee:

> The committee will survey the field of space- and ground-based astronomy and astrophysics, recommending priorities for the most important new initiatives of the decade 2000-2010. The principal goal of the study will be an assessment of proposed activities in astronomy and astrophysics and the preparation of a concise response addressed to the agencies supporting the field, the congressional committees with jurisdiction over these agencies, and the scientific community. The study will restrict its scope to experimental and theoretical aspects of subfields involving remote observations from the Earth and space and analysis of astronomical objects. Missions to make in situ studies of the Earth and planets, which have been treated by other National Research Council and Academy reports, will be excluded. Attention will be given to effective implementation of proposed and existing programs and to the organizational infrastructure and the human aspects of the field involving demography and education. Promising areas for the development of new technologies will be suggested.
>
> A brief review of the initiatives of other nations will be given together with a discussion of the possibilities of joint ventures and other forms of

international cooperation. Prospects for combining resources—private, state, federal, and international—to build the strongest program possible for U.S. astronomy will be explored. Recommendations for new initiatives will be presented in priority order within different categories.

The committee will also address two questions posed by the House Science Committee staff: Have NASA and NSF mission objectives resulted in a balanced, broad-based, robust science program for astronomy? That is, NASA's mission is to fund research that supports flight programs and focused campaigns such as Origins, whereas NSF's mission is to support basic research. Have these overall missions been adequately coordinated and has this resulted in an optimum science program from a productivity standpoint? What special strategies are needed for strategic cooperation between NASA and NSF? Should these be included in agency strategic plans? How do NASA and NSF determine the relative priority of new technological opportunities (including new facilities) compared to providing long-term support for associated research grants and facility operations?

The committee will consult widely within the astronomical and astrophysical community and make a concerted effort to disseminate its recommendations promptly and effectively.

The two major questions posed by the House Science Committee staff (detailed above) were accompanied by several other questions that were treated in a report entitled *Federal Funding of Astronomical Research*.[1] That report was submitted to the survey committee as input to its deliberations.

The National Research Council established the survey under the auspices of the BPA, which oversaw the study in close consultation with the Space Studies Board. After consultations with members of the National Academy of Sciences Astronomy Section, members of astronomy departments in U.S. universities, and other leading astronomers, the BPA presented a slate of nominees for membership on the survey committee to the chair of the National Research Council. The NRC chair subsequently appointed the 15-member Astronomy and Astrophysics Survey Committee (AASC), with Joseph H. Taylor, Jr., and Christopher F. McKee as co-chairs, to carry out the study.

To provide detailed input to the AASC on the wavelength-based subdisciplines of astronomy and other areas, nine panels were established. Each panel's vice chair was selected from the membership of the

---

[1]National Research Council. 2000. Federal Funding of Astronomical Research (Washington, D.C.: National Academy Press).

AASC. The panel vice chairs were thus able to serve as liaisons between the panels and the main committee and to articulate the priorities of the subdisciplines within the AASC in the process of setting priorities. The panels included more than 100 people, who together were able to encompass the enormous intellectual breadth of modern astronomy and astrophysics.

Each panel met three times and also held two open "town meeting" sessions at the January and June 1999 meetings of the American Astronomical Society. Many of the panel members also held sessions at other professional gatherings, as well as at astronomical departments and centers throughout the United States.

The seven science panels were charged with preparing reports that identified the most important scientific goals in their respective areas, prioritizing the new initiatives needed to achieve these goals, recommending proposals for technology development, considering the possibilities for international collaboration, and discussing any policy issues relevant to their charge. The science panels were as follows:

- High-energy Astrophysics from Space;
- Optical and Infrared Astronomy from the Ground;
- Particle, Nuclear, and Gravitational-Wave Astrophysics;
- Radio and Submillimeter-Wave Astronomy;
- Solar Astronomy;
- Theory, Computation, and Data Exploration; and
- Ultraviolet, Optical, and Infrared Astronomy from Space.

Their reports are contained in this volume.

The reports of the other two panels—Astronomy Education and Policy, and Benefits to the Nation from Astronomy and Astrophysics— were revised and incorporated into the AASC main report.[2] As mentioned above, the AASC also drew on the report *Federal Funding of Astronomical Research* as well as other NRC reports cited in the text. Further valuable input to the AASC and its panels was provided by four ad hoc cross-panel working groups: Astronomical Surveys (T. Prince, Chair), Extrasolar Planets (D. Jewitt, Chair), Laboratory Astrophysics (C. Alcock, Chair), and NSF-Funded National Observatories (F. Bash, Chair).

---

[2]Astronomy and Astrophysics Survey Committee, National Research Council. 2001. *Astronomy and Astrophysics in the New Millennium* (Washington, D.C.: National Academy Press).

Members of the survey committee and the panels consulted widely with their colleagues to solicit advice and to inform other members of the astronomical community of the main issues facing the committee. This consultation process provided useful input for the panel reports and also gave the survey committee a good picture of the community consensus on the various initiatives under consideration for inclusion among the priorities of the main report.

At the final AASC meeting in late 1999, the panel chairs participated with members of the survey committee to develop the new decadal survey's recommendations. The committee based its final recommendations and priorities in significant part on the panel reports and on the discussions with the panel chairs. The overall priorities are presented in *Astronomy and Astrophysics for the New Millennium*, the report of the AASC. The AASC's priorities take precedence over those of the panel reports in the present volume. The panel reports contain, in addition to more detailed discussion of these priorities, further projects and topics that were not selected by the AASC for inclusion among the overall priorities that are viewed as having importance for the field as a whole. They also contain cost estimates,[3] which formed the basis for the cost estimates in the AASC report. The panel reports were reviewed by the National Research Council together with the AASC report.

---

[3]The size categories for new initiatives are based on the capital cost for ground-based projects and on the total cost, excluding technology development, for space-based projects. Only costs to be borne by the federal government are included. The AASC's cost estimates for these initiatives are based on discussions with agency personnel and on presentations to the panels; they are given in FY2000 dollars. For ground-based projects, small projects have capital costs of up to $5 million; moderate, from $5 million to $50 million; and major, above $50 million. In contrast to the practice in previous decadal surveys, the tabulated costs for ground-based capital projects include operations and new instrumentation for 5 years at rates of 7 percent and 3 percent, respectively, of the capital cost per year. In addition, grants for data analysis and associated theory are included at a rate of 3 percent of the capital cost per year for major projects, 5 percent for moderate projects, and 0 percent for small projects. The total costs that were used in the survey committee report for ground-based initiatives are thus typically 1.65, 1.75, and 1.50 times the capital costs for major, moderate, and small initiatives, respectively.

There are several exceptions to these general rules, however. Square Kilometer Array (SKA) technology development includes only funds for a theory challenge, budgeted at $200,000 per year for the decade. The Telescope System Instrumentation Program (TSIP) does not require operations or instrumentation funds and is too fragmented to have a grants program. The National Virtual Observatory (NVO), the National Astrophysics Theory Postdoctoral Program, and the Laboratory Astrophysics Program are not capital projects and therefore have no added costs. The Large-aperture Synoptic Survey Telescope (LSST) is ex-

The AASC is grateful to the many astronomers, both in the United States and from abroad, who provided written advice or participated in organized discussions. We thank the National Science Foundation, the National Aeronautics and Space Administration, and the Keck Foundation for providing support for the project. We are grateful to Robert Milkey and Kevin Marvel and to the American Astronomical Society for assistance in the community outreach and town meeting sessions. The committee also acknowledges the assistance of NRC staff members, particularly the outstanding work of Joel Parriott and Roc Riemer, who provided support for the entire project, Susan Maurizi and Liz Fikre, who edited the reports, and the National Academy Press, which published the reports. We are also indebted to Robert Sokol and Ken Van Pool of Design@Large for their innovative design of the booklet that gives an overview of and popularizes the results of the survey. The timely completion of this report would not have been possible without the unstinting efforts of David Hollenbach, who served both as a member of the committee and as Executive Officer. Many other people too numerous to cite individually assisted in various aspects of the survey. We thank them all for their assistance.

Christopher F. McKee and Joseph H. Taylor, Jr., *Co-chairs*
Astronomy and Astrophysics Survey Committee

---

pected to have significant expenses for data analysis, so the total operations cost estimated by the Panel on Optical and Infrared Astronomy from the Ground has been used for this project.

The cost estimates for space-based initiatives do not include technology development. NASA has adopted a policy of deferring the construction of new missions until all major technological problems have been solved, a policy the committee endorses. These costs are typically about 30 percent of the construction costs of a mission. In some cases, entire missions will serve as precursors for other missions, such as the Space Interferometry Mission for the Terrestrial Planet Finder. The Explorer and Discovery missions are regarded as small initiatives. Since they are peer-reviewed, the committee did not prioritize them. Moderate missions are those with construction, launch, and operations costs between the $140 million cap on Explorer missions and $500 million; major missions have estimated costs above $500 million.

The cost estimates for ground-based projects listed in the reports of the panels are different from those listed in the Executive Summary and the survey committee report, because the costs for projects described by the panels have not been inflated using the calculations described above. The total cost estimates for space-based projects listed in the reports of the panels may differ from those listed in the Executive Summary and the survey committee report in some cases. The Next Generation Space Telescope, discussed in the report of the Panel on Ultraviolet, Optical, and Infrared Astronomy from Space, is an example. Technology development costs were included in some cases and the numbers in the panel reports were not rounded.

Acknowledgment of Reviewers

This report has been reviewed in draft form by individuals chosen for their diverse perspectives and technical expertise, in accordance with procedures approved by the NRC's Report Review Committee. The purpose of this independent review is to provide candid and critical comments that will assist the institution in making its published report as sound as possible and to ensure that the report meets institutional standards for objectivity, evidence, and responsiveness to the study charge. The review comments and draft manuscript remain confidential to protect the integrity of the deliberative process. We wish to thank the following individuals for their review of the survey committee report and/or one or more of the panel reports:

W. David Arnett, Steward Observatory, University of Arizona,
Peter Banks, ERIM International, Inc. (retired),
Gordon A. Baym, University of Illinois at Urbana-Champaign,
Roger Chevalier, University of Virginia,
Anita L. Cochran, University of Texas at Austin,
Marshall H. Cohen, California Institute of Technology,
Anne P. Cowley, Arizona State University,
Val L. Fitch, Princeton University,
Bill Green, former Congressman, New York,
Karen L. Harvey, Solar Physics Research Group,
John P. Huchra, Harvard-Smithsonian Center for Astrophysics,
Robert P. Kirshner, Harvard-Smithsonian Center for Astrophysics,
Chryssa Kouveliotou, NASA Marshall Space Flight Center,
Richard G. Kron, Yerkes Observatory,
Jeffrey Linsky, University of Colorado/JILA,

Richard McCray, University of Colorado/JILA,
Melissa McGrath, Space Telescope Science Institute,
Mark Morris, University of California, Los Angeles,
Martin J. Rees, Institute of Astronomy, Cambridge University, U.K.,
Morton S. Roberts, National Radio Astronomy Observatory–
    Charlottesville,
Patrick Thaddeus, Harvard-Smithsonian Center for Astrophysics,
J. Anthony Tyson, Lucent Technologies, and
David T. Wilkinson, Princeton University.

Although the reviewers listed above have provided many constructive comments and suggestions, they were not asked to endorse the conclusions or recommendations, nor did they see the final draft of the report before its release. The review of the survey committee report and of the panel reports was overseen by Nicholas P. Samios, Brookhaven National Laboratory, and Lewis M. Branscomb, John F. Kennedy School of Government, Harvard University. Appointed by the National Research Council, they were responsible for making certain that an independent examination of the reports was carried out in accordance with institutional procedures and that all review comments were carefully considered. Responsibility for the final content of this report and the panel reports rests entirely with the authoring committee and the institution.

Contents

EXECUTIVE SUMMARY   1

1   REPORT OF THE PANEL ON HIGH-ENERGY ASTROPHYSICS
    FROM SPACE   17
    Summary, 18
    A Decade of Opportunity, 20
        Emergence of Structure, 21
        Gravity Power, 25
        Origin of the Elements, 33
        Surprises, 39
    A New Beginning, 39
        Chandra, 39
        XMM-Newton, 41
        HETE-2 and Swift, 41
        INTEGRAL, 42
    The Next Steps, 42
        Proposed Major Mission: Constellation-X, 42
        First-Priority Proposed Intermediate Mission: GLAST, 50
        Second-Priority Proposed Intermediate Mission: EXIST, 54
    Investing for the Future, 56
        MAXIM (E, F), 56
        Generation-X (A, I), 58
        MeV Spectroscopy Mission (J, L), 58
    Smaller Programs, 58
        Potential Explorer Research, 59
        Ultralong-Duration Ballooning, 60

Laboratory Astrophysics, 60
Theoretical Challenges, 61
Policy Issues, 61
Long-Term Scientific Support for Observers, 61
Junior Faculty Instrumentation Program, 62
Education and Public Outreach, 62
Acronyms and Abbreviations, 63

2   REPORT OF THE PANEL ON OPTICAL AND INFRARED
ASTRONOMY FROM THE GROUND   65
Summary, 66
Major Initiative, Priority One: GSMT, 67
Major Initiative, Priority Two: LSST, 68
Moderate Initiative, Priority One: TSIP, 68
Science Opportunities, 69
Answering Fundamental Questions, 69
Exploiting the Diverse, Unique Facilities of U.S. Ground-based
O/IR Astronomy, 71
Major Initiative, Priority One: Develop and Build a Next-Generation
Ground-Based Telescope (GSMT), 73
Mission Description, 73
Science with the GSMT, 75
Theory Challenge for GSMT, 88
Technology Basis, 88
Key Technology Issues, 89
Cost Issues, 91
Context Issues, 93
Ancillary Benefits, 94
Major Initiative, Priority Two: A Large-Aperture Synoptic
Survey Telescope (LSST), 94
Mission Description, 94
Science with LSST: The Wide Area Variability Experiment, 95
Theory Challenge for LSST, 102
Data Flow and Information Distribution, 102
Multiplicative Advantages and Discovery Space Potential, 104
Technology and Cost Issues, 104
Context Issues, 106
Ancillary Benefits, 106

Moderate Initiative, Priority One: Telescope System
     Instrumentation Program: Leveraging Nonfederal Investment
     and Increasing Public Access, 107
  Definition, 107
  Science Drivers for 8-M Telescopes with Advanced
     Instrumentation, 108
  Guidelines for the Telescope System Instrumentation
     Program, 110
  Technology Issues, 116
  Cost Issues, 116
  Context Issues, 116
  Other Issues, 117
Acronyms and Abbreviations, 119

3   REPORT OF THE PANEL ON PARTICLE, NUCLEAR, AND
    GRAVITATIONAL-WAVE ASTROPHYSICS   123
    Summary, 124
    Science Opportunities, 125
      Gravitational-Wave Astrophysics, 126
      Cosmic Particle Acceleration, 128
      Neutrino and Nuclear Astrophysics, 133
      Search for Dark Matter, 137
    Existing Programs, 138
      Gravitational Waves, 138
      Very-high-energy Gamma Rays, 139
      Galactic Cosmic Rays, 139
      Highest-Energy Cosmic Rays, 140
      Neutrino Astronomy, 141
      Solar Neutrinos, 141
      Dark Matter Searches, 142
    Recommended New Initiatives, 142
      Gravitational-Wave Astronomy (LISA), 143
      Ground-Based Gamma-Ray Astrophysics (VERITAS), 146
      Program in Particle Astrophysics, 149
    Technology for the Future, 158
      Policy Issues, 158
      Facilities, 160
      Recommendations for the Funding Agencies, 160
    Acronyms and Abbreviations, 163

4   REPORT OF THE PANEL ON RADIO AND
     SUBMILLIMETER-WAVE ASTRONOMY   167
     Summary, 168
     Science Opportunities, 172
        The Large-Scale Structure of the Universe, 173
        The Formation and Evolution of Galaxies, 178
        The Formation and Evolution of Stars, 190
        The Formation and Evolution of Planets, 194
        The Origin and Evolution of Life, 198
     Existing Programs, 199
        National Centers, 199
        University Radio Facilities, 200
     Recommended New Initiatives, 202
        Expansion of the VLA, 203
        Square Kilometer Array, 205
        Combined Array for Millimeter Astronomy, 207
        Advanced Radio Interferometry Between Space and Earth, 207
        South Pole Submillimeter Telescope, 208
        Other High-Priority Projects, 209
     Technology for the Future, 211
        Ground-Based Needs and Opportunities, 212
        Space-Based Needs and Opportunities, 213
     Policy Issues, 213
        Open Skies Policy, 213
        Radio Spectrum Management, 214
        The National Radio Astronomy Observatory and the
           Atacama Large Millimeter Array, 214
        Agency Funding and Management Policies, 214
     Acknowledgments, 215
     Acronyms and Abbreviations, 216

5   REPORT OF THE PANEL ON SOLAR ASTRONOMY   221
     Summary, 222
        Strategy for the Decade 2001 to 2010, 222
        Observational Efforts, 223
        Theory and Data Mining, 224
        New Technologies, 224
        Policy Issues, 225
     Why Do Solar Physics Research?, 225
        Key to the Magnetodynamic Universe, 225

Solar-Terrestrial Physics, 227
Origin and Evolution of Life on Planets, 228
The Most Significant Advances in the Last Decade, 228
Goals Achieved, 228
The Solar-Stellar Connection, 230
A Systems Approach to Solar Physics—Toward a Decade
    of Understanding, 231
The Concept Behind the Solar Magnetism Initiative, 233
Global Solar Databases, 234
Operational Forecasting, 234
International Cooperation, 234
Existing Programs, 236
Ground-Based Observational Efforts, 236
Space-Based Observational Efforts, 238
New Initiatives, 244
From the Ground, 245
In Space, 257
Theory and Data Mining: The Solar Magnetism Initiative, 264
Technologies for the Future, 266
Adaptive Optics, 266
Solar-Lite, 267
High-Resolution Vector Magnetometry of UV Lines, 267
Connection to Laboratory Astrophysics, 268
Atomic/Molecular/Nuclear Physics, 268
Plasma Physics, 268
Policy and Educational Aspects, 269
The University-Based Solar Physics Community in the
    United States, 269
Funding Aspects, 270
The National Solar Observatory, 270
Education, 270
Acronyms and Abbreviations, 271

6   REPORT OF THE PANEL ON THEORY, COMPUTATION,
    AND DATA EXPLORATION   275
Summary, 276
The Scope of Theoretical Astrophysics, 276
Theory Initiatives Proposed by This Panel, 277
Data Exploration Initiative Proposed by This Panel:
    The National Virtual Observatory, 280

Summary of Panel Findings and Recommendations, 282
Description of Theoretical Astrophysics, 285
    The New Theorist, 285
    Successes of the Previous Decade, 287
Theory Challenges Tied to Priority Missions and Projects, 293
    Introduction, 293
    Examples of Theory Challenges, 294
    Computational Threads in Theory Challenges, 301
The National Virtual Observatory, 303
    Motivation for the NVO, 303
    Major Aspects of the Virtual Observatory, 306
    Project Scope, Structure, and Time Line, 310
National Postdoctoral Fellowships in Theoretical Astrophysics, 312
Right-Sizing Theory Support, 314
Institutional Issues for Theoretical Astrophysics, 317
    Unique Role for the Department of Energy, 317
    Institutes for Visiting Theorists, 319
    High-Performance Computing, 320
Acronyms and Abbreviations, 322

7   REPORT OF THE PANEL ON ULTRAVIOLET, OPTICAL,
    AND INFRARED ASTRONOMY FROM SPACE   327
Summary, 328
    Major Missions, 328
    Moderate Missions, 329
    Small Missions, 330
    Technology Development, 331
Science Opportunities, 332
Assumed Facilities, 336
    The Hubble Space Telescope, 336
    The Space Interferometry Mission, 337
Recommended New Initiatives, 339
    Major Missions, 339
    Moderate Missions, 352
    Small Missions, 367
Technology for the Future, 369
    Energy-Sensitive UV/Optical Detectors, 369
    Refrigerators, 370
    Spacecraft Communications, 371
    Ultralightweight ("Gossamer") Optics, 372
Acronyms and Abbreviations, 372

Executive
Summary

This Executive Summary is reprinted from the Astronomy and Astrophysics Survey Committee report, *Astronomy and Astrophysics in the New Millennium* (National Academy Press, Washington, D.C., 2001). It should be read in conjunction with the preface to this volume of panel reports; the preface explains the relationship between the recommendations of the survey committee and the recommendations of the panels.

ASTRONOMY AND ASTROPHYSICS IN THE NEW MILLENNIUM

In the first decade of the new millennium, we are poised to take a giant step forward in understanding the universe and our place within it. The decade of the 1990s saw an enormous number of exciting discoveries in astronomy and astrophysics. For example, humanity's centuries-long quest for evidence of the existence of planets around other stars resulted in the discovery of extrasolar planets, and the number of planets known continues to grow. Astronomers peered far back in time, to only a few hundred thousand years after the Big Bang, and found the seeds from which all galaxies, such as our own Milky Way, were formed. At the end of the decade came evidence for a new form of energy that may pervade the universe. Nearby galaxies were found to harbor extremely massive black holes in their centers. Distant galaxies were discovered near the edge of the visible universe. In our own solar system, the discovery of Kuiper Belt objects—some of which lie beyond the orbit of Pluto—opens a new window onto the history of the solar system. This report presents a comprehensive and prioritized plan for the new decade that builds on these and other discoveries to pursue the goal of understanding the universe, a goal that unites astronomers and astrophysicists with scientists from many other disciplines.

The Astronomy and Astrophysics Survey Committee was charged with surveying both ground- and space-based astronomy and recommending priorities for new initiatives in the decade 2000 to 2010. In addition, the committee was asked to consider the effective implementation of both the proposed initiatives and the existing programs. The committee's charge excludes in situ studies of Earth and the planets, which are covered by other National Research Council committees: the Committee on Planetary and Lunar Exploration and the Committee on Solar and Space Physics. To carry out its mandate, the committee established nine panels with more than 100 distinguished members of the astronomical community. Broad input was sought through the panels, in forums held by the American Astronomical Society, and in meetings with representatives of the international astronomical community. The committee's recommendations build on those of four previous decadal surveys (NRC, 1964, 1972, 1982, 1991), in particular the report of the 1991 Astronomy and Astrophysics Survey Committee, *The Decade of*

Discovery in Astronomy and Astrophysics (referred to in this report as the 1991 survey; also known as the Bahcall report).

The fundamental goal of astronomy and astrophysics is to understand how the universe and its constituent galaxies, stars, and planets formed, how they evolved, and what their destiny will be. To achieve this goal, researchers must pursue a strategy with several elements:

- Survey the universe and its constituents, including galaxies as they evolve through cosmic time, stars and planets as they form out of collapsing interstellar clouds in our galaxy, interstellar and intergalactic gas as it accumulates the elements created in stars and supernovae, and the mysterious dark matter and perhaps dark energy that so strongly influence the large-scale structure and dynamics of the universe.
- Use the universe as a unique laboratory for probing the laws of physics in regimes not accessible on Earth, such as the very early universe or near the event horizon of a black hole.
- Search for life beyond Earth, and if it is found, determine its nature and its distribution.
- Develop a conceptual framework that accounts for all that astronomers have observed.

Several key problems are particularly ripe for advances in this decade:

- Determine the large-scale properties of the universe: the amount, distribution, and nature of its matter and energy, its age, and the history of its expansion.
- Study the dawn of the modern universe, when the first stars and galaxies formed.
- Understand the formation and evolution of black holes of all sizes.
- Study the formation of stars and their planetary systems, and the birth and evolution of giant and terrestrial planets.
- Understand how the astronomical environment affects Earth.

These scientific themes, all of which now appear to offer particular promise for immediate progress, are only part of the much larger tapestry that is modern astronomy and astrophysics. For example, scientists cannot hope to understand the formation of black holes without understanding the late stages of stellar evolution, and the full significance of observations of the galaxies in the very early universe will not be clear until it is clear how these galaxies have evolved since that time. Although the new initiatives that the committee recommends will advance knowledge in many other areas as well, they were selected explicitly to address one or more of the important themes listed above.

In addition, the committee believes that astronomers can make important contributions to education. Building on widespread interest in astronomical discoveries, astronomers should:

- Use astronomy as a gateway to enhance the public's understanding of science and as a catalyst to improve teachers' education in science and to advance interdisciplinary training of the technical workforce.

OPTIMIZING THE RETURN ON THE NATION'S INVESTMENT IN ASTRONOMY AND ASTROPHYSICS

The United States has been generous in its support of astronomy and astrophysics and as a result enjoys a leading role in almost all areas of astronomy and astrophysics. So that the nation can continue to obtain maximum scientific return on its investment, the committee makes several recommendations to optimize the system of support for astronomical research.

BALANCING NEW INITIATIVES WITH THE ONGOING PROGRAM

An effective program of astronomy and astrophysics research must balance the need for initiatives to address new opportunites with completion of projects accorded high scientific priority in previous surveys.

- **The committee reaffirms the recommendations of the 1991 Astronomy and Astrophysics Survey Committee (NRC, 1991) by endorsing the completion of the Space Infrared Telescope Facility (SIRTF), the Millimeter Array (MMA; now part of the Atacama Large Millimeter Array, or ALMA), the Stratospheric Observatory for Infrared Astronomy (SOFIA), and the Astrometric Interferometry Mission (now called the Space Interferometry Mission, or SIM). Consis-**

tent with the recommendations of the Task Group on Space Astronomy and Astrophysics (NRC, 1997), the committee stresses the importance of studying the cosmic microwave background with the Microwave Anisotropy Probe (MAP) mission, the European Planck Surveyor mission, and ground-based and balloon programs.

The committee endorses U.S. participation in the European Far Infrared Space Telescope (FIRST), and it endorses the planned continuation of the operation of the Hubble Space Telescope (HST) at a reduced cost until the end of the decade.

- **To achieve the full scientific potential of a new facility, it is essential that, prior to construction, funds be identified for operation of the facility, for renewal of its instrumentation, and for grants for data analysis and the development of associated theory.**

NASA already follows this recommendation in large part by including Mission Operations and Data Analysis (MO&DA) in its budgeting for new missions. The committee recommends that funds for associated theory be included in MO&DA as well. It recommends further that the National Science Foundation include funds for facility operation, renewal of instrumentation, and grants for data analysis and theory along with the construction costs in the budgets for all new federally funded, ground-based facilities. These recommendations are consistent with those of the 1991 survey. For the purpose of total project budget estimation, the committee adopted a model in which operation amounts to 7 percent of the capital cost per year and instrumentation amounts to 3 percent per year for the first 5 years of operation. The committee recommends that total project budgets provide for grants for data analysis and associated theory at the rate of 3 percent of the capital cost per year for major facilities and 5 percent per year for moderate ones. On the basis of this model, the committee has included funds for operations, instrumentation, and grants for a period of 5 years in the cost estimates provided in this report for most ground-based initiatives.

- **Adequate funding for unrestricted grants that provide broad support for research, students, and postdoctoral associates is required to ensure the future vitality of the field; therefore new initiatives should not be undertaken at the expense of the unrestricted grants program.**

Grants not tied to a facility or program—unrestricted grants—often drive the future directions of astronomy.

STRENGTHENING GROUND-BASED ASTRONOMY AND ASTROPHYSICS

The committee addresses several structural issues in ground-based astronomy and astrophysics.

- **U.S. ground-based optical and infrared facilities, radio facilities, and solar facilities should each be viewed by the National Science Foundation (NSF) and the astronomical community as a single integrated system drawing on both federal and nonfederal funding sources. Effective national organizations are essential to coordinate, and to ensure the success and efficiency of, these systems. Universities and independent observatories should work with the national organizations to ensure the success of these systems.**

- **Cross-disciplinary competitive reviews should be held about every 5 years for all NSF astronomy facilities. In these reviews, it should be standard policy to set priorities and consider possible closure or privatization.**

The National Radio Astronomy Observatory (NRAO) and the National Astronomy and Ionosphere Center (NAIC) currently serve as effective national organizations for radio astronomy, and the National Solar Observatory (NSO) does so for solar physics. The National Optical Astronomy Observatories (NOAO) as currently functioning and overseen does not fulfill this role for ground-based optical and infrared astronomy. A plan for the transition of NOAO to an effective national organization for ground-based optical and infrared astronomy should be developed, and a high-level external review, based on appropriate, explicit criteria, should be initiated.

The Department of Energy (DOE) supports a broad range of programs in particle and nuclear astrophysics and in cosmology. The scientific payoff of this effort would be even stronger with a clearly articulated strategic plan for DOE's programs that involve astrophysics.

- **Given the increasing involvement of the Department of Energy in projects that involve astrophysics, the committee**

recommends that DOE develop a strategic plan for astro-physics that would lend programmatic coherence and facilitate coordination and cooperation with other agencies on science of mutual interest.

ENSURING THE DIVERSITY OF NASA MISSIONS

NASA's Great Observatories have revolutionized understanding of the cosmos, while the extremely successful Explorer program provides targeted small-mission opportunities for advances in many areas of astronomy and astrophysics. The committee endorses the continuation of a vigorous Explorer program. There are now fewer opportunities for missions of moderate size, however, despite the enormous role such missions have played in the past.

- **NASA should continue to encourage the development of a diverse range of mission sizes, including small, moderate, and major, to ensure the most effective returns from the U.S. space program.**

INTEGRATING THEORY CHALLENGES INTO THE NEW INITIATIVES

The new initiatives recommended below are motivated in large part by theory, which is also key to interpreting the results. Adequate support for theory, including numerical simulation, is a cost-effective means for maximizing the impact of the nation's capital investment in science facilities. The committee therefore recommends that

- **To encourage theorists to contribute to the planning of missions and facilities and to the interpretation and under-standing of the results, one or more explicitly funded theory challenges should be integrated with most moderate or major new initiatives.**

COORDINATING PROGRAMS AMONG FEDERAL AGENCIES

Because of the enormous scale of contemporary astronomical projects and the need for investigations that cross wavelength and discipline boundaries, cooperation among the federal agencies that support astronomical research often has benefits. To determine when interagency collaboration would be fruitful, each agency should have in place a strategic plan for astronomy and astrophysics and should also have cross-disciplinary committees (such as DOE and NSF's Scientific Assessment Group for Experiments in Non-Accelerator Physics [SAGENAP] and NASA's Space Science Advisory Committee [SSAC]) available to evaluate proposed collaborations. The Office of Science and Technology Policy could play a useful role in facilitating such interagency cooperation.

COLLABORATING WITH INTERNATIONAL PARTNERS

International collaboration enables projects that are too costly for the United States alone and enhances the scientific return on projects by bringing in the scientific and technical expertise of international partners. In many cases, international collaboration provides opportunities for U.S. astronomers to participate in major international projects for a fraction of the total cost, as in the case of the European Solar and Heliospheric Observatory (SOHO), XMM-Newton, Planck Surveyor, and FIRST missions, and the Japanese Advanced Satellite for Cosmology and Astrophysics mission. Valuable opportunities for international collaboration exist for smaller missions as well. Collaborations on major projects require the full support of the participating scientific communities, which can be ensured if the projects are among the very highest priorities of the participants, as is the case with ALMA.

The committee affirms the value of international collaboration for ground- and space-based projects of all sizes. International collaboration plays a crucial role in a number of this committee's recommended initiatives, including the Next Generation Space Telescope, the Expanded Very Large Array, the Gamma-ray Large Area Space Telescope, the Laser Interferometer Space Antenna, the Advanced Solar Telescope, and the Square Kilometer Array technology development, and it could play a significant role in other recommended initiatives as well.

NEW INVESTMENTS IN ASTRONOMY AND ASTROPHYSICS

Many mysteries confront us in the quest to understand our place in the universe. How did the universe begin? What is the nature of the dark matter and the dark energy that pervade the universe? How did the first stars and galaxies form? Researchers infer the existence of stellar mass black holes in our galaxy and supermassive ones in the nuclei of galaxies. How did they form? The discovery of extrasolar planets has opened an entirely new chapter in astronomy, bringing a host of unresolved questions. How do planetary systems form and evolve? Are planetary systems like our solar system common in the universe? Do any extrasolar planetary systems harbor life? Even a familiar object like the Sun poses many mysteries. What causes the small variations in the Sun's luminosity that can affect Earth's climate? What is the origin of the eruptions on the solar surface that cause "space weather"?

To seek the answers to these questions and many others described in this report, the committee recommends a set of new initiatives for this decade that will substantially advance the frontiers of human knowledge. Table ES.1 presents these initiatives, combined for both ground- and space-based astronomy, in order of priority. The committee set the priorities primarily on the basis of scientific merit, but it also considered technical readiness, cost-effectiveness, impact on education and public outreach, and the relation to other projects. The initiatives were divided into three categories—major, moderate, and small—that were defined separately for ground- and space-based projects based on estimated cost (see Chapter 1). The estimated cost of the recommended program for the decade 2000 to 2010 is $4.7 billion in FY2000 dollars, about 20 percent greater than the $3.9 billion inflation-adjusted cost of the recommendations of the 1991 survey. Two of the recommended projects, the Terrestrial Planet Finder (TPF) and the Single Aperture Far Infrared (SAFIR) Observatory, could start near the end of this decade or at the beginning of the next. The committee has assumed that about 15 percent of the total estimated cost for these two projects will fall in this decade.

MAJOR INITIATIVES

- The **Next Generation Space Telescope (NGST),** the committee's top-priority recommendation, is designed to detect light from the first stars and to trace the evolution of galaxies from their formation to the present. It will revolutionize understanding of how stars and planets form in our galaxy today. NGST is an 8-m-class infrared space telescope with 100 times the sensitivity and 10 times the image sharpness of the Hubble Space Telescope in the

TABLE ES.1 Prioritized Initiatives (Combined Ground and Space) and Estimated Federal Costs for the Decade 2000 to 2010[a,b]

Initiative	Cost[c] ($M)
Major Initiatives	
Next Generation Space Telescope (NGST)[d]	1,000
Giant Segmented Mirror Telescope (GSMT)[d]	350
Constellation-X Observatory (Con-X)	800
Expanded Very Large Array (EVLA)[d]	140
Large-aperture Synoptic Survey Telescope (LSST)	170
Terrestrial Planet Finder (TPF)[e]	200
Single Aperture Far Infrared (SAFIR) Observatory[e]	100
Subtotal for major initiatives	2,760
Moderate Initiatives	
Telescope System Instrumentation Program (TSIP)	50
Gamma-ray Large Area Space Telescope (GLAST)[d]	300
Laser Interferometer Space Antenna (LISA)[d]	250
Advanced Solar Telescope (AST)[d]	60
Square Kilometer Array (SKA) technology development	22
Solar Dynamics Observatory (SDO)	300
Combined Array for Research in Millimeter-wave Astronomy (CARMA)[d]	11
Energetic X-ray Imaging Survey Telescope (EXIST)	150
Very Energetic Radiation Imaging Telescope Array System (VERITAS)	35
Advanced Radio Interferometry between Space and Earth (ARISE)	350
Frequency Agile Solar Radio telescope (FASR)	26
South Pole Submillimeter-wave Telescope (SPST)	50
Subtotal for moderate initiatives	1,604
Small Initiatives	
National Virtual Observatory (NVO)	60
Other small initiatives[f]	246
Subtotal for small initiatives	306
DECADE TOTAL	4,670

[a]Cost estimates for ground-based capital projects include technology development plus funds for operations, new instrumentation, and facility grants for 5 years.

[b]Cost estimates for space-based projects exclude technology development.

[c]Best available estimated costs to U.S. government agencies in millions of FY2000 dollars and rounded. Full costs are given for all initiatives except TPF and the SAFIR Observatory.

[d]Cost estimate for this initiative assumes significant additional funding to be provided by international or private partner; see *Astronomy and Astrophysics in the New Millennium: Panel Reports* (NRC, 2001) for details.

[e]These missions could start at the turn of the decade. The committee attributes $200 million of the $1,700 million total estimated cost of TPF to the current decade and $100 million of the $600 million total estimated cost of the SAFIR Observatory to the current decade.

[f]See Chapter 1 for details.

infrared. Having NGST's sensitivity extend to 27 μm would add significantly to its scientific return. Technology development for this program is well under way. The European Space Agency and the Canadian Space Agency plan to make substantial contributions to the instrumentation for NGST.

- The **Giant Segmented Mirror Telescope (GSMT),** the committee's top ground-based recommendation and second priority overall, is a 30-m-class ground-based telescope that will be a powerful complement to NGST in tracing the evolution of galaxies and the formation of stars and planets. It will have unique capabilities in studying the evolution of the intergalactic medium and the history of star formation in our galaxy and its nearest neighbors. GSMT will use adaptive optics to achieve diffraction-limited imaging in the atmospheric windows between 1 and 25 μm and unprecedented light-gathering power between 0.3 and 1 μm. The committee recommends that the technology development for GSMT begin immediately and that construction start within the decade. Half the total cost should come from private and/or international partners. Open access to GSMT by the U.S. astronomical community should be directly proportional to the investment by the NSF.

- The **Constellation-X Observatory** is a suite of four powerful x-ray telescopes in space that will become the premier instrument for studying the formation and evolution of black holes of all sizes. Each telescope will have high spectral resolution over a broad energy range, enabling it to study quasars near the edge of the visible universe and to trace the evolution of the chemical elements. The technology issues are well in hand for a start in the middle of this decade.

- The **Expanded Very Large Array (EVLA)**—the revitalization of the VLA, the world's foremost centimeter-wave radio telescope—will take advantage of modern technology to attain unprecedented image quality with 10 times the sensitivity and 1,000 times the spectroscopic capability of the existing VLA. The addition of eight new antennas will provide an order-of-magnitude increase in angular resolution. With resolution comparable to that of ALMA and NGST, but operating at much longer wavelengths, the EVLA will be a powerful complement to these instruments for studying the formation of protoplanetary disks and the earliest stages of galaxy formation.

- The **Large-aperture Synoptic Survey Telescope (LSST)** is a 6.5-m-class optical telescope designed to survey the visible sky every week down to a much fainter level than that reached by existing surveys. It will catalog 90 percent of the near-Earth objects larger than 300 m and assess the threat they pose to life on Earth. It will find some 10,000 primitive objects in the Kuiper Belt, which contains a fossil record of the formation of the solar system. It will also contribute to the study of the structure of the universe by observing thousands of supernovae, both nearby and at large redshift, and by measuring the distribution of dark matter through gravitational lensing. All the data will be available through the National Virtual Observatory (see below under "Small Initiatives"), providing access for astronomers and the public to very deep images of the changing night sky.

- The **Terrestrial Planet Finder (TPF)** is the most ambitious science mission ever attempted by NASA. It is currently envisaged as a free-flying infrared interferometer designed to study terrestrial planets around nearby stars—to find them, characterize their atmospheres, and search for evidence of life—and to obtain images of star-forming regions and distant galaxies with unprecedented resolution. The committee's recommendation of this mission is predicated on the assumptions that TPF will revolutionize major areas of both planetary and nonplanetary science and that, prior to the start of TPF, ground- and space-based searches will confirm the expectation that terrestrial planets are common around solar-type stars. Both NGST and SIM lie on the technology path necessary to achieve TPF.

- The **Single Aperture Far Infrared (SAFIR) Observatory** is an 8-m-class space-based telescope that will study the important and relatively unexplored spectral region between 30 and 300 μm. It will enable the study of galaxy formation and the earliest stage of star formation by revealing regions too enshrouded by dust to be studied by NGST, and too warm to be studied effectively with ALMA. As a follow-on to NGST, SAFIR could start toward the end of the decade, and it could form the basis for developing a far-infrared interferometer in the succeeding decade.

MODERATE INITIATIVES

GROUND-BASED PROGRAMS

The committee's recommended highest-priority moderate initiative overall is the Telescope System Instrumentation Program (TSIP), which would substantially increase NSF funding for instrumentation at large telescopes owned by independent observatories and provide new observing opportunities for the entire U.S. astronomical community. Its second priority among ground-based initiatives is the Advanced Solar Telescope (AST), which offers the prospect of revolutionizing understanding of magnetic phenomena in the Sun and in the rest of the universe. The committee's next recommendation is that a program be established to plan and develop technology for the Square Kilometer Array, an international centimeter-wave radio telescope for the second decade of the century. In order of priority, the other recommended moderate initiatives are the following: The Combined Array for Research in Millimeter-wave Astronomy (CARMA) will be a powerful millimeter-wave array in the Northern Hemisphere. The study of very-high-energy gamma rays will take a major step forward with the construction of the Very Energetic Radiation Imaging Telescope Array System (VERITAS). The Frequency Agile Solar Radio telescope (FASR) will apply modern technology to provide unique data on the Sun at radio wavelengths. The South Pole Submillimeter-wave Telescope (SPST) will take advantage of the extremely low opacity of the Antarctic atmosphere to carry out surveys at submillimeter wavelengths that are possible nowhere else on Earth.

SPACE-BASED PROGRAMS

The committee's top recommendation for a moderate space-based mission is the Gamma-ray Large Area Space Telescope (GLAST). This joint NASA-DOE mission will provide observations of gamma rays from 10 MeV to 300 GeV with six times the effective area, six times the field of view, and substantially better angular resolution than the Energetic Gamma Ray Experiment aboard the Compton Gamma Ray Observatory. The committee's second-priority space-based project is the Laser Interferometer Space Antenna (LISA), which will be able to detect gravity waves from merging supermassive black holes throughout the visible universe and from close binary stars throughout our galaxy. The committee has assumed that LISA's cost will be shared with the European Space

Agency. Four additional space-based missions have priority. The Solar Dynamics Observer (SDO), a successor to the pathbreaking SOHO mission, will study the outer convective zone of the Sun and the structure of the solar corona. The highly variable hard-x-ray sky will be mapped by the Energetic X-ray Imaging Survey Telescope (EXIST), which will be attached to the International Space Station. The Advanced Radio Interferometry between Space and Earth (ARISE) mission is an orbiting antenna that will combine with the ground-based VLBA to provide an order-of-magnitude increase in resolution for studying the regions near supermassive black holes in active galactic nuclei.

SMALL INITIATIVES

Several small initiatives recommended by the committee span both ground and space. The first among them—the National Virtual Observatory (NVO)—is the committee's top priority among the small initiatives. The NVO will provide a "virtual sky" based on the enormous data sets being created now and the even larger ones proposed for the future. It will enable a new mode of research for professional astronomers and will provide to the public an unparalleled opportunity for education and discovery.

The remaining recommendations for small initiatives are not prioritized. The committee recommends establishing a laboratory astrophysics program and a national astrophysical theory postdoctoral program for both ground- and space-based endeavors. Augmentation of NASA's Astrophysics Theory Program will help restore a balance between the acquisition of data and the theory needed to interpret it. Ultralong-duration balloon flights offer the prospect of carrying out small space-based experiments at a small fraction of the cost of satellites. The Low Frequency Array (LOFAR), a joint Dutch-U.S. initiative, will dramatically increase knowledge of the universe at radio wavelengths longer than 2 m. The Advanced Cosmic-ray Composition Experiment for the Space Station (ACCESS) will address fundamental questions about the origin of cosmic rays. Expansion of the Synoptic Optical Long-term Investigation of the Sun (SOLIS) will permit investigation of the solar magnetic field over an entire solar cycle.

TECHNOLOGY

Technological innovation has often enabled astronomical discovery.

Advances in technology in this decade are a prerequisite for many of the initiatives recommended in this report as well as for initiatives in the next decade. For the recommended space-based initiatives, technology investment as specified in the existing NASA technology road map is an assumed prerequisite for the cost estimates given in Table ES.1. It is essential to maintain funding for these initiatives if NASA is to keep these missions on schedule and within budget. The committee endorses NASA's policy of completing a mission's technological development before starting the mission. The committee similarly endorses such a policy as the NSF is applying it to the design and development of ALMA.

For possible ground-based initiatives in the decade 2010 to 2020, investment is required in very large, high-speed digital correlators; in infrared interferometry; and in specialized dark-matter detectors. Future space-based initiatives require investment in spacecraft communication and x-ray interferometry, as well as technology for the next-generation observatories. Such technology will include energy-resolving array detectors for optical, ultraviolet, and x-ray wavelengths; far-infrared array detectors; refrigerators; large, lightweight optics; and gamma-ray detectors.

ASTRONOMY'S ROLE IN EDUCATION

Because of its broad public appeal, astronomy has a unique role to play in education and public outreach. The committee recommends that the following steps be taken to exploit the potential of astronomy for enhancing education and public understanding of science:

• Expand and improve the opportunities for astronomers to engage in outreach to the K-12 community.
• Establish more pilot partnerships between departments of astronomy and education at a few universities to develop exemplary science courses for preservice teachers.
• Improve communication, planning, and coordination among federal programs that fund educational initiatives in astronomy.
• Increase investment toward improving public understanding of the achievements of all NSF-funded science and facilities, especially in the area of astronomy.

1

Report of the Panel on High-Energy Astrophysics from Space

SUMMARY

X rays and gamma rays are emitted by the hottest gases and the most energetic events in the universe. Because of their penetrating power, they enable us to see into regions that are inaccessible in other wave bands, and because of their energy, they probe matter under the most extreme conditions. They also allow us to see out to large distances, observing the universe when it was much younger than it is today. X rays and gamma rays can only be observed from space, so their use for astronomy is young compared with other wavelengths. Still, dramatic discoveries of cosmological gamma-ray bursts, magnetars, baryon-rich clusters of galaxies, iron lines from accretion disks, and microquasars have led to a better understanding of these energetic environments and have taken us closer to a number of long-range scientific quests: finding the first light of the modern universe, elucidating relativistic gravity by directly imaging black holes, and understanding the origin of the elements that are critical for forming planets and life.

The technological capability is at hand to take the next steps toward these goals. Accordingly, the Panel on High-Energy Astrophysics from Space of the Astronomy and Astrophysics Survey Committee recommends a program for the coming decade that will require the building of three new telescopes:

• The Constellation-X Observatory (Con-X) is a major, high-spectral-resolution, broad-bandpass, x-ray spectroscopy mission. It is proposed as a launch of four telescopes on two rockets well away from Earth. Their combined sensitivity will improve upon that of existing and imminent x-ray missions by factors of 20 to 100, depending on wavelength.

• The top-priority, intermediate-class mission is the Gamma-ray Large Area Space Telescope (GLAST), which will use technology developed for particle physics experiments to detect high-energy gamma rays from quasars, pulsars, and gamma-ray bursts.

• The second-priority, intermediate-class mission is the Energetic X-ray Imaging Survey Telescope (EXIST), which will be attached to the International Space Station. It will monitor the whole sky at hard x-ray energies every 90 minutes.

In addition, the panel proposes a prioritized, advanced technology program that will comprise three missions: the Microarcsecond X-ray

Imaging Mission (MAXIM), Generation-X, and the MeV Spectroscopy Mission. Such a program would do three things:

• Lay the foundation for ultimately resolving black holes using x-ray interferometry in space.
• Develop the mirror and detector technology required to create a 10-m-diameter, focusing x-ray telescope that can detect emission from the first galaxies and stars in the universe.
• Develop instruments sensitive enough to perform extensive gamma-ray spectroscopy of the sites of element formation.

The six missions are listed in Table 1.1. A healthy high-energy astrophysics program should also embody a balance between larger and smaller projects. Accordingly, the panel presents four unprioritized recommendations:

• Maintain the Explorer program, which offers timely opportunities for opening up fresh territory, including nuclear line spectroscopy, x-ray surveys to map the "missing baryons," and continuous x-ray monitoring of the entire sky.
• Develop ultralong-duration ballooning, a cost-effective approach to hard x-ray and gamma-ray astronomy.
• Increase the investment in laboratory astrophysics so as to be ready to interpret the results anticipated from the observing program.

TABLE 1.1 New Major and Intermediate Missions Considered in Chapter 1

Mission	Specialty	Recommendation
Con-X	X-ray spectroscopy	2008 launch
EXIST	Hard x-ray survey	2005 deployment on ISS
Generation-X	Large-aperture x-ray telescope	Technology development
GLAST	Hard gamma-ray survey	2005 launch
MAXIM	X-ray interferometry	Technology development
MeV Spectroscopy Mission	Gamma-ray spectroscopy	Technology development

- Support focused theoretical challenges centered on the principal targets of observation. This would also enhance the scientific return from Con-X, GLAST, and EXIST.

Finally, the panel advocates three unprioritized policy actions:

- Provide sustained support for data analysis groups.
- Support instrumentalists in junior faculty positions.
- Maintain the exemplary record of public outreach.

A DECADE OF OPPORTUNITY

Our universe is an astounding place, and we are privileged to be alive when its scope, contents, and history are being revealed. From the almost-perfect microwave background radiation to immense clusters of galaxies; from the first quasars to the sleeping, giant black holes that they leave behind; from dense hydrogen gas clouds, where stars and their scalding planets are discreetly born, to the life-giving elements that these stars spawn—we have discovered worlds more wondrous than our boldest prophecies and more subtle than our most careful predictions. This flow of enduring discovery has been sustained by applying ingenious technology to increasingly sensitive telescopes operating throughout the electromagnetic spectrum as well as by exploring the universe using cosmic rays, neutrinos, and—soon, it is hoped—gravitational radiation.

The panel was charged with surveying x-ray and gamma-ray astronomy and recommending new initiatives for the coming decade at a particularly exciting time. As astronomers absorb the momentous discoveries of the U.S.-led Compton Gamma-Ray Observatory (CGRO) and Rossi X-ray Timing Explorer (RXTE), the Japanese-led spectroscopic satellite ASCA, the German-led low-energy survey satellite ROSAT, and the Italian-led broadband x-ray to gamma-ray mission Beppo-SAX, they are starting to make fundamental discoveries using the recently launched Chandra X-ray Observatory and the European-led X-ray Multi-Mirror Observatory (XMM-Newton). In addition, the European gamma-ray observatory INTEGRAL and the U.S. missions Hete-2 and Swift will also be launched and are expected to make major advances in gamma-ray astronomy.

However, these current missions have nominal lifetimes of 5 years, and the scientific opportunity and technology are already in hand to go well beyond their capabilities. It is therefore imperative to plan now for their successors. To focus its conclusions, the panel has organized its report around three long-term scientific "quests":

- To see the first light at the end of the universe's dark age and comprehend our cosmic origin;
  - To image black holes and elucidate relativistic gravity; and
- To understand the origin of the elements essential for forming Earth-like planets and life.

These quests will probably take several decades and will involve observations over the whole electromagnetic spectrum. For this reason, the panel has selected 12 associated near- to mid-term challenges (items A through L in Table 1.2) that are specific to high-energy astrophysics and that could be met over the next 10 to 15 years.

EMERGENCE OF STRUCTURE

The central task in contemporary observational cosmology is to reconcile the ancient and the modern universe. By detecting tiny fluctuations in the microwave background radiation, astronomers expect that they will soon have comprehensive measurements of the minor irregularities in the expanding universe from a time when it was less than a million years old. These irregularities grew, under gravity, to form the structure that we see around us now. Measurements such as these will also allow us to estimate the size and shape of the ancient universe. Meanwhile, observations of nearby stars and galaxies reveal the size and shape of the modern universe as it ages from roughly 1 to 13 billion years. Although the full story is not yet in, there is confidence that, within a few years, it will be possible to link these two views, using cosmological theory, for a universe containing cold dark matter (CDM) and, perhaps, dark energy. This will give us a description of the overall expansion of the universe—the stage upon which great cosmic dramas are enacted.

However, even if this endeavor is brilliantly successful, it will not tell us how, when, or where the first stars and galaxies formed. Indeed, we still do not know if the first luminous objects are stars in developing dwarf galaxies, as most theory predicts, stars in normal galaxies like our own, or accreting black holes in galactic nuclei. Although there has been impressive progress in recent years using optical observations of very distant galaxies and quasars, these observations are proving difficult to interpret,

TABLE 1.2 High-Energy Astrophysics Challenges to 2015

Designation	Challenge	Sections in Which Discussed
A	Find and weigh the first clusters of galaxies	Emergence of Structure, Hot Intergalactic Medium (Con-X), Generation-X
B	Detect local intergalactic gas and measure its density and temperature	Emergence of Structure, Hot Intergalactic Medium (Con-X)
C	Observe the first generation of gamma-ray bursts, perhaps associated with the first massive stars	Emergence of Structure, Cosmic Rays (GLAST), Gamma-ray Bursts (GLAST), Gamma-Ray Bursts (EXIST)
D	Find the first active galactic nuclei (AGN)	Emergence of Structure, Cosmic Rays (GLAST), Obscured AGN and the X-ray Background (EXIST)
E	Form an indirect image of the flow of gas around a black hole	Gravity Power, Black Holes and Neutron Stars (Con-X), Cosmic Rays (GLAST), Obscured AGN and the X-ray Background (EXIST), Galactic Survey (EXIST), MAXIM, All-Sky Monitors
F	Understand how jets are created and collimated	Gravity Power, Black Holes and Neutron Stars (Con-X), Blazars (GLAST), Cosmic Rays (GLAST), Gamma-ray Bursts (EXIST), MAXIM, All-Sky Monitors
G	Measure accurately the variation of neutron star radii with mass	Gravity Power, Black Holes and Neutron Stars (Con-X), Galactic Survey (EXIST)
H	Solve the mystery of gamma-ray bursts	Gravity Power, Gamma-ray Bursts (GLAST), Gamma-ray Bursts (EXIST), All-Sky Monitors
I	Balance the cosmic energy budget of galaxies and their active nuclei	Gravity Power, Black Holes and Neutron Stars (Con-X), Blazars (GLAST), Gamma-ray Bursts (GLAST), Obscured AGN and the X-ray Background (EXIST), Generation-X

TABLE 1.2 Continued

Designation	Challenge	Sections in Which Discussed
J	Use x-ray and gamma-ray observations to associate evolving stars with their post-supernova remnants and the elements they form	Origin of the Elements, Nucleosynthesis (Con-X), Galactic Survey (EXIST), Gamma-ray Bursts (EXIST), MeV Spectroscopy Mission, Nuclear Line X-ray Spectroscopy
K	Determine accurately the relative abundances and distribution in the interstellar medium of the 20 most common elements	Origin of the Elements, Nucleosynthesis (Con-X)
L	Understand the cosmic history of element production and dispersal	Origin of the Elements, Hot Intergalactic Medium (Con-X), Nucleosynthesis (Con-X), MeV Spectroscopy Mission, Nuclear Line X-ray Spectroscopy, Soft X-ray Surveys

for three reasons. First, we do not understand how the distribution of luminous galaxies relates to that of the dark matter. Second, the first stars create heavy elements that quickly condense into dust grains and lead to variable and uncertain obscuration of optical light. Third, the formation of galaxies involves much complex physics that is difficult to quantify. As a result, future progress is believed to depend on observations at both longer (infrared) and shorter (x-ray) wavelengths.

The practical approach to the first light quest is to investigate how structure emerges on all scales. The pivotal discovery—that the largest collections of galaxies, called "clusters," are luminous x-ray sources— gave us a probe of large-scale structure that avoids all three of the above problems. This is because the penetrating x-ray photons allow us to measure the mass of both the gas and the dark matter and because the formation of clusters is believed to be simpler than the formation of individual galaxies. X-ray observations of local clusters of galaxies provided the first indication—and, in many respects, the strongest argument yet—that the universe contains too little dark matter to arrest its expansion.

A central tenet of the CDM theory is that large structures were formed from the merging of smaller structures. In other words, clusters of galaxies should be relatively young. Some support for this view comes from the observation that many local clusters appear to comprise colliding subclusters, which may be visible by virtue of the strong x-ray-emitting shock waves that they develop. However, the similarity of distant x-ray clusters to those that we see around us now and the discovery of an apparently very dense x-ray cluster, which must have formed when the universe was less than 8 billion years old, suggest that much of the large-scale structure was in place earlier than had once been thought. To understand what really happened, we need to know the size of these young clusters, when they were formed, and how they themselves congregated. These considerations naturally motivate our first challenge, namely *to find and weigh the first clusters of galaxies (A)*.

Another way to understand the development of structure is to find the gas that does not condense into stars, galaxies, and clusters. Optical astronomers have probably found much of this gas—at epochs when the universe was only a few billion years old and temperatures were around 30,000 K—through its absorption of quasar light. However, they also know that most of this gas is no longer in this form and that it is necessary to understand what has become of it. We already know, from microwave background observations, that its current temperature must be less than 30 million K and that the recently launched Far Ultraviolet Spectroscopic Explorer (FUSE) will detect any gas with a temperature around 300,000 K. However, numerical simulations suggest that the temperature ought to be closer to a few million kelvin, which is well suited to x-ray observation. Therefore, in order to describe most of the matter in the expanding universe, we need *to detect local intergalactic gas and measure its density and temperature (B)*.

A more direct approach to finding the first light of the universe comes from observing gamma-ray bursts. These are now known to be ultraluminous explosions, probably associated with massive stars and already seen from when the universe was only 2 billion years old. They should be visible from much earlier times and may turn out to be a signature of the very first stars. If so, we should be able to use observations of gamma-ray bursts to study the history of formation of these stars and to determine whether or not they are localized in normal galaxies. Consequently, we desire *to observe the first generation of gamma-ray bursts, perhaps associated with the first massive stars (C)*. Like supernovae, gamma-ray bursts can serve as invaluable cosmological probes even if we are unsure how they work.

We now know that most normal galaxies contain nuclei that, although mostly dormant now, were very active in the past. The most luminous of these are the quasars, and they can be recognized using x-ray and gamma-ray observations. Quasars have already been seen from when the universe was only a billion years old and should be observable from much earlier epochs. There is no sign yet from the x-ray observations (in contrast to the optical searches) that we are seeing the onset of quasar activity. It is even possible that quasars formed before stars. X-ray and gamma-ray observations are particularly important for finding distant quasars because they are much less susceptible to absorption than optical emission. The final challenge associated with the first quest is, then, *to find the first active galactic nuclei (AGN) (D)*.

GRAVITY POWER

The study of black holes began as a theoretical consequence of Einstein's General Theory of Relativity more than 80 years ago. During the past decade, the evidence that they exist in abundance—in the nuclei of most normal galaxies as massive (a million to a few billion solar masses) black holes and as 5- to 30-solar-mass products of stellar evolution in close binary systems—has become overwhelming, and their masses have been confidently measured in roughly 10 cases. (More recently, there have also been reports that black holes with masses several thousand times that of the Sun may have been detected, and these might have a cosmological origin.)

An astrophysical black hole is described by just two parameters, a mass (which also measures its size) and a rate of rotation. Black holes have to be observed indirectly through their effects on nearby matter. As relativistic objects, they accelerate gas near their surfaces to speeds close to that of light and so can convert mass into radiant energy with an efficiency a hundred times greater than nuclear reactions. This happens in quasars, which sometimes outshine their host galaxies by a factor as large as 1000 to 10,000. It also happens in the nuclei of "normal" galaxies like our own, which contains a 2.6-million-solar-mass black hole (Figure 1.1). And it happens in binary stars, where the gas swirling around the black hole can be much brighter than the regular stellar companion. Indeed, far from being seen as an end point, the formation of a stellar black hole is now regarded as a beginning—the start of a new phase when it can convert matter into radiant energy with far greater efficiency than was possible for it as a star. We are deeply curious about how they function. This is our second quest.

Black holes were predicted to swallow their gas via accretion disks,

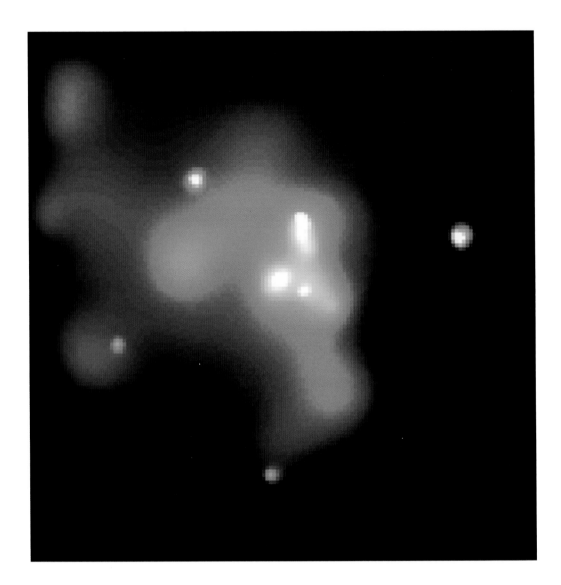

FIGURE 1.1 Chandra x-ray image of the galactic center. Infrared observations have demonstrated that there is a 2.6-million-solar-mass black hole at the dynamical center of our galaxy. This massive black hole accretes gas from its surroundings and heats it to x-ray-emitting temperatures. It is spatially coincident with the x-ray point source at the center of this image. The intensity of this source is surprisingly small. The other sources in the image are mostly associated with gas and stars near the black hole. Courtesy of NASA/Massachusetts Institute of Technology/Pennsylvania State University (PSU).

which orbit the hole rather like the rings that encircle Saturn (Figure 1.2). The existence of disks has been substantiated by several observations, most notably the measurement, by ASCA, of relativistically broadened iron line profiles (combined with narrow absorption features) from nearby galactic nuclei. These measurements have also been used to argue that the black holes must be spinning very rapidly. The gas in accretion disks is believed to sink toward the central hole and be heated by the frictional effect of a magnetic field. This magnetic field can also sustain an active corona rather like that observed around the Sun and from which extremely energetic x-ray photons are produced. Roughly half of the coronal radiation is reprocessed by the accretion disk, and much of the x-ray spectrum of nearby Seyfert galaxies (which resemble low-power quasars) has been interpreted in this manner. However, the arrangement and thermal states of the absorbing and the emitting gas are unknown.

A new diagnostic of the hole–disk–corona connection has been studied in great detail in accreting stellar black holes using RXTE. It has been found that these sources often exhibit quasi-periodic oscillations (QPOs), which are probably derived from the excitations of waves in the accretion disk. Interestingly, the x-ray photons are so energetic that they must have been created in the corona. Some QPOs are associated with bursts that happen on neutron star surfaces. These provide good measures of the (high) spin frequencies of the neutron stars and test our understanding of their properties.

To understand how gas accretes onto a black hole, we must use observations from binary stars and AGN and combine these with laboratory atomic astrophysics investigations and three-dimensional, numerical magnetohydrodynamical simulations. In essence, what we are trying to do is *to form an indirect image of the flow of gas around a black hole (E)*, in much the same way that a geophysicist might analyze seismic waves, gravity data, surface geology, and so on to create an image of Earth's interior. The x-ray counterparts of these diagnostics include the variable iron line profiles and the QPOs, which ought to be characteristic of the mass and spin of the hole and the rate and manner by which gas is supplied to it.

In addition to disks, many accreting black holes form a pair of jets that appear to flow with speeds close to that of light along opposite directions that are perpendicular to the disk (Figure 1.3). These jets were first found associated with galactic nuclei using radio astronomy. However, they have also been seen emanating from stellar black holes and

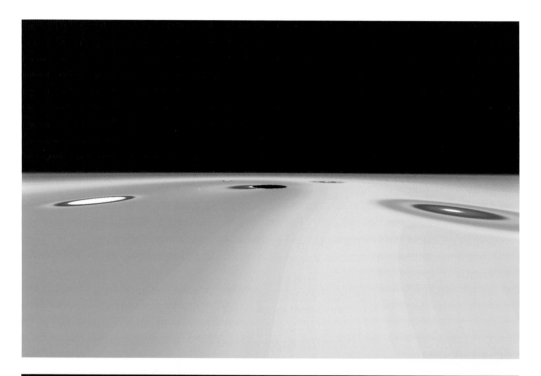

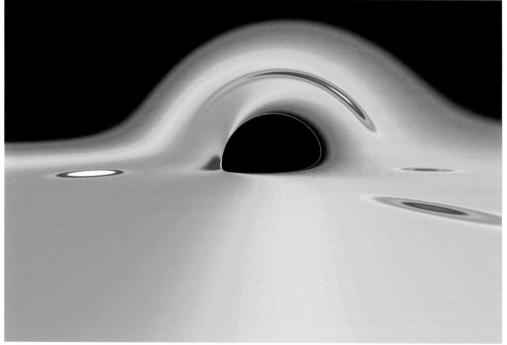

FIGURE 1.2 Numerical simulation of the appearance of a flat accretion disk orbiting a rapidly spinning black hole. The disk is endowed with four orbiting bright features to aid visualization. The brightest features are shown as white, the faintest as red. The observer is at rest and hovering just above the disk at a distance equal to 10 times the radius of the black hole. The top frame shows the appearance of a disk to an observer when the gravitational bending of the light is ignored. The bottom frame exhibits the strong distortions caused by general relativity. The observer sees the disk behind the black hole apparently elevated because of the bending of light rays. The strong effect of the Doppler shift, the gravitational redshift, and the "dragging" of inertial frames are also apparent. The faint red curve that apparently delineates the surface of the hole is actually formed by rays from the underside of the accretion disk. One long-term goal of the MAXIM technology development program is to make images of real accretion disks that would exhibit these physical effects. Courtesy of Kevin Rauch, Johns Hopkins University.

are particularly well-suited to x- and gamma-ray observations. It is a consequence of the near-light speed of these jets that they are overwhelmingly intense when we observe directly along one of them, in which case they are known as blazars. Even allowing for this beaming, the gamma-ray power can still outshine all of the other emission. Because relativistic jets can carry away such a large fraction of the energy released by the accreting gas, their formation is now seen as an integral part of accretion onto a black hole. Nearly a hundred gamma-ray blazars have been found by CGRO at GeV[1] energies and as many unidentified and possibly related sources have also been detected. The challenge now is *to understand how jets are created and collimated (F)*. In particular, it will be necessary to determine if they derive their power from the spin energy of the black hole or the gravitational energy of the accreting gas. This is a second area where there are thought to be unusually good opportunities for close collaboration between observers, theorists, and laboratory astrophysicists, particularly in exploring the behavior of ultrarelativistic plasma.

Neutron stars are also relativistic objects, and they permit quantitative tests of strong gravity. They contain matter with density greater than that found in atomic nuclei and are usually born with magnetic fields millions of times stronger than we can sustain in the laboratory. Conse-

---

[1]In this report, the panel refers to the energies of individual photons in electron volts (eV), with 1 eV = 1.6×10^{-12} erg (1 keV is a thousand eV, 1 MeV is a million eV, 1 GeV is a billion eV, and 1 TeV is a trillion eV).

FIGURE 1.3 Image of the jet in the nearby active elliptical galaxy Centaurus A superimposed on the optical image of the galaxy. The x-ray image was obtained with the high-resolution camera on Chandra (scale bar = 1 arcmin). The jet is created by a massive black hole in the center of the galaxy and powers a pair of giant radio sources. Several point x-ray sources, probably neutron stars and black holes that accrete gas from their companion stars, can also be discerned. Jets like those in Centaurus A are called blazars when they are pointed toward us and are also prodigious gamma-ray sources. Note the prominent dust lanes in the optical image. These absorb soft x rays and allow only the hard x rays to escape from the nucleus. Courtesy of NASA/CXC/Smithsonian Astrophysical Observatory (SAO) and NSF/Association of Universities for Research in Astronomy, Inc. (AURA)/National Optical Astronomy Observatories.

quently, neutron stars form unique cosmic laboratories, allowing us to study matter under physical conditions that cannot be reproduced on Earth. In particular, the rate at which neutron stars cool can be measured and can provide a unique test of the properties of matter at densities greater than that of atomic nuclei.

All neutron stars spin, and—in contrast to black holes—this spin causes regular pulsations with periods that can be as short as ~1.5 ms. Most neutron stars are isolated and powered by their spin energy, so they slow down with time. They are most luminous in gamma-ray photons. Other neutron stars attract gas from companion stars, and the gravitational energy that is released powers their emission, which is mostly observed at x-ray energies. This gas spins up the neutron star, and it appears that an equilibrium period of ~3 ms is commonly attained. These rapidly spinning neutron stars may be one of the most promising sources of gravitational radiation.

Even more remarkably, it has recently been discovered that a minority of neutron stars—called magnetars—are born with magnetic fields up to a billion times stronger than we can create and so strong that the uncertainty principle requires that electrons have ultrarelativistic energies independent of their densities. We observe magnetars through bursts of "soft" gamma-ray photons. The high magnetic fields are discovered by measuring the very rapid rate of their spin period change, which is much faster than that of radio pulsars; this requires a third energy source, magnetic energy. There is much fundamental astrophysics to explore here. However, perhaps the most significant use of neutron stars will be *to measure accurately the variation of neutron star radii with mass (G)*, which tells us how neutron star matter responds to gravity and produces pressure. This will enable us to understand in much more detail what happens in a supernova explosion and will be an invaluable contribution to fundamental physics, complementing the results that are coming from heavy-ion colliders.

In recent years, it has become clear from high-energy observations that gamma-ray bursts (GRBs) are cosmologically distant and consequently so powerful that they must transform a fraction of a solar rest mass into pure energy within seconds. The only objects thought likely to create GRBs are black holes and neutron stars. These explosions may create relativistic jets moving even closer to the speed of light than blazar jets. As with supernovae, their ejecta are decelerated by the surrounding gas, and they create afterglows at essentially all frequencies that can be followed for up to a year. Surprisingly, CGRO has detected four bursts at

~1 GeV energies, including one that persisted for an hour (one burst was reported at TeV energies). Also, as with supernovae, there appear to be many different types: Some GRBs are relatively low power and may be by-products of supernovae associated with massive stars; others show features like those of AGN. In spite of these breakthroughs and many creative suggestions, there is still no widely accepted explanation for GRBs. GRBs have stimulated study of the behavior of matter under completely new physical conditions. In particular, understanding the coalescence of neutron stars requires detailed study of the behavior of nuclear matter using general relativity. It is a prime challenge to observers, laboratory astrophysicists, and theorists combined *to solve the mystery of gamma-ray bursts (H)*. Success in this endeavor will surely also clarify our understanding of advanced stellar evolution and AGN.

We can also consider the global properties of AGN to understand how they operate on the average. By measuring the masses of black holes in nearby galactic nuclei, it is possible to estimate how much radiant energy was produced in forming them. This quantity turns out to be significantly larger than we actually observe at optical, ultraviolet, and soft x-ray energies. However, many AGN are quite overluminous when observed in hard (in contrast to soft) x-ray photons (Figure 1.4). This suggests that most of the light they emit is absorbed by cold gas and dust and then reradiated as infrared radiation. Only the penetrating hard x-ray photons and gamma-ray photons can show these "obscured" AGN as they truly are. Furthermore, because this high-energy emission is likely to vary, we are unlikely to confuse its black-hole origin with stars, which is a problem with similar observations at infrared wavelengths. High-energy observations are necessary to tell us the true powers of AGN and to understand how efficiently they convert mass into energy.

There is an additional consideration. We already know that most of the soft x-ray background comes from discrete sources (Figure 1.5). The same is probably true of the hard (~10 to 100 keV) x-ray background. Observing the x-ray background allows us to measure the time-averaged luminosities of accreting black holes in AGN and to see how they have changed over cosmic time. It is already clear that the average AGN power is at least several percent of that associated with stars (and might even be comparable). AGN power may also account for a significant fraction of the recently measured far-infrared background. The inevitable consequence of all of this is that AGN are likely to have played an important role in galaxy formation. To place these ideas on a more quantitative footing, we need *to balance the cosmic energy budget of*

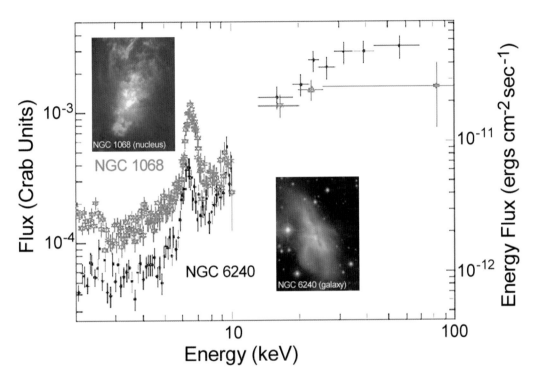

FIGURE 1.4 X-ray spectra recorded by Beppo-SAX of two highly absorbed Seyfert galaxies, NGC 1068 and NGC 6240. Note the fluorescent iron lines around 7 keV and the increasing fluxes to well beyond 10 keV, which are indicative of strong absorption by heavy elements in intervening gas, partially visible in the accompanying galaxy images from the Hubble Space Telescope (HST) and the European Southern Observatory (ESO), respectively. EXIST could discover thousands of AGN like this that probably dominate the hard x-ray background. Courtesy of Beppo-SAX collaboration.

galaxies and their active nuclei (I). This connects strongly to our first quest, which involved understanding the role of massive black holes in the birth and evolution of galaxies.

ORIGIN OF THE ELEMENTS

One of the most important discoveries in recent years has been of large numbers of extrasolar planets. These are typically "Jupiters," comprising mostly hydrogen and helium gas; often located extremely close to their stars, they are, consequently, very hot. Earth-like planets, by contrast, are made out of the heavier elements and are solid. By their

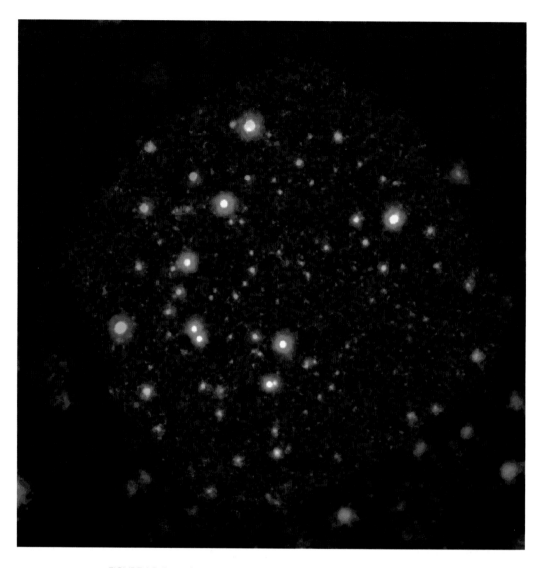

FIGURE 1.5 Deep fields from ROSAT, Chandra, and Hubble. The left-hand image of the x-ray sky was taken as a long exposure by the German-led x-ray satellite ROSAT. The image, which measures 25 arcmin on a side, contains a variety of sources, including about 40 distant AGN. These sources are representative of those that form the x-ray background. (The colors indicate the "hardness" of the x-ray spectrum, with blue being hard—relatively more high-energy photons—and red, soft.) Con-X will be able to obtain high-resolution spectra for essentially all of these x-ray sources. Courtesy of ROSAT/G. Hasinger. The bottom right image is of the famous Hubble Deep Field (HDF) North as observed at optical and x-ray wavelengths. The image size is roughly 2.5 arcmin. Chandra (top right) can detect x-ray sources that are more than 10 times fainter than ROSAT and

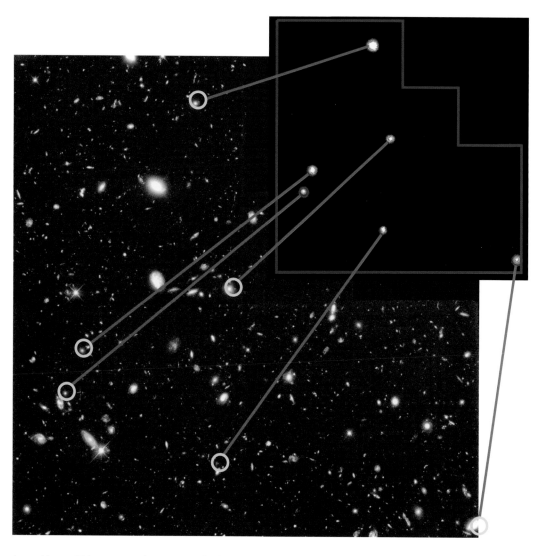

is sensitive to higher-energy photons as well. Of the six x-ray sources found so far in the HDF North, one is identified with the bright active nucleus of a galaxy (an AGN), three are associated with elliptical galaxies (some of which appear to contain weak AGN), one is associated with an optically faint galaxy that shows relatively strong emission in the infrared (probably a hidden AGN), and one appears to lie in the outer parts of a nearby spiral galaxy. Some AGN are seen only through their penetrating high-energy x rays, suggesting that optical, ultraviolet, and low-energy x-ray photons are absorbed by gas and dust. Optical surveys for quasars and AGN have underestimated their density and importance in the evolution of galaxies. Top right courtesy of NASA/PSU/G.P. Garmire, A.E. Hornschemeier, W.N. Brandt, and the ACIS team, and NASA/CXC. Bottom right courtesy of R. Williams and the Hubble Deep Field team, NASA/STScI.

nature, high-energy observations are not useful for finding these potentially habitable environments; they are only good for assessing those
factors that are inimical to life! However, where high-energy astrophysics
can make a contribution is by determining where and when the heavy
elements—the raw materials of life—were made in the first place.

With the possible exception of carbon and nitrogen, most of the
heavy elements in nature were made in supernovae. These exploding
stars can be divided into two basic classes—those of type II (including
also those of types Ib and Ic), where the explosion is preceded by gravitational collapse and a neutron star or black hole is left behind, and those
of type Ia, where there is a thermonuclear explosion of a white dwarf and
no remnant. For both classes, the sudden release of energy raises the
temperature to the point where violent nuclear reactions occur. The
ejection of both kinds of elements—those made explosively and those
made prior to the supernova—is responsible for producing most of the
atoms heavier than helium in our universe. This basic picture has been
corroborated by many observations. However, important details of the
explosion mechanism are not yet understood, and these have a crucial
bearing on the yield of freshly synthesized elements from different types
of supernovae. For example, we still do not understand how the fusion
flame propagates in a type Ia supernova—faster or slower than sound—
nor do we know the mass of the progenitor star. In type II supernovae,
we do not understand exactly how the collapsing star gets converted to
an explosion and, in particular, which elements fall into the neutron star
and which are ejected.

We can address these issues most directly by observing the gamma-
ray lines that are made by radioactive nuclei created in the explosion. In
fact, we can match the lifetimes of these unstable nuclei to the timescales
over which we need to average. For example, the supernova that was
observed in 1987 in the Large Magellanic Cloud was detected in lines
created by a cobalt isotope that decays on a timescale of several months.
Supernovae in the Galaxy that are part of the historical record have been
observed in lines formed by a titanium isotope, which decays in about a
century. Lines from aluminum and iron isotopes, which last a million
years and have half-lives of a few million years, may chart the average
rate of formation of massive stars in the Galaxy long after these stars and
their remnants have vanished from view. In addition, gamma-ray
photons, made by decaying pions, tell us about cosmic-ray-induced
production of the light elements lithium, beryllium, and boron within the
expanding remnants. Gamma-ray nucleosynthetic studies are in their

infancy, because the poor sensitivity of gamma-ray telescopes has so far limited us to detecting just a few lines from nearby supernovae.

The debris from the explosion expands rapidly, creating a strong shock wave and heating the interstellar gas to x-ray temperatures. This, in turn, decelerates and heats the debris as well. Both types of hot gas radiate strong atomic and ionic emission lines (Figure 1.6). Many of these lines have already been measured and, to some extent, mapped, using ASCA. However, neither the sensitivity nor the angular resolution

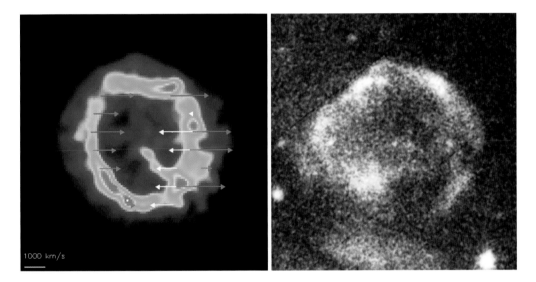

FIGURE 1.6 On the left is a Chandra x-ray image of a nearby supernova remnant, E0102-72.3, using an emission line created by ionized neon in the hot gas formed by the explosion. The arrows measure the velocity along the line of sight (toward us when directed to the right), which can be determined using the Doppler shift of the wavelength of the line. Images like these can be used to test our understanding of how the chemical elements are formed in supernova explosions. Courtesy of NASA/Massachusetts Institute of Technology (MIT)/C. Canizares and J. Houck. On the right is a detail from an image of the Large Magellanic Cloud made using the EPIC detector on the XMM-Newton x-ray observatory, launched in late 1999 by the European Space Agency (ESA), showing a new supernova remnant. The bright source in the bottom right-hand corner is the remnant of the famous supernova explosion that was observed in 1979. XMM-Newton complements the Chandra Observatory by emphasizing collecting area and spectroscopy over image quality. Con-X is designed to use mirrors similar to, although much larger than, the XMM mirrors. Courtesy of ESA/XMM/G. Hasinger.

has so far been adequate to develop a quantitative understanding of the different types of explosion and the yield of trace elements. It will be necessary *to use x-ray and gamma-ray observations to associate evolving stars with their postsupernova remnants and the elements they form (J)*. A very good example is provided by the well-studied supernova remnant Cassiopeia A, which now appears to contain a compact object, either a neutron star or a black hole. Another good example is provided by the Crab Nebula, the remnant of a supernova that occurred in 1054. The Crab Nebula is mostly powered by a central pulsar—a rapidly spinning, magnetized neutron star. (Both Cassiopeia A and the Crab Nebula were formed by type II explosions. The study of type Ia supernovae has been given added impetus in recent years by their role in cosmic distance determination.)

Eventually, the supernova debris mixes with the ambient interstellar gas, so we also need to understand how the mix of the elements has evolved in different environments. One approach is to measure the strengths of absorption lines in gas that lies between us and bright x-ray sources like quasars. This technique is particularly robust because, unlike with optical and ultraviolet studies, it is not sensitive to the ionization state of the absorbing gas and the fraction of the elements that has been incorporated into dust grains. It should be possible *to determine accurately the relative abundances and distribution in the interstellar medium of the 20 most common elements (K)*. This is the material that is incorporated into new stars, planets, and—occasionally—living organisms.

The dispersal of the elements does not stop in the interstellar medium. Observations of nearby galaxies with high rates of star formation have already found hot, outflowing gas that has been recently enriched. For those galaxies that reside within the rich clusters, this gas will be trapped and will augment the hot intracluster gas. All other galaxies should enrich the general intergalactic medium, which, as mentioned above, is expected to be at x-ray temperatures and must now account for the bulk of normal matter. In addition to locating intergalactic gas, it is of prime importance *to understand the cosmic history of element production and dispersal (L)*. This relates directly to the history of star formation and should help us understand how much of the formation happened outside normal galaxies. It also relates to our first two quests, because quasars and gamma-ray bursts provide ideal probes of the build-up of heavy elements in the early universe.

SURPRISES

As must be clear from the above, if history is any guide, the most significant discoveries of the next decade in high-energy astrophysics will be unanticipated and will not fit into some preconceived pattern. It is therefore important to structure a research program so as to permit a broad investigation of "discovery space"—new areas of investigation that open as new observational technologies and techniques are used. Each of the facilities now described by the panel fulfills this requirement.

A NEW BEGINNING

This report is being written as astronomers start to use the Chandra and XMM-Newton observatories. (A third x-ray observatory, Japan's ASTRO-E, was lost following a launch failure. It was to have complemented Chandra and XMM-Newton by emphasizing high-energy x-ray observations.) In addition, three gamma-ray missions are awaiting launch.

CHANDRA

Launched on July 23, 1999, and performing as planned, the Advanced X-ray Astrophysics Facility of the National Aeronautics and Space Administration (NASA), renamed the Chandra X-ray Observatory,[2] is poised to make major discoveries in x-ray astronomy. It is designed primarily as an imaging x-ray telescope that uses grazing incidence reflectors to achieve a 0.5 arcsec angular resolution over a 7-arcmin (diameter) field of view, comparable to what is obtained on the ground using optical telescopes. Chandra can detect x-ray photons with energy in the range ~0.1 to 10 keV, with an effective area for imaging of 600 cm^2 at ~1 keV; armed with charge-coupled devices (CCDs) and microchannel plate detectors, it has over a hundred times better sensitivity than the Einstein Observatory, which was launched in 1978, and 10 times the angular resolution of ROSAT. It also has a strong spectroscopic capability using transmission gratings and will provide energy resolution of up to one part in a thousand (depending upon the energy) (Figure 1.7). As it is

---

[2]See <http://chandra.harvard.edu/>.

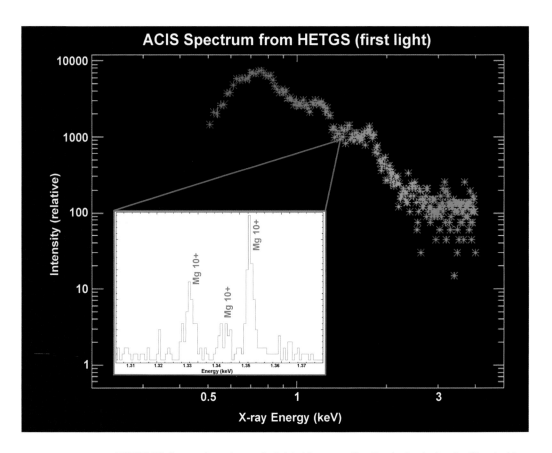

FIGURE 1.7 Spectra from the nearby bright binary star Capella, obtained using the Chandra X-ray Observatory. The x rays are produced by hot gas in a corona. The moderate-resolution spectrum measured using the Advanced CCD Imaging Spectrometer shows several broad features; those in the ~1- to 2-keV interval are contributed mainly by Si and Mg. The inset shows a small detail of the high-resolution spectrum obtained with the High-Energy Transmission Grating Spectrometer close to 1.3 keV, where a magnesium triplet is clearly resolved. These spectra can be used to deduce the physical conditions and abundance in the source. Con-X will be able to take spectra like this for sources that are a hundred times fainter than Capella. Courtesy of NASA/Chandra X-ray Center/ MIT, SAO.

in high Earth orbit, it will allow essentially uninterrupted observation of faint and variable sources.

Among Chandra's many scientific objectives are the following:

• Performing x-ray spectroscopy of normal stars so as to detect their coronae and stellar winds and compare them with the Sun;

- Mapping supernova remnants using emission lines from the most common elements at moderate spectral resolution;
- Observing binary star x-ray sources from as far away as the Virgo cluster of galaxies and performing a census so as to understand more clearly their role in stellar evolution;
- Observing black hole accretion disks using their iron emission lines; and
- Imaging galaxies, AGN (including jets), and clusters of galaxies to much greater distances and earlier times than before.

XMM-NEWTON

By emphasizing field of view (30 arcmin) and effective area ($\sim$6000 cm^2 for imaging at 1 keV) at the expense of angular resolution (15 arcsec), the European-led XMM-Newton[3] (launched in December 1999) nicely complements the Chandra X-ray Observatory. It has an onboard optical-UV monitor and a high-resolution x-ray spectrometer using reflection gratings, which provides simultaneous observations with the imaging detectors. Many of XMM-Newton's overall science objectives are similar to those of Chandra. Its special strengths are sensitivity to extended, low-surface-brightness sources, notably the x-ray background, and spectroscopy of faint point sources. Like Chandra, XMM-Newton is performing well and producing a steady stream of remarkable astronomical discoveries.

HETE-2 AND SWIFT

There are two missions poised to capitalize on recent developments in our understanding of gamma-ray bursts. HETE-2[4] was launched in October 2000. It is expected to locate roughly 30 gamma-ray bursts per year to better than $\sim$10 arcsec and communicate these positions to waiting ground stations in about 5 s. This will allow immediate optical follow-up. Swift,[5] which is planned for launch in 2003, is expected to improve upon this capability by obtaining immediate arcminute positions for roughly one burst per day and then using its onboard optical, UV, and

---

[3]See <http://sci.esa.int/xmm/>.
[4]See <http://space.mit.edu/HETE/>.
[5]See <http://swift.gsfc.nasa.gov/>.

x-ray telescopes to obtain arcsecond accuracy after a few minutes have elapsed, facilitating a determination of the distances and powers of the bursts. It will also produce a 10 to 100 keV survey of the sky with sensitivity at least 10 times better than that of the best existing survey.

INTEGRAL

The European gamma-ray mission, INTEGRAL[6] (2001 launch), will break new ground in gamma-ray spectroscopy and imaging. It will be sensitive up to ~10 MeV and have an energy resolution of ~2 keV at the crucial electron-positron annihilation line at ~0.5 MeV. Its solid-state germanium detectors will give it a narrow line sensitivity, which is a 10-fold improvement over CGRO. INTEGRAL also carries a "coded mask" imager to reconstruct source positions to an accuracy of ~2 arcmin. The key scientific objectives include the study of explosive nucleosynthesis in type Ia supernovae out to a distance of about 50 million light-years through the detection of cobalt lines and the survey of recent and past Galactic supernovae through the detection and mapping of titanium and aluminum lines.

THE NEXT STEPS

After having considered a variety of proposals, the panel unanimously recommends that NASA start and launch three missions over the coming decade. The capabilities of these missions would represent major advances over the capabilities of current missions, and the three would complement each other in addressing the quests and challenges described above.

PROPOSED MAJOR MISSION: CONSTELLATION-X

MISSION DEFINITION

Con-X[7] (Figure 1.8) is planned as a high-throughput, x-ray facility emphasizing observations with unprecedented energy resolution, E/DE, of between about 300 and 5000 over a broad energy range, ~0.25 to

---

[6]See <http://sci.esa.int/integral/>.
[7]See <http://constellation.gsfc.nasa.gov>.

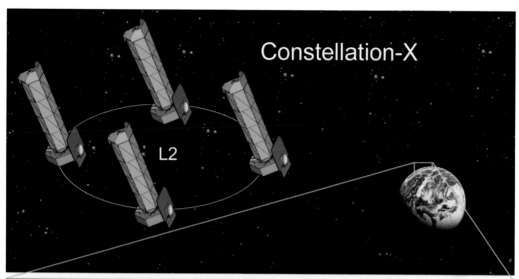

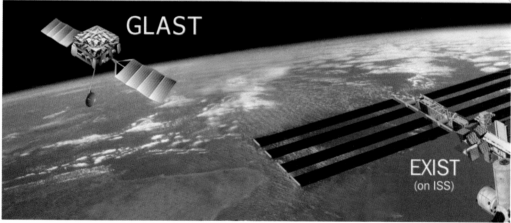

FIGURE 1.8 Montage showing the three proposed missions: Con-X, GLAST, and EXIST. Con-X will operate on the far side of Earth from the Sun at the second Lagrange point, designated L2. GLAST and EXIST will both operate in low Earth orbit. Courtesy of NASA/Goddard Space Flight Center (GSFC).

40 keV. As such, it will complement Chandra in much the same way that the Keck and Gemini optical telescopes complement HST. Con-X's large effective area in spectroscopic mode ($\sim$1.5 m^2) will exceed the spectroscopic effective areas of Chandra and XMM-Newton by factors of $\sim$20 to 100. Con-X will be able to time individual photons to $\sim$100 μs.

As currently configured, Con-X's high sensitivity will be attained using four telescopes launched by two Atlas/Delta rockets to the second Lagrange point, on the far side of Earth from the Sun. The main mirror design is a larger and lighter version of the XMM mirror, while the CCDs will benefit from experience gained with Chandra. The main spectroscopic detector takes the form of a cooled microcalorimeter array. Each telescope will also carry a separate multilayered mirror for the hard x-ray observations. Fabricating these components constitutes the main technology challenge for Con-X.

Con-X has been under study for over 5 years, and a broad-based facility science team has been formed to refine its scientific objectives and match them to attainable instrument and spacecraft capability. Technology development for Con-X is part of the 1997 strategic plan of NASA's Office of Space Science (OSS). The Con-X time line envisages a phase C/D start in 2005, leading quickly to a 2008 launch. The cost is expected to be $450 million for phase C/D development, $189 million for the two launches, and $133 million for 5 years of mission operations and data analysis (MO&DA), giving a total cost of $772 million.

HOT INTERGALACTIC MEDIUM (A, B, L)

As discussed above, some giant clusters of galaxies appear to have formed sooner in the life of the universe than had been expected. Provided that all clusters formed after the time when the universe was about 4 billion years old, they should be detectable by Chandra and XMM-Newton. If they are, Con-X will be able to measure the composition and temperature of the x-ray-emitting gas and convert these measurements into mass distributions for the gas, the stars, and the dark matter. These observations will provide a quantitative measure of the growth of large-scale structure, especially when combined with radio observations that measure the effect that clusters have on the microwave background radiation.

In addition, Con-X will perform detailed analyses of the closer and brighter clusters, measuring their gas flow speeds with accuracies of ~ 20 km s^{-1}, somewhat better than the formal resolution. It should also locate the shock waves expected to surround these clusters. By measuring the abundances of all the common ions, the thermal state of the gas can be specified and its recent history reconstructed. This is essential if, as we believe, clusters merge with one another. The combination of all of these x-ray measurements with optical studies, microwave background observations, and numerical simulations should allow us to determine when these individual clusters were assembled and how they subsequently grew. Measuring the steady increase of metals inside clusters as the universe ages also traces the star formation, which, in turn, relates to the evolutionary history of the galaxies inside the cluster.

If, as numerical simulations strongly suggest, most of the mass of the intergalactic medium is now at million-kelvin temperatures, then Con-X should be able to measure its distribution and follow its dynamics by seeing absorption lines (notably those formed by oxygen ions) in the spectra of hundreds of bright, background quasars in much the same way that optical astronomers have been able to detect 30,000-K intergalactic gas from when the universe was only a few billion years old using hydrogen and carbon atoms (Figure 1.9). Indeed, Con-X should also be able to see emission from this hot gas associated with nearby groups of galaxies and draw conclusions parallel to those for the clusters (Figure 1.10).

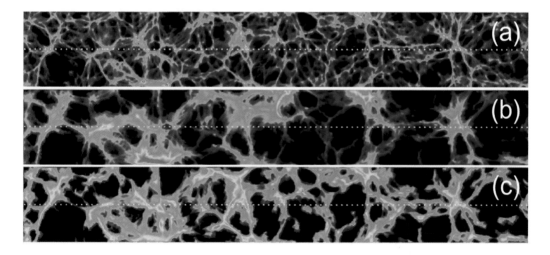

FIGURE 1.9 Numerical simulations of the gas density in the local intergalactic medium. These simulations show small sections of the expanding universe and can be used to produce estimates of the column densities of different ionization states and elements as a function of recession velocity along a line of sight like the dotted lines shown in the figures: (a) density of oxygen ions with two electrons, (b) density of oxygen ions with one electron, and (c) overall density of all heavy elements. Courtesy of N.Y. Gnedin, U. Hellsten, and J. Miralda-Escudé, *Astrophysical Journal* 509 (1998): 56-61.

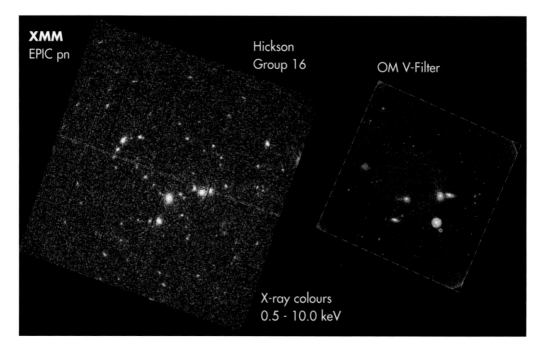

FIGURE 1.10 X-ray image of a nearby, compact group of galaxies known as Hickson 16. The right-hand frame is a corresponding optical image. Both images were acquired with XMM-Newton. Con-X will be able to measure the composition, distribution, and temperature of the gas in groups like Hickson 16. Courtesy of ESA/XMM-Newton.

NUCLEOSYNTHESIS (J, K, L)

Similarly, soft x-ray absorption studies of our own interstellar medium will provide total abundances of the common elements C, N, O, Fe, Ne, Mg, and Si. This should be a great advance in sensitivity and reliability over optical observations and the capability of XMM-Newton. Observations of supernova remnants promise an even more dramatic demonstration of Con-X's spectroscopic prowess. Trace elements like P, K, Cl, Cr, and Mn can also be measured. Detecting these elements, and their ionization states, soon after they are formed will serve as a quantitative test of our understanding of how radiation escapes from these remnants as well as of the theory of nucleosynthesis inside supernova explosions. The most abundant elements can be assayed from supernovae occurring as far away as the Virgo cluster, which should be sufficient to derive a fair average rate of element production.

Another major design goal is to be able to map the distribution of the hot gas that has already been detected in the ~100 kpc dark matter halos of galaxies. Here, spectroscopy has a dual role. It will provide gas densities and abundances and, using the Doppler shift and Newton's laws, it will be able to determine the distribution of the dark matter. Used in conjunction with optical/ultraviolet observations of nearby galaxies, Con-X observations should be able to measure the abundances of elements created by massive stars and so provide a measurement of the distribution of star formation within nearby galaxies, which, in turn, provides a record of when these stars formed.

Perhaps the most spectacular demonstrations of Con-X's spectroscopic capability will come from observations of the coronas of nearby bright stars. These will measure thousands of emission lines with resolving powers greater than several hundreds and will allow the physical conditions and abundances within the coronas to be studied, putting our rapidly improving understanding of the solar corona to the test. In addition to allowing astronomers to detect trace elements essential to life, understanding star–planet relationships is highly relevant to the third quest, because stellar coronas are an important source of x rays and low-energy cosmic rays, whose intensity is a strong factor determining the habitability of planets.

BLACK HOLES AND NEUTRON STARS (E, F, G, I)

Con-X should greatly advance our understanding of black hole astrophysics because the accreting gas naturally emits x-ray photons just before it falls into the hole. As outlined above, astronomers have learned much over the past 5 years and anticipate further enlightenment over the next 5. A key capability is Con-X's large aperture and fine energy resolution, which will permit rapid measurement of the shape of the variable, strong iron emission and absorption lines that have already been detected from some Seyfert galaxies. These line features are imprinted upon the spectrum of the x-ray photons that the disk reflects from its own hot corona. Its shape is dictated by the rotation speed, the orientation of the disk, the bending of light rays by the strong gravitational field, the gravitational redshift, and the spin of the hole. Con-X will measure line profiles for a large sample of AGN and should determine the black hole spin rates as well as verify that space-time geometry is as predicted by Einstein's general theory of relativity (Figure 1.11).

More detailed information can be derived by observing the time variation of the iron line in response to changes in the strength of the x-

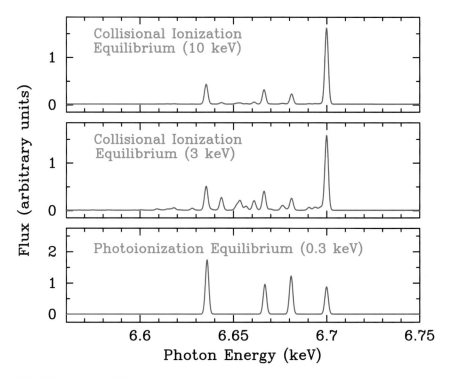

FIGURE 1.11 Detailed high-resolution simulation of an x-ray source that shows helium-like iron. Two cases of collisional excitation and one of photoionization are presented with associated temperatures. Note that the ratios of the lines are quite different in each case. In this way it should be possible to infer the physical conditions in the source. Con-X should be able to perform time-resolved spectroscopy of the inner regions of accretion disks orbiting massive black holes and to form indirect "imaging" of the gas flow. Courtesy of M. Sako, Columbia University.

ray photons from the corona. These variations are delayed by the time it takes light to cross the disk region (typically minutes to hours), and analyzing these variations offers another promising approach to making an indirect image of the flow of gas around the black hole. As it will be possible to carry out these studies on distant quasars as well as nearby Seyfert galaxies, astronomers hope to learn why only a minority of AGN form powerful radio jets.

These observations can then be related to the gas at larger radii, which is probed by measuring its emission and the absorption features imprinted upon the spectrum of the escaping x-ray photons. In particular, the hard x-ray capability of Con-X will be invaluable for finding heavily obscured AGN. Some of this gas will be associated with intense bursts of star formation occurring within the nucleus, and astronomers hope to clarify the relationship of these "starbursts" to the emission coming from directly around the black hole. In another approach, it will be possible to infer whether or not our own galactic center, which is currently dormant, was active over the past few centuries by observing the x-ray photons scattered back towards us by more distant molecular gas.

Analogous accretion disk studies will be possible for the stellar black holes. In addition, it will be possible to see x-ray line photons reflected off the companion stars and thereby obtain reliable mass estimates for many of the black holes that have so far stubbornly resisted this analysis. Another opportunity is to seek QPO variation in the emission lines, which, if detected, should greatly clarify their physical origin.

Isolated young neutron stars can be thought of as giant nuclei whose observable properties depend crucially on the basic principles of nuclear physics, general relativity, quantum electrodynamics, superconductivity, superfluidity, and so on. They have been studied assiduously over the past 30 years and yet astronomers are still discovering completely unexpected properties. Perhaps the largest impediment to understanding neutron stars in a quantitative manner is that astronomers still do not know how much pressure matter exerts when the density exceeds that of nuclear matter. It is this that determines the radius of a neutron star of a given mass. By measuring the spectrum of the x-ray emission from the surfaces of hot neutron stars, Con-X will be able to determine the gravitational redshift with high accuracy and, consequently, their radii and masses. These observations will also identify the surface composition, which is important for understanding how neutron stars interact with their environments and how they cool. Specifically, by accurately measuring the surface temperature and composition of a sample of neutron stars of known age, it will be possible to measure how fast they cool. This, in turn, sheds light on the detailed physics of dense matter. In addition, it should be possible to use these observations to map the surface magnetic field by, for example, measuring cyclotron lines.

FIRST-PRIORITY PROPOSED INTERMEDIATE MISSION: GLAST

MISSION DEFINITION

The Gamma Ray Large Area Space Telescope (GLAST)[8] is conceived as a successor to the highly successful Energetic Gamma-Ray Experiment (EGRET) on CGRO. It will observe gamma-ray photons with energies from 10 MeV to 300 GeV and an energy resolution, $E/\Delta E$, of ~50. Like EGRET, GLAST is a pair production telescope, but with six times the effective area, and by observing six times as much sky at any given time, it will be over 40 times as sensitive. It will position individual photons to within 2 deg at 100 MeV and 10 arcmin at 10 GeV as well as to ~2 μs in time. Images of bright, extended sources should have angular resolution of ~10 arcmin. Bright point sources can be located to 30 arcsec, which should allow the identification of some of the large number of sources that could not be located from EGRET observations.

Unlike EGRET, GLAST will rely on modern solid-state detectors, similar to those used successfully in particle physics. Silicon-based technology has been chosen for gamma-ray tracking.

GLAST is currently in phase A and is planned to enter phase C in 2002 in time for a 2005 launch. The phase C/D cost is $205 million, of which ~$50 million is expected to be contributed by DOE and the French and Japanese space agencies. A Delta launch into low Earth orbit costs $53 million and MO&DA for 10 years will be $48 million, for a combined total cost to the United States of $286 million.

BLAZARS (F, I)

GLAST will be able to detect several thousand AGN and observe them when the universe was roughly a billion years old (Figure 1.12). It will have the sensitivity to see rapid variations in a large number of these sources. Comparing these variations with those that occur at lower frequency is the best way to understand how jets are formed and what they contain. Already, large, multiwavelength campaigns have made it possible to construct a tentative model of how jets emit. Relativistic

---

[8]See <http://glast.gsfc.nasa.gov/>.

1 Year All-Sky Survey (E$_\gamma$ > 100 MeV)

FIGURE 1.12 Simulation of a 100-MeV all-sky survey by the proposed mission GLAST after 1 year's observation. Several thousand sources should be detected. The inset shows the region around 3C279 observed above ~1 GeV. Note that the positional accuracy improves with increasing energy. Courtesy of NASA/GLAST/GSFC.

electrons are accelerated at shock fronts formed inside the jet, and these electrons radiate x-ray photons by the synchrotron process. These x-ray photons are subsequently scattered by the same relativistic electrons, producing hard gamma-ray photons. As these jets propagate outward, they form, successively, optical, infrared, and radio synchrotron sources.

GLAST is designed to answer three outstanding questions: whether the jet plasma contains protons or positrons, how they are confined, and what the role is of the radiation field through which the jet propagates. Answering these questions should lead to a much better understanding of how jets are powered at their sources, especially whether this is due to spin of the black hole or the energy liberated by the accreting gas. Furthermore, by determining the evolution of the strengths of these jets over cosmic time, it should also be possible to check whether or not they dominate the gamma-ray background, as is widely suspected.

A comprehensive understanding of blazars requires a knowledge of the total gamma-ray spectrum extending to energies beyond the GLAST limit of 300 GeV. Here the detectors are ground-based and involve detecting atmospheric Cherenkov emission. Although it is not part of its charge, the panel regards the interplay between space-based and ground-based detectors as so strong that it endorses the proposal to construct the Very Energetic Imaging Telescope Array System (VERITAS).

COSMIC RAYS (J, K, L)

Hard gamma-ray photons should also be emitted by the cosmic rays that are believed to be accelerated within supernova remnants. In particular, GLAST should observe the gamma-ray photons produced through the decay of neutral pions around ~70 MeV and demonstrate, conclusively, that the majority of the cosmic rays with energies in the range ~1 GeV to ~100 TeV are accelerated in these sites. GLAST will also have the angular resolution to determine where this acceleration is taking place, in particular to determine if the ions are accelerated mainly at strong shock fronts or throughout the body of the remnant. Comparison with radio and x-ray observations will enable a fairly comprehensive description of particle acceleration to be developed. Because cosmic rays account for such a large fraction of the energy in a supernova remnant, they must be included in the dynamical analysis that underpins the nucleosynthetic investigations. On a larger scale, it will be possible to use these types of observation to observe the overall consequences of particle acceleration within nearby galaxies and even clusters of galaxies.

A quite different type of cosmic-ray decay might be detectable. One of the strongest candidates for dark matter is a new elementary particle called a neutralino—the lightest "supersymmetric" partner to the normal particles. These particles may not be stable and, according to some theories, may decay over cosmological timescales into gamma-ray photons with a particular energy in the GLAST range. If this gamma-ray line were detected, it would have major implications for physics as well as astrophysics.

Gamma-ray Bursts (C, H, I)

The discovery by Beppo-SAX in 1997 of afterglows from GRBs has revolutionized our understanding of these cataclysmic events. Their incredible power plus the multiwavelength nature of the afterglows has captured the public imagination and led more astronomers to work on understanding what mechanisms are responsible for GRBs.

GLAST has a central role in the future of this research because GeV photons are strongly susceptible to absorption—they create an electron-positron pair within the source, just as they do in the detector. As a result, observing GeV gamma-ray photons from nearby bursts (in contrast to the ≤ 1 MeV photons detectable by Swift) places strong constraints on the physical conditions within the source region. There are already indications that the gamma-ray burst emission originates from relativistic jets that move even faster than those associated with blazars. Based on the EGRET detections, it is projected that GLAST, with its greater sensitivity and much shorter dead time, will see a GRB about once every 2 days, quite possibly including bursts from the first generation of stars, although these can only be seen at lower energy. GLAST seems to be a particularly good candidate to produce the key insight that will identify the predominant source of GRBs.

Above a certain energy, gamma rays from GRBs and blazars are also subject to absorption outside their sources as they propagate through the infrared cosmological background radiation. By measuring this absorption and (in the case of blazars) combining it with hard x-ray spectra that EXIST (see below) can measure, GLAST should be able to measure the infrared background when the universe was quite young. This infrared background is believed to be dominated by reprocessed stellar light (although, as discussed above, AGN may contribute as much as one-half of it), so these gamma-ray observations can monitor the history of star formation in the universe. An understanding of the absorption on the infrared background is also needed to measure the intrinsic high-energy spectrum of gamma-ray bursts.

SECOND-PRIORITY PROPOSED INTERMEDIATE MISSION: EXIST

MISSION DEFINITION

An imaging survey of the hard x-ray sky is our second candidate for a moderate mission. There has been no hard x-ray survey of the whole sky to match the existing ROSAT soft x-ray survey since that performed by the High-Energy Astronomical Observatory (HEAO-1) satellite. The International Space Station-attached EXIST[9] would carry out this survey using eight wide-field (~40 deg), coded-aperture, hard x-ray (~5 to 600 keV) telescopes with good energy resolution ($E/\Delta E \sim 100$) that image the whole sky every 90-min orbit. (This is a particularly important feature because the hard x-ray sky is so variable and is enabled by having eight telescopes.) The final survey limit would be roughly 100 to 1000 times fainter than the HEAO-1 limit and roughly 10 times fainter than the anticipated Swift hard x-ray survey (with a much broader energy range and superior angular resolution). EXIST would provide ~30 arcsec source localization for bright sources and angular resolution of ~5 arcmin for bright extended sources as well as ~1 μs photon timing. The International Space Station provides a nearly ideal platform for this fixed-pointing, scanning telescope (although it could also be a free flyer). A single telescope could be prototyped on an ultralong-duration balloon flight.

This project is enabled by recent advances in hard x-ray Cd-Zn-Te detectors (which are also used for medical imaging). The most difficult technology challenge is to construct ~1 m² arrays of these detectors for each telescope. Other challenges concern data handling and integration with the International Space Station.

EXIST was selected as a New Mission Concept in 1994, and the proposal has developed considerably since that time. An EXIST science

---

[9]See <http://exist.gsfc.nasa.gov>.

working group has been formed. Because of its strong synergy with Con-X and GLAST, a 2005 launch would be optimal. The projected phase C/D cost is $120 million, and the MO&DA for a 2-year mission is $30 million, for a combined estimated cost of $150 million.

Obscured AGN and the X-ray Background (D, E, I)

Perhaps the most compelling survey science is the first deep survey of AGN above ~20 keV. There are already strong indications from ASCA and Beppo-SAX that many AGN are heavily obscured. EXIST should discover more than 3000 such self-absorbed AGN, which can then be subjected to deep follow-up study. (These sources are probably quite variable, and EXIST is well suited to monitor them.) This will be necessary to understand the source composition of the hard x-ray background (at around 40 keV, where it is most luminous). In addition, EXIST observations will facilitate a secure calculation of the contribution of accreting black holes in AGN to the "luminosity density" of the universe and a direct comparison with the galaxy luminosity density that should be measured by the Next Generation Space Telescope (NGST). Further comparison with the distribution of black hole masses in dormant galactic nuclei will lead to a quantitative understanding of the evolution of black holes in different types of galaxy.

Galactic Survey (E, G, J)

With its very large field of view, EXIST is well suited to detect the "soft x-ray transients" that are usually associated with black holes. Studying these objects and understanding why they behave as they do will lead to a census of stellar mass black holes within our galaxy. In addition, its good energy resolution should allow EXIST to perform the first high-sensitivity, high-energy galactic search for supernovae that are hidden inside molecular hydrogen clouds by seeking gamma-ray emission lines from radioactive titanium. If these lines are seen, they should help us to understand the overall supernova rate and relate this to the theory of advanced stellar evolution.

Sensitive, hard x-ray observations of accreting neutron stars should measure their magnetic fields. In addition, it will be possible to observe QPOs from the disk coronas around neutron stars and stellar black holes at the high energies where they are most prominent.

Gamma-ray Bursts (C, F, H, J)

EXIST is also well matched to the study of GRBs. With a projected sensitivity 20 times better than CGRO and 4 times better than Swift, EXIST should detect GRBs every 4 hours and furnish arcminute positions for all and ~10 arcsec positions for the brightest cases. If there is a large population of low-power GRBs associated with supernovae, like the recent example 1998bw, then EXIST should find them. It also has the sensitivity to detect bursts from when the universe was less than a billion years old, and this will provide a direct probe of early star and galaxy formation. EXIST can time tag each photon with microsecond accuracy, which will be a useful diagnostic tool for exploring the kinematics of the expanding, ultrarelativistic blast waves and jets.

INVESTING FOR THE FUTURE

The missions that have just been described are large steps towards the long-term goals with which this report began. To go further requires investing in selected technologies that will be needed by missions that could be launched after 2010. The panel has identified three important areas where there has recently been considerable progress and where the prospects for future advances seem particularly good. In order of priority, they are MAXIM, Generation-X, and the MeV Spectroscopy Mission.

MAXIM (E, F)

The second quest, imaging a black hole, could succeed, in principle, using x-ray interferometry, as proposed for MAXIM[10] (Figure 1.13).

Success would require ~0.1 µarcsec angular resolution, a seven-order-of-magnitude improvement over Chandra. At first sight, this seems unrealistically ambitious. However, new interferometer designs, using grazing incidence mirrors, suggest a clever way of using widely separated spacecraft to form an interferometer in a manner that will combine photons with different energies and accommodate source variability. One spacecraft holds the mirrors, which are separated by a fixed

---

[10]See <http://maxim.gsfc.nasa.gov/>.

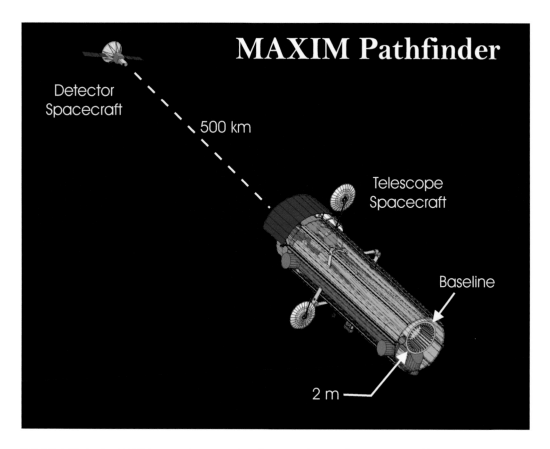

FIGURE 1.13 In the MAXIM approach to x-ray interferometry, beams of x rays separated by several meters are formed by the "optics spacecraft." The x-ray photons in these beams are then detected by a second "detector spacecraft" roughly 500 to 1000 km away. Ultimately, it is proposed to use x-ray interferometry to form images of gas flow around a black hole. Courtesy of NASA/MAXIM/GSFC.

"baseline" as large as ~100 m. These mirrors reflect the x rays onto a second spacecraft up to 1000 km away, where the interference fringes are formed.

One possible intermediate goal is a pathfinder mission designed to demonstrate ~100 µarcsec resolution at ~1 keV, comparable to what is achieved at radio wavelengths using very long baseline interferometry (VLBI). In order to have an adequate flux of x-ray photons, a collecting area of 100 cm^2 and a baseline of ~2 m are required. The detector would have to be ~500 km away. The biggest technology challenges for such a pathfinder mission include developing x-ray optics based on large optical flats and fabricating large, two-dimensional cryogenic detector arrays with sufficient energy resolution. The pointing ($\sim$30 μarcsec) and metrology ($\sim$0.1 nm) demands are similar in character to those of the Space Interferometry Mission (SIM) and the Laser Interferometer Space Antenna (LISA), and much of the technology should be transferable.

GENERATION-X (A, I)

Attaining the first goal of seeing directly the very faint first galaxies and stars requires much larger collecting areas to capture enough photons to form an image or attempt spectroscopy; x-ray detectors are already approaching 100 percent efficiency. Furthermore, to make the identifications, it will be necessary to localize sources to within an arcsecond. These requirements motivate the fabrication and deployment of x-ray mirrors with effective areas exceeding $\sim$100 m^2, a hundred times larger than Con-X and comparable to a single Keck telescope. Large-format detectors capable of sub-eV energy resolution will also be necessary.

A paced program of mirror and detector technology development directed toward these long-term goals is recommended.

MeV SPECTROSCOPY MISSION (J, L)

As discussed above, the measurement of nuclear gamma-ray lines provides a direct probe of the formation of the elements in supernovae. Most of the lines have MeV energies, and measuring them poses an unusual instrumental challenge because it is not possible to use focusing optics. The INTEGRAL mission will be an important pathfinder for this field. It appears that in order to attain its scientific goals, it will be necessary to surpass INTEGRAL in sensitivity by a factor of roughly 30. The panel recommends support for technology leading to a future mission with sensitivity matched to our scientific goals for reasonable cost.

SMALLER PROGRAMS

A healthy high-energy astrophysics program must embody a balance between larger and smaller projects. Accordingly, the panel also endorses four smaller programs.

POTENTIAL EXPLORER RESEARCH

There is strong support in the high-energy astrophysics community for the Explorer program, as demonstrated by the large number of exciting proposals submitted by high-energy astrophysicists. As the central purpose of this program is to encourage innovation and rapid responses to emerging scientific opportunities, the panel does not endorse any specific mission proposals. However, three exciting research areas appear to be particularly well suited to Explorer missions.

NUCLEAR LINE X-RAY SPECTROSCOPY (J, L)

Recent advances in multilayer coatings used at grazing incidence make it possible to concentrate soft gamma-rays onto a small detector, greatly reducing the background and allowing unprecedented sensitivity with a much more modest instrument than previously imagined. Several important nuclear lines occur in the accessible energy range, below ~200 keV, including three isotopes of Ti, Co, and Ni that are stringent diagnostics of the explosion mechanism for type II supernovae. Many supernovae could be observed in this manner, providing unique diagnostics of how elements are created and disseminated.

SOFT X-RAY SURVEYS (L)

A systematic low-energy survey of a substantial area of sky to depths significantly beyond that investigated by the all-sky survey ROSAT could map distant clusters and the hot structures surrounding them to probe the missing baryons. Spectroscopic surveys of diffuse emission offer another approach to observing the intergalactic medium that would complement absorption line measurements by the large observatory missions, and they could also help determine the nature of the hot interstellar medium in our own galaxy.

ALL-SKY MONITORS (E, F, H)

Soft x-ray transients, microquasars, and AGN are all highly variable. Continuous monitoring of large numbers of these sources at low and intermediate x-ray energies will complement the high-energy studies proposed above using EXIST and help produce an indirect image of accretion disks and their coronas. Continuous monitoring of a large fraction of the sky should also discover new and important rare phenomena. "Lobster eye" optics promises an order of magnitude gain in sensitivity, enabling thousands of AGN, binary x-ray sources, variable stars, and so on to be monitored simultaneously.

ULTRALONG-DURATION BALLOONING

There is a good history of performing high-energy astrophysics from balloons. A particularly fine example was the rapid response to supernova 1987a, which exploded in the Large Magellanic Cloud in 1987. Hard x-ray and gamma-ray missions were successfully mounted from the Southern Hemisphere and were able to confirm directly the production of radioactive nuclei within the expanding supernova remnant. Since that time the capabilities of balloons have increased considerably. The ultralong-duration balloon (ULDB) program offers the prospect of carrying several ton payloads to ~40 km altitudes for flights of several months' duration. The panel recognizes that this capability offers relatively cheap access to near space for certain classes of hard x-ray and gamma-ray payloads and urges that this capability be developed further. It is noted that ballooning is particularly well matched to the needs of the best younger scientists, who require experience in building and flying instruments with relatively rapid turnaround. Additional support of roughly $5 million per year is needed for the development and operation of the balloons.

LABORATORY ASTROPHYSICS

High-energy astrophysics missions will return spectroscopic data over the next decade with unprecedented breadth and detail. This will take us into virgin territory largely unexplored by theoretical chemists and experimental astrophysicists. In addition, the laboratory study of high-energy-density fluid dynamics, magnetohydrodynamics, and plasma physics appropriate to the interpretation of supernovae, jets, and GRBs is in its infancy. There is need for an increased, though still comparatively small, investment (the panel estimates roughly $2 million per year for high-energy astrophysical studies) in both computational modeling and experimental facilities in order to make the best scientific use of the new observatories. A cross-agency initiative involving the Department of Energy (DOE), NASA, and the National Science Foundation (NSF) is recommended.

THEORETICAL CHALLENGES

The contributions of theory to the development of high-energy astrophysics are legion. Black holes, neutron stars, cosmological GRBs and relativistic blast waves, supernova nucleosynthesis, gamma-ray jets, cosmic-ray acceleration in supernova remnants, and the hot intergalactic medium were all widely discussed in the theoretical literature before observations established their reality. However, what was found was not usually exactly what had been predicted, so the theory had to be modified. To continue this symbiotic relationship, the panel proposes that each of its three recommended missions sponsor a particular theory challenge (see further discussion in Chapter 6) in order to refine mission planning and to obtain the best scientific return. The expenditure on this program should be matched to the perceived benefit, which may vary from mission to mission. Candidate challenges for Con-X, GLAST, and EXIST are those designated E, F, and G, respectively, and elaborated upon in Chapter 6.

POLICY ISSUES

The panel identified three policy issues for which it makes focused recommendations.

LONG-TERM SCIENTIFIC SUPPORT FOR OBSERVERS

The NASA data support centers serve the critical function of writing and maintaining the data analysis software needed for the dissemination, interpretation, and archiving of high-energy data. The concomitant large pool of scientists creates a critical mass for broad scientific investigations. The role of academic researchers in high-energy astrophysics has historically been to propose single observations, and they receive incremental funding to carry out the specific observations at hand. However, the lack of long-term, stable funding makes it difficult to create critical masses of researchers in universities similar to those in other countries, and it is also hindering the training of graduate students, a potential loss for the field. Accordingly, the panel recommends that NASA invest in a small number of focused, high-energy astrophysics data analysis groups within universities and research institutes. In particular, computing facilities and an appropriate mix of graduate students and postdoctoral scientists needs to be supported.

JUNIOR FACULTY INSTRUMENTATION PROGRAM

Space missions are becoming increasingly complex and their organization concentrated in a few centers. This has the unfortunate side effect that instrument and spacecraft development is becoming remote from most universities, in particular from graduate students, who are eager to enter the field at a time when there are great opportunities and a need for new blood. As evidence for this, the panel cites the apparent shortage of U.S.-trained postdoctoral scientists with experience of high-energy instrumentation. To alleviate this problem, the panel proposes that NASA initiate a modest program specifically directed at supporting a small number of carefully selected junior faculty working in high-energy instrumentation.

EDUCATION AND PUBLIC OUTREACH

In recent years, high-energy astrophysicists have established an enviable record for responding to a strong public interest in their discipline. Black holes, supernovae, and gamma-ray bursts have captured the public imagination like few other topics in the physical sciences and are at least as firmly established as cosmology and the search for extraterrestrial life. High-energy astrophysicists have experimented successfully with a variety of new education and outreach initiatives, including the High-Energy Astrophysics Learning Center and the Astronomy Picture of the Day. Individual missions such as Chandra, Swift, GLAST, and Con-X maintain interactive Web sites and have distributed compact disks and brochures widely. In recent months the early-release images from Chandra have been broadly accessed and disseminated in the news media. The panel recognizes the importance of these outreach activities and recommends that they continue to be encouraged, financially supported, and rewarded in a manner that is described in more detail in the survey committee report.

ACRONYMS AND ABBREVIATIONS

ACIS—Advanced X-ray Astrophysics Facility (now Chandra) CCD
   Imaging Spectrometer
AGN—active galactic nuclei
ASCA—Advanced Satellite for Cosmology and Astrophysics mission
   (Japan)
ASTRO-E—fifth in a series of Japanese x-ray astronomy satellites; with
   help from the United States
AURA—Association of Universities for Research in Astronomy, Inc.
Beppo-SAX—Satellite per Astronomia X, a European collaboration x-ray
   mission
CCD—charge-coupled device
CDM—cold dark matter
CGRO—Compton Gamma-Ray Observatory
Chandra—Chandra X-ray Observatory (NASA, launched in 1999)
CXC—Chandra X-ray Center at the Smithsonian Astrophysical
   Observatory
DOE—Department of Energy
EGRET—Energetic Gamma Ray Experiment aboard CGRO
EPIC—European Photon Imaging Camera (on XMM-Newton)
ESA—European Space Agency
ESO—European Southern Observatory
EXIST—Energetic X-ray Imaging Survey Telescope
FUSE—Far Ultraviolet Spectroscopic Explorer
GLAST—Gamma-ray Large Area Space Telescope
GRBs—gamma-ray bursts
HDF—Hubble Deep Field
HEAO-1—High-Energy Astronomical Observatory
HETE-2—High-Energy Transient Explorer (launched in 2000)
HST—Hubble Space Telescope
INTEGRAL—International Gamma-Ray Astrophysics Laboratory
ISS—International Space Station
LISA—Laser Interferometer Space Antenna
MAXIM—Microarcsecond X-ray Imaging Mission
MO&DA—mission operations and data analysis (NASA)
NASA—National Aeronautics and Space Administration
NGST—Next Generation Space Telescope
NSF—National Science Foundation
OSS—Office of Space Science (NASA)

QPOs—quasi-periodic oscillations, x-ray pulses from compact objects
ROSAT—Roentgen Satellite (German-U.S.-U.K. collaboration)
RXTE—Rossi X-ray Timing Explorer
SAO—Smithsonian Astrophysical Observatory
SIM—Space Interferometry Mission
STScI—Space Telescope Science Institute
ULDB—ultralong-duration balloon
VERITAS—Very Energetic Radiation Imaging Telescope Array System
VLBI—very long baseline interferometry
XMM-Newton—X-ray Multi-Mirror Observatory, a European
    collaboration x-ray space mission

2

Report of the Panel on Optical and Infrared Astronomy from the Ground

SUMMARY

As we cross the threshold of the new millennium, astronomy with ground-based optical and infrared (O/IR) telescopes will continue to play its fundamental role in shaping our understanding of the workings of the universe, enriching the golden era of discovery that astronomy has enjoyed in the last decades. As a result of past investments in astronomical facilities, the United States led the world in observational research throughout the 20th century. Our nation has the talent, the knowledge, and the resources to carry this great tradition of leadership into the 21st century, building on a generation of powerful 8-m-class telescopes and anticipating future telescope facilities of unprecedented power and resolution.

However, state-of-the-art, ground-based O/IR facilities have grown in scale and complexity so that a new paradigm is needed that balances diversity and coordination. This paradigm focuses effort on unique and complementary capabilities and will enable the efficient development and operation of the next generation of facilities together with the effective use of existing ones. Establishing a common vision within the astronomy community of how these facilities should evolve is the foundation of the recommendations of the Panel on Optical and Infrared Astronomy from the Ground for the coming decade. In this context, the panel proposes three initiatives to encourage the evolution of U.S. O/IR ground-based facilities as a system, by combining and coordinating the assets and efforts of federally funded and independent observatories:

  • A next-generation, giant-aperture, adaptive-optics-equipped telescope whose spatial and spectral resolution will enable the unraveling of complex physical processes in the first galaxies, in nearby planetary systems, and in newborn stars. A unique opportunity exists to bring together federal and independent observatories to build and operate this facility.

  • A large-aperture, very-wide-field synoptic survey telescope that will search the solar system for its ancient materials and open a new time window on astronomical phenomena. This facility has particular resonance with the new role envisioned for the National Optical Astronomy Observatories (NOAO).

  • An enhanced instrumentation program for independent observatories that capitalizes on and encourages the significant investment of nonfederal funds, both to maximize the operation of U.S. facilities as an efficient system and to increase public access to these facilities.

The panel has translated these generalized initiatives into concrete recommendations and prioritized them.

MAJOR INITIATIVE, PRIORITY ONE: GSMT

Develop the technology to build a giant (30-m class) segmented-mirror, adaptive-optics-equipped, ground-based O/IR telescope (GSMT) and begin its construction within the decade.   With diffraction-limited performance down to at least 1 μm, an order-of-magnitude increase in light-gathering power, and a factor-of-4 gain in spatial resolution, GSMT will enable breakthrough science in studies of star and planet formation, stellar populations, and early galaxy evolution. The GSMT's spatial resolution of 14 milliarcsec at 2 μm and its high spectral resolution in the near-infrared region will significantly exceed the performance of the Next Generation Space Telescope (NGST, scheduled for launch in 2008), providing an important complementarity such as that developed between the Hubble Space Telescope and the Keck telescopes. Furthermore, with the ability to add new instrumentation to GSMT, its capabilities can evolve in response to scientific advances in the early NGST-GSMT era, making it more productive and developing the scientific case for even more advanced facilities on the ground and in space. The GSMT will push relevant technology such as adaptive optics (AO) to its limits, toward what could be the ultimate ground-based telescope in the decades to come. The panel recognizes that, even with anticipated innovation in design and technology, construction of this facility requires an enormous investment of resources, perhaps exceeding $400 million; the operating costs will be similarly large over the facility's lifetime.[1] In response to the challenge of garnering such huge resources, the panel emphasizes its belief that GSMT offers a golden opportunity for partnership between national and independent observatories. To assure the maximum science return, it is essential that a broad scientific community have

---

[1]The estimated costs for ground-based initiatives that appear in the survey committee report (Astronomy and Astrophysics Survey Committee, National Research Council. 2001. *Astronomy and Astrophysics in the New Millennium*, Washington, D.C.: National Academy Press) include instrumentation, grants, and operations, as described in the preface. These costs for the GSMT are estimated be about $200 million.

access to GSMT; the panel believes that public access should be maximized within the constraints of available funding and that the partnership between the public and private components of U.S. O/IR ground-based astronomy should be strengthened.

MAJOR INITIATIVE, PRIORITY TWO: LSST

Build a large-aperture (6.5-m class), very-wide-field (~3 deg) synoptic survey telescope (LSST) to produce a periodic digital map of the sky. Its unique combination of large aperture and wide field (~ 10 deg^2) will allow LSST to map the entire sky down to 24th magnitude in a few days. Such capability will enable a wide-area variability experiment (WAVE), a finite-duration project that will accomplish many important scientific goals through a small number of simple survey modes. For example, WAVE on LSST will do the following:

- Discover and track 10,000 objects in the Kuiper Belt, a largely unexplored, primordial component of our solar system.
- Locate potentially threatening near-Earth objects (NEOs) down to 300 m in size.
- Discover and monitor many kinds of variable objects, including supernovae, active galactic nuclei (AGN), and microlensed stars.
- Produce extremely deep images over hundreds of square degrees for studying the distribution of dark matter through weak gravitational lensing.

More than just a telescope, LSST with WAVE will make important strides in data processing, data mining tools, and archiving components and will play a key role in the National Virtual Observatory (NVO) described in the survey committee report. Its data product will have widespread application to all fields of astrophysics and will have enormous educational potential by virtue of its ability to produce a "living sky" that can be downloaded in the classroom. While this unique, state-of-the-art facility could capitalize on the complementary strengths of independent observatories, its mode of operation and data product would make LSST an ideal undertaking for NOAO in its developing new role.

MODERATE INITIATIVE, PRIORITY ONE: TSIP

Support the Telescope System Instrumentation Program (TSIP) to foster the more coherent development of public and independent telescope facilities and to increase public access. By substantially increasing its funding of instrumentation for the new generation of large-aperture telescopes at independent observatories, the NSF would encourage the continuation of substantial nonfederal investments, leverage their scientific productivity, and open up new observing opportunities for the entire U.S. astronomy community. The panel therefore proposes TSIP, which would fulfill this critical need and encourage the evolution of U.S. ground-based O/IR facilities as a coherent system. Particularly in an era of enormous investment by the European Southern Observatory, the systemization of all U.S. resources is essential to maintain leadership in the field. Leadership in astronomy is important not only to the discipline itself, but also to the vital role that astronomy plays in improving the scientific literacy of the public.

SCIENCE OPPORTUNITIES

ANSWERING FUNDAMENTAL QUESTIONS

The world's astronomy community has built powerful tools with which to answer fundamental questions about the birth of galaxies, stars, and planets and to explore the most exotic phenomena in the universe. These tools include (1) a new generation of ground-based O/IR telescopes, (2) the powerful new millimeter-wave arrays—the Combined Array for Research in Millimeter-wave Astronomy (CARMA) and the Submillimeter Array (SMA), with the Atacama Large Millimeter Array (ALMA) to come later in the decade, and (3) the Hubble Space Telescope (HST) and Chandra X-ray Observatory, now in operation, the Space Infrared Telescope Facility (SIRTF) and the Space Interferometry Mission (SIM), on the way, and the Next Generation Space Telescope (NGST), to come. Properly instrumented and supported, these facilities will provide unprecedented opportunities to solve many of the mysteries that 20th century astronomical exploration uncovered. In this chapter, the panel recommends two additional ground-based O/IR facilities—a 30-m-class telescope (GSMT) and a large-aperture synoptic survey telescope (LSST)—which, in concert with the above facilities, will enable astronomers to accomplish the following goals over the next decade or two:

• Describe the complete cosmological state of the universe with better than 10 percent accuracy. Using type Ia supernovae and other standard candles, gravitational lenses, and the Sunyaev-Zel'dovich effect, test the Friedmann-Robertson-Walker model back to $z \sim 3$. Is the universe accelerating as a result of dark energy?

• Follow the history of star formation and chemical evolution over all of cosmic time. Find and characterize the first generation of stars in the early universe and relate these to the oldest stars in our galaxy and its neighbors. Chart chemical evolution in stars and the interstellar medium through star-by-star studies in nearby galaxies and integrated galaxy-light measurements back to the earliest galaxies. Balance the baryon budget: describe the location and nature of all the baryons through cosmic time.

• Test the hierarchical model of galaxy formation: observe and analyze the assembly of galaxies from the earliest star systems. Follow the buildup of large-scale structure from reionization to the present day and connect the evolution of individual galaxies to their environment within large-scale structure. Map the dark matter and galaxy distributions with sufficient precision to understand the role of dark matter in galaxy formation and obtain clues to its nature.

• Connect the formation of supermassive black holes to galaxy formation and evolution. Discover the epoch of black hole formation, the dynamics of the process, and the relation to star formation in the nuclei of galaxies. Understand better the physical processes in active galactic nuclei and solve the riddle of gamma-ray bursts.

• Examine in detail the processes of star and planet formation. Describe the accretion process that forms a star, including the physical properties of the raw material and the dynamic processes from molecular cloud formation to nuclear ignition. Study the evolution of planetary disks and observe the building of planets around nearby stars, from Kuiper Belt objects (KBOs) at the 100-AU zone to the 1-AU zone, where Earth-like worlds could be formed.

• Analyze the surviving building blocks of the solar system. Map the Kuiper Belt in our solar system and measure the physical characteristics of KBOs as examples of the primordial material that built the planets. Identify all NEOs whose impact with Earth could have catastrophic consequences.

• Take a census of the planetary populations around other stars. Understand the relationship between brown dwarfs and planets and the distribution of giant planets in neighboring systems. Search for worlds that could, or already do, support life as we know it.

These are ambitious goals, and ground-based O/IR facilities will play a substantial, often crucial, part in achieving them. Despite the large foreign investments around the world in ground-based astronomy, the United States is well positioned to maintain its leadership role in this field.

EXPLOITING THE DIVERSE, UNIQUE FACILITIES OF U.S. GROUND-BASED O/IR ASTRONOMY

Astronomers now probe the physics of exotic and extreme environments using observations across the electromagnetic spectrum, ranging from radio waves to gamma rays and everything in between. Ground-based O/IR research remains at the heart of this endeavor. Ground-based spectroscopic observations have been crucial to research with the HST for such diverse topics as high-redshift galaxies, the extragalactic distance scale and cosmological parameters, and stellar populations in the Milky Way and other nearby galaxies. The study of radio galaxies relies on a combination of radio, optical, and infrared (IR) data, as does the goal of linking the evolution of the interstellar medium to stellar evolution. The x-ray halos of galaxies and the hot interstellar gas in rich clusters are being understood thanks to both space-based observations and ground-based optical data, and the recent discovery that gamma-ray bursts take place in distant galaxies, making them the most energetic known phenomenon, is the result of hard-won spectroscopic data from the Keck telescopes and rapid-response imaging from a variety of ground-based instruments.

The ambitious scientific program outlined above requires a broad suite of telescopes with a range of aperture sizes and powerful state-of-the-art instrumentation employing the latest array detectors. Ground-based O/IR facilities available to U.S. astronomers include both national and independent installations in a unique combination that has kept U.S. astronomy strong. Remarkably, most of the glass resides at independent observatories, but the National Science Foundation's (NSF's) role is nevertheless vital to the health of the entire enterprise: NSF not only supports NOAO and Gemini but also plays a crucial role at the independent observatories by virtue of its grant support for research and instrumentation.

Strong competition for leadership in astronomy is now coming from the European Southern Observatory (ESO), which has made huge investments at two Chilean observatories. The most recent of the investments was for the Very Large Telescope (VLT), four 8-m telescopes at

Cerro Paranal costing approximately $700 million. Eight first-generation instruments are under construction at a cost of approximately $90 million. An ongoing instrumentation program funded at $10 million per year is anticipated to provide state-of-the art facilities throughout the decade. A world-class interferometer is also under construction for the VLT. Early results from the VLT are impressive. In addition, ESO has announced its intention to build a truly enormous telescope, currently planned to have a 100-m aperture; this billion-dollar-plus initiative is unprecedented in the history of ground-based astronomy.

To meet this challenge, the United States must use its unique combination of federal, state, and private resources to best advantage. The spirit of independence and individual initiative that has characterized U.S. ground-based O/IR astronomy should continue, for it has had highly productive and creative results. But federal resources are sufficiently scarce, and the need at both national and independent observatories so acute that a greater degree of cooperation is urgently needed. Put simply, the suite of U.S. observatories should function as a coherent system to ensure that U.S. astronomers will have the means to participate fully in pursuing the fundamental goals outlined above. Although the recommended new facilities and programs described below are completely justified by scientific arguments alone, the panel sees as equally important the additional overarching goal of strengthening the system and creating a process for its development. To reach this goal, NSF and other agencies that fund O/IR astronomy should appreciate all the elements of the U.S. O/IR system and implement policies that guide the system's development so that federal funding will achieve maximum science and maximum opportunity.

Implicit in this new goal is a changing role for NOAO. The McCray report, *A Strategy for Ground-Based Optical and Infrared Astronomy*,[2] and the AURA-sponsored study on the future of NOAO[3] emphasized that to be a more effective component of the U.S. ground-based O/IR system, NOAO must have as its first priority to represent the entire U.S. astronomy community, carrying on activities that benefit all. These activi-

---

[2]Panel on Ground-Based Optical and Infrared Astronomy, National Research Council. 1995. *A Strategy for Ground-Based Optical and Infrared Astronomy* (Washington, D.C.: National Academy Press); also known as the McCray report after the panel's chair, Richard McCray.

[3]Association of Universities for Research in Astronomy, Inc./NOAO. 2000. *Building the Future: NOAO Long Range Plan: 2001-2005* (Washington, D.C.: AURA).

ties will include providing the scientific leadership and technical expertise needed for building the largest facilities, identifying and providing complementary capabilities that support the suite of large telescopes, and representing U.S. interests in efforts that develop as collaborations, either between public and private institutions or with international partners. In the future, NOAO will probably run unique facilities in preference to nonunique ones; as well, it could play an important, multifaceted role in coordinating the entire suite of U.S. ground-based O/IR facilities.

MAJOR INITIATIVE, PRIORITY ONE: DEVELOP AND BUILD A NEXT-GENERATION GROUND-BASED TELESCOPE (GSMT)

MISSION DESCRIPTION

The panel recommends that highest priority be given to the design of a giant (30-m class) segmented-mirror, AO-equipped, ground-based O/IR telescope (GSMT), with the goal of beginning construction before the end of the decade. The GSMT will be a filled-aperture, diffraction-limited telescope with atmospheric correction by AO down to at least 1 μm. It will achieve order-of-magnitude gains over any extant ground-based O/IR telescope and will provide substantial gains in spatial resolution and near-IR high-resolution spectroscopy even over NGST, for which it will be an essential complement (Figure 2.1). The facility will achieve substantial breakthroughs in the science covered in the NASA Origins theme. Especially in view of the ESO initiative to build even larger, more ambitious ground-based telescopes over a substantially longer timescale than this decade, the successful development of a next-generation ground-based telescope, accessible to all U.S. astronomers and contemporaneous with NGST, is essential for maintaining the tradition of U.S. leadership in astronomy. The ESO proposal for a 100-m-class telescope would offer even more spectacular gains for many kinds of observations, but it is the opinion of the panel that the proposal is too ambitious for the current decade and that an intermediate step, to a 30-m telescope, would be optimal in terms of science, technology, and allocation of resources. Of particular importance, the panel believes, is that a next-generation, ground-based facility be available during the lifetime of NGST.

The GSMT will operate over the wavelength range from 0.3 to

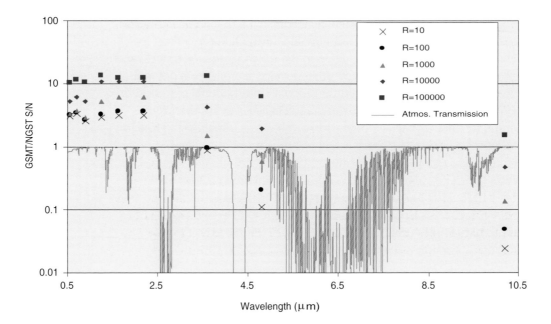

FIGURE 2.1 The performance of a 30-m ground-based telescope (GSMT) is compared with that of NGST for point sources for a number of spectral resolutions over a range of wavelengths. The vertical axis is the ratio between the S/N achieved for an observation of a given duration (for an object much fainter than the sky) by GSMT and the S/N achieved by NGST. Diffraction-limited image quality at all wavelengths and an emissivity of 10 percent are assumed for GSMT. NGST's advantage beyond 4 μm is likely to be even greater owing to improvements in detector dark currents and read noise. Superposed is a line showing atmospheric transmission. The comparison shows that GSMT is substantially more effective than NGST in obtaining moderate- to high-resolution spectra of faint compact objects, especially below 2.5 μm. NGST is more effective at longer wavelengths, at wavelengths blocked by the atmosphere, and for observations done at low spectral resolution. It should also be noted that no spectroscopic capability is planned for NGST below 1.0 μm, nor is GSMT expected to deliver diffraction-limited images below 1.0 μm. Courtesy of L. Ramsey, Pennsylvania State University, 1999.

25 μm, with a field of view (FOV) of ~20 arcmin and expected diffraction-limited images over a ~1 arcmin field ranging from ~8 milliarcsec at 1 μm to 0.2 arcsec at 25 μm and with seeing-limited performance of ~0.5 arcsec in the UV. The telescope will provide diffraction-limited performance with AO for λ ≥ 1.0 μm. Its high spatial resolution and powerful spectroscopic capability will be a true quantum leap over any other existing or planned U.S. facility. At the same time, the immense aperture

of the telescope will, in and of itself, make the GSMT a uniquely powerful facility for partially corrected or uncorrected observations.[4]

SCIENCE WITH THE GSMT

The need for very high spatial resolution and moderate to high spectral resolution, with dramatically increased sensitivity, drives the development of GSMT. When steps of the size proposed are taken, it is often the case that the greatest impact will come from discoveries that cannot be predicted—such is the nature of exploration and discovery in astronomy—and that are a major source of the excitement that surrounds a powerful new facility of this kind. Nevertheless, the panel sees many unique opportunities for GSMT to help answer *today's* leading scientific questions. Summarized below are a few of the exciting possibilities.

STAR AND PLANET FORMATION

The development of the theory of stellar structure and evolution was one of the greatest achievements of 20th century science. Yet this elegant theory, which explains the life cycle of stars, is incomplete in one critical aspect: it does not predict or account for the formation of stars. Despite its key role in processes as diverse as the origin of planetary systems and the evolution of galaxies, star formation is probably the least-understood aspect of the fundamental processes. Nonetheless, over the last quarter of the 20th century impressive advances in our understanding of star formation came from the continued development of new technological observational capabilities from both the ground and space. During this period astronomers learned the following:

• Stars form continually in our galaxy within the dense cores of giant molecular clouds.

---

[4]In addition to the ESO project, called OWL, there are three other programs in the early planning stages: MAXAT, a 30- to 50-m telescope (New Initiatives Office at NOAO), the 30-m-class CELT (Caltech and the University of California), and ELT, a 25-m scale-up of the HET (Pennsylvania State University and the University of Texas). The GSMT described here corresponds closely with CELT or MAXAT. Although it is too early to judge the future direction of those projects, the panel believes that GSMT could evolve directly from either of them, one from the private, the other from the public sector, or from a joint project created by merging the two.

- The process of star formation is almost always accompanied by the formation of a circumstellar disk. Systems of planets like our own appear to be a natural by-product of the star formation process and are therefore common, as the first observations of extrasolar planets also suggest.
- Star formation is a complex and dynamic process dominated by gravitational collapse, which is accompanied by the energetic ejection of spectacular bipolar jets and outflows.
- Stars tend to form in pairs, groups, or clusters but rarely in isolation.

Existing theories cannot simultaneously account for all these facts, and there are additional mysteries—for example, the form of the initial mass function (IMF) and the efficiency of star formation both need to be understood in order to construct a credible theory of star formation. The physical process of star formation spans an enormous range in both spatial scale (about eight orders of magnitude) and density (about 20 orders of magnitude). However, despite the progress of the past two decades, direct observation of the key stages remains a formidable challenge. For example, researchers know little about the crucial processes that occur on relatively small physical scales (less than 200 AU), such as the development of energetic bipolar jets, the growth of a protostar through accretion and infall, and the formation of planets from a circumstellar disk.

The GSMT, working in concert with NGST and the millimeter-wave arrays CARMA and ALMA, will have a profound impact on our understanding of these matters. Probing the 1- to 200-AU scales with a variety of wavelengths will provide a more detailed and comprehensive picture of the earliest stages of star and planet formation than was previously possible. In particular, GSMT will have the angular resolution and sensitivity to study regions as small as 1 AU (at 1 μm) in the nearest protostellar clouds, vastly increasing our knowledge of many phenomena:

- *The origin and nature of bipolar jets.* High-angular- and high-spectral-resolution observations should be able to determine how close to the central protostar the jets are collimated and whether jets form as disk winds or are instead driven from close to the stellar surface. Such knowledge could tell us whether such ejections regulate the mass of the star and the form of the IMF.
- *The structure and nature of protostars.* High-resolution spectros-

copy at near-infrared wavelengths will probe the velocity and density structure of protostellar environments on scales from a few astronomical units down to the stellar surface (even in seeing-limited mode). Protostars gain mass through infall and disk accretion, which are believed to dominate in the inner regions (see Figure 2.2). GSMT observations of the protostellar disk and envelope will show whether material accretes directly from the disk or along dipole field lines from a truncated disk and whether the accretion is steady or episodic. The GSMT's greater sensitivity will also enable measurement of the photospheric absorption lines from protostellar atmospheres, which are too heavily veiled to be easily detected with smaller telescopes. Such observations will critically constrain the theory of stellar evolution through direct measurements of

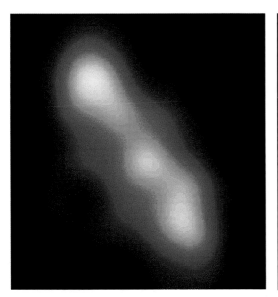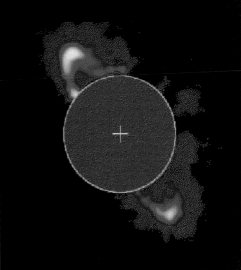

FIGURE 2.2 Direct images of the debris disk around the main sequence A star HR 4796 show the advantage of high spatial resolution and the interplay between ground- and space-based facilities. On the left is a mid-IR image (24.5 μm) from the Keck telescope of the thermal emission from the disk. On the right is a near-IR image (1.1 μm) taken with NICMOS aboard the Hubble Space Telescope, which detects the disk in light scattered from the central star. (The large gray disk is due to the coronagraphic mask, which greatly reduces the light of the central star.) With GSMT, much higher resolution images could be obtained of these disks around A stars as well as images of these disks around lower mass and younger stars, which are currently unobtainable. Courtesy of D. Koerner, University of Pennsylvania.

effective temperatures, surface gravities, rotation rates, and even the accretion energy of protostars.

• *Protostellar companions and masses.* High-angular-resolution imaging and spectroscopy will permit measurement of the frequency, separations, and orbital motions of binary companions to protostars and more evolved young stellar objects, such as T Tauri stars, on scales of 1 to 5 AU in the nearest star-forming regions. Such measurements would yield the first direct determinations of protostellar masses, crucial to the development of a complete theory of star formation, and indicate the survivability of protoplanetary disks in multiple-star systems.

• *Disk structure and chemistry.* The bulk of the mid-infrared emission from a protoplanetary disk is confined to the inner circumstellar regions with a radius of less than 20 AU. The improvement in angular resolution with GSMT will allow the first spatially resolved mapping of the dust structure and chemistry of young disks in the region where planetary systems are thought to form. For both these disks and the older debris disks (see Figure 2.2), maps of the thermal emission at mid-infrared wavelengths or of scattered light at near-infrared wavelengths could reveal gaps and spiral arms in the surface density caused by gravitational interaction with embedded protoplanets. For instance, Jupiter would have formed a gap of ~1 AU in width in the primitive solar nebula, a feature that would be detectable in GSMT images of the nearest star-formation regions (at a distance of 140 pc).

• *Outer solar system.* A more thorough understanding of the early development of our own solar system is needed to interpret observations of extrasolar planetary systems. Near-infrared spectroscopy of KBOs to determine which ices and minerals are present on their surfaces is possible now only for the brightest objects (see Figure 2.3). GSMT spectroscopy will greatly increase the sample so as to cover more of the compounds that could characterize this large population. Also, high-spatial-resolution GSMT images of KBOs will show their sizes and shapes and the homogeneity of their surface compositions, providing constraints on their formation and collisional history. Also, high-spatial-resolution GSMT images of KBOs will begin to resolve their surfaces. Diffraction-limited images at 1 μm with GSMT will provide a spatial resolution of 200 km for an object at 40 AU, corresponding to about 16 resolution elements on the largest known KBOs.

• *Direct planet detection.* Simulations indicate that with its current AO system, the Keck telescope could detect reflected starlight from mature Jupiter-sized planets in 10- to 40-AU orbits around six bright stars

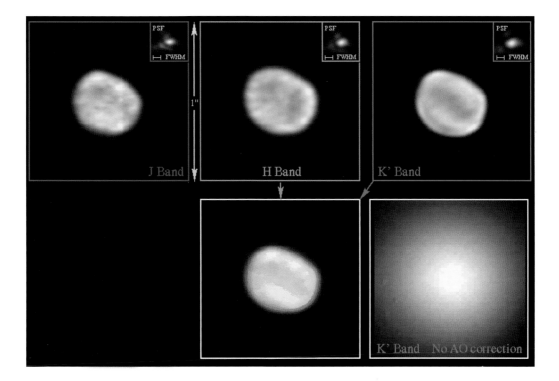

FIGURE 2.3 Vesta, one of the brightest main belt asteroids, as imaged with the Keck AO system on its second night of operation. Images in three bands (J = 1.25 µm, H = 1.65 µm, and K' = 2.1 µm) were used to produce a true-color image. The open loop (no AO correction) image is shown in the bottom right-hand frame. The top row of images shows the results of deconvolution (i.e., the estimate of the object that, when convolved by the point spread function—shown in the inset—most accurately reproduces the data). In the true-color image, blue represents J; green, H; and red, K. The albedo at 1.2 µm is dominated by the reflection of pyroxene and at 2.1 µm, mostly by that of olivine. Thus, a very blue area shows a concentration of surface pyroxene, and a red one shows a concentration of olivine. Courtesy of R.P. Binzel, Massachusetts Institute of Technology, C. Dumas, NASA Jet Propulsion Laboratory, and the W.M. Keck Observatory Adaptive Optics Team.

within 2 to 4 pc of Earth. The time required to detect a planet scales with the fourth power of the telescope aperture diameter (D), and the minimum separation scales as D, so GSMT will be much more powerful. Young (<10-million-year-old) planets will be detected by GSMT through their thermal emission at 2 µm at separations down to ~5 AU out to distances of ~140 pc.

• *Comparison with brown dwarfs.* Detailed spectroscopy of brown dwarfs shows that their atmospheres are unlike those of any known star and more like that of Jupiter. Finding extrasolar planets allows us to ask whether all giant planets have similar structures and compositions and how these compare with the properties of brown dwarfs, which are intermediate in mass between giant planets and the coolest stars. The GSMT will be able to characterize atmospheres of extrasolar planets through near-IR spectroscopy.

UNRAVELING THE EPOCH OF GALAXY FORMATION

Within the next decade the panel expects existing 8-m-class telescopes to make substantial progress in assessing the global statistics of galaxies and quasi-stellar objects (QSOs) as a function of look-back time. However, to place these objects in a meaningful cosmological context will require much more demanding physical measurements of small-spatial-scale internal kinematics, chemical abundances and gradients, gas-phase physical conditions, stellar content, and subkiloparsec morphology. These quantities should all be measured as a function of large-scale environment and of look-back time. The observations will be necessary to understand galaxy formation and evolution and how they tie into the development of the large-scale structure of the universe.

Such contextual observations require high angular resolution and sensitivity that cannot be obtained with telescopes of the 8-m-diameter class. The NGST has similar goals in the area of galaxy evolution, but GSMT can both complement and support NGST with its powerful capabilities. The NGST will open up the currently unexplored dawn of the universe—the end of the "dark age" left by the cooling Big Bang—by observing galaxies at $z > 5$. For such objects the most important diagnostic features of the spectra have been redshifted into the thermal IR, where NGST's extraordinary sensitivity will be unrivaled. However, GSMT will be a powerful tool for studying galaxies with more modest redshifts during the period of cosmic history when most of today's stars and metals were formed (see Figure 2.4).

To give a specific example, while the measurement of redshifts to $z \sim 2.5$ to 4.5 is now almost routine, only the very brightest objects can be observed with sufficient spectroscopic precision to delve into the astrophysics. A spectral resolution of $R \geq 5000$—higher than what is possible for very faint objects observed with 8-m-class telescopes—is required to resolve, both spatially and in terms of wavelength, the

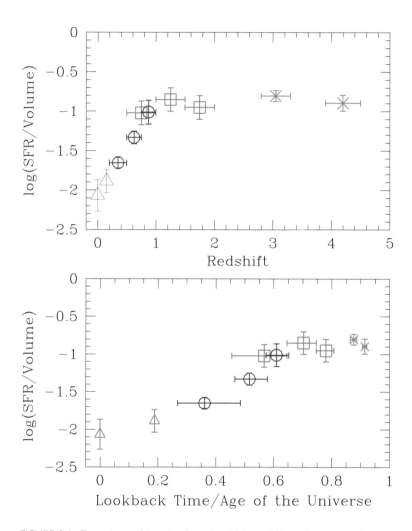

FIGURE 2.4 Two views of the star formation history of the universe, based on current data. Plotted in each case is the logarithm of the star formation rate (SFR), the rate at which gas in galaxies is turned into stars, per unit volume. In the top panel, this quantity is plotted against redshift. In the bottom panel, the redshifts have been converted to lookback times, given as a fractional age of the universe. Note that the cosmic time before $z \sim 5$ is relatively brief—NGST will be the crucial facility for exploring this epoch. Over the next decade, large surveys for galaxies in the redshift range $1 < z < 5$, when most of the stars in the present-day universe formed, will be carried out using SIRTF and 4- to 8-m-class ground-based survey telescopes. NGST and 8- to 10-m telescopes will provide follow-up spectroscopy for measuring redshift and determining star formation rates. The GSMT will provide unique access to the chemical and dynamical history of the $1 < z < 5$ universe through its powerful spectroscopic and imaging capabilities. Courtesy of C. Steidel, California Institute of Technology.

rotation curve or velocity dispersion of a (potentially) low-mass galaxy,[5] and the typical half-light radius of such objects at high redshift is only ~0.2 to 0.3 arcsec. Not surprisingly, then, astronomers do not even know such rudimentary information as whether these galaxies are supported by rotation or dispersion. Without the physical measurements allowed by high-dispersion spectroscopy, they are unable to tie one observed epoch to another and are unable to connect theory to observation.

The GSMT will alter this situation dramatically by giving fourfold improvement in (diffraction-limited) resolution as well as a 16-fold gain in light-gathering power, relative to an 8-m-class telescope. Multiobject $R = 5000$ spectroscopic capability will allow us to establish the relationship between luminosity and mass and to measure the chemical properties of $1 < z < 5$ galaxies through nebular line diagnostics (the rich interstellar absorption line spectrum in the rest-frame UV) and the integrated stellar light. GMST will achieve the same ~50 pc spatial resolution for $1 < z < 5$ galaxies as ground-based observations of galaxies at the distance of the Virgo cluster. This would place up to ~50 resolution elements across typical compact galaxies at high redshift, thereby converting them from "fuzzballs," as currently observed, to resolved objects rich in morphological complexity (see Figure 2.5). GSMT spectroscopic determinations will measure the chemical abundances in individual star clusters and giant HII regions and will trace the kinematics of large-scale outflows across the face of each galaxy by observing interstellar absorption lines against the UV continuum from massive stars. These are the essential measurements for understanding the assembly of baryons into galaxies in the early universe.

With such data astronomers will be able to answer the fundamental questions of galaxy evolution: When did galactic bulges form? Are distant galaxies rotationally supported? What controls the decline in the global star formation rate for $z < 1$? What is the mass function (as opposed to the luminosity function) of distant galaxies? How much metal mass is ejected from galaxies during their robust star-forming phase, polluting the intergalactic medium (IGM)? Are chaotic morphologies really indicative of mergers, or are they a natural consequence of rapid star formation? (See Figure 2.6.)

---

[5]At this resolution, the terrestrial background in the nonabsorbed bands in the 0.6- to 2.5-μm range approaches the low background accessible from space—the bulk of the background comes from very narrow OH airglow lines and not from thermal emission.

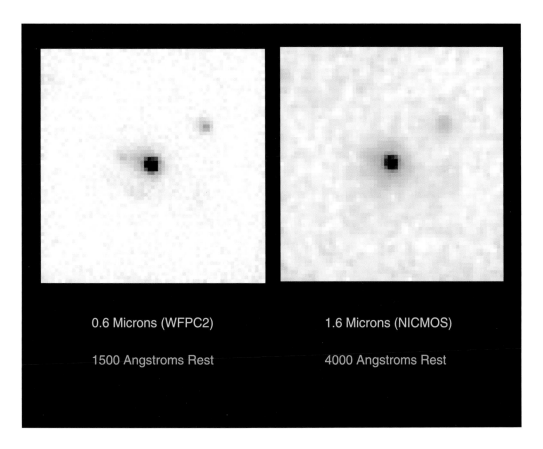

0.6 Microns (WFPC2)   1.6 Microns (NICMOS)

1500 Angstroms Rest   4000 Angstroms Rest

FIGURE 2.5 Hubble Space Telescope images, in the rest-frame far-UV and rest-frame optical bandpasses, of one of the brightest known $z \sim 3$ galaxies. The galaxy has a morphology typical of high-redshift, star-forming galaxies, with a half-light radius of only ~0.25 arcsec, surrounded by more extended and diffuse nebulosity. The core of the galaxy is barely resolved at 1.6 μm with the HST. Higher spatial resolution, such as that of a diffraction-limited GSMT, will be required to resolve such galaxies into individual star-forming knots and to delve into the detailed kinematics that will allow measurements of, for example, dynamical mass, chemical abundance gradients, and the distribution of outflowing metal-enriched gas. Courtesy of C. Steidel, California Institute of Technology, 1999.

Astronomers believe that the diffuse IGM contained more than 90 percent of the baryons during the $2 < z < 5$ epoch. Theoretical models suggest that the Lyman-alpha forest may be a more reliable tracer of the overall matter distribution (because the gas exists in regions close to the mean density of the universe) than galaxies, which are highly biased

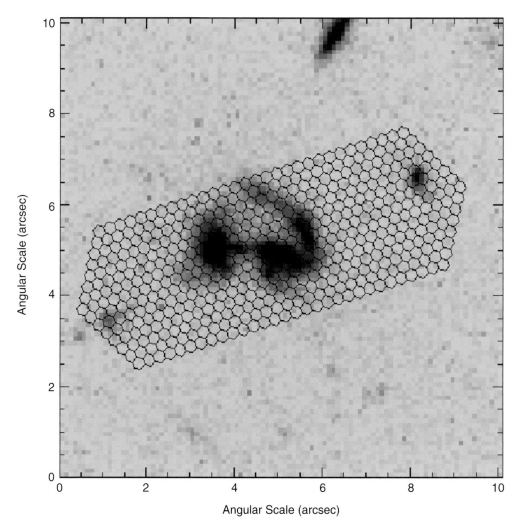

FIGURE 2.6 A merger of two galaxies at $z = 0.831$ in a 10 arcsec × 10 arcsec patch of the Hubble Deep Field as it would be sampled by an integral field spectrograph, which produces spectra over a two-dimensional format for pixels approximately 0.1 arcsec². The GSMT will sample such an object with seven times the spatial resolution, corresponding to ~100 pc, the size of a large star-forming region at high redshift. Such a spectrograph would be ideal for GSMT's study of metal abundance and the kinematics of complex objects at $1 < z < 4$. Courtesy of I. Parry, Institute of Astronomy, Cambridge, United Kingdom, and the Cambridge IR Panoramic Survey Spectrograph group.

tracers. With $R \geq$ 30,000 spectroscopy of background QSOs, measurements have been made of the temperature, metal content, and ionization state of the IGM, all on scales of a few kilometers per second. This technique has already produced the most reliable determinations of the cosmic deuterium/hydrogen ratio, the metal content of the diffuse IGM, and the chemical evolution history of high-column-density gas. What is lacking is the three-dimensional information necessary to place the observations into the context of the galaxy distribution at the same cosmic epochs, allowing detailed comparisons with increasingly sophisticated simulations of structure formation. The problem is that QSOs are rare objects—a sufficiently dense sampling requires measuring intergalactic absorption in the spectra of faint galaxies (see Figure 2.7). The GSMT will extend the magnitude limit for high-precision spectroscopy to $R \sim 22$, increasing the surface density of suitable probes of the high-redshift

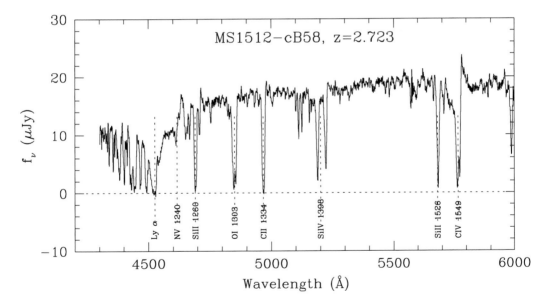

FIGURE 2.7 Keck/LRIS spectrum (3.5-Å resolution) of a gravitationally lensed (by a factor of 30 to 40 in apparent luminosity), star-forming galaxy at $z \sim 3$. In addition to the strong interstellar absorption features, the spectrum is of high enough quality to identify weak stellar photospheric lines from O-stars and to separate the P Cygni profiles from stellar winds for high-ionization features like CIV $\lambda 1549$ and NV $\lambda 1240$. An analysis of this spectrum indicates chemical abundances slightly less than solar and a galactic-scale outflow of gas at a velocity of more than 200 km s^{-1}. The GSMT will obtain such measurements routinely for common, unlensed galaxies. Courtesy of C. Steidel, California Institute of Technology.

universe to an (estimated) 150 deg^{-2} in the redshift range $2 < z < 4$. This would allow mapping both the galaxy distribution and the diffuse gas distribution over the same cosmological volumes at high redshift, providing for the first time a full accounting of the distribution of baryons. Because of the required high spectral resolution, such crucial observations are beyond the planned capabilities of NGST.

THE STAR FORMATION HISTORY OF NEARBY GALAXIES

Did star formation begin in all galaxies at the same time, and did it proceed from the outside in or inside out? Was it episodic or continuous, and does this depend on galaxy mass or environment? Such fundamental questions are best answered by dissecting stellar populations directly, that is, by determining the age and metal abundances of stars as a function of epoch and position within our galaxy and its neighbors. NGST will probably be able to determine ages for old stellar systems at the distance of M31 (turn-off magnitudes near $R = 29$) in low-surface-brightness regions, but GSMT will resolve the main sequence turnoffs in higher-density regions, to within about 2 kpc of the nucleus of M31 (even closer to the center of M32), and will enable study of stars as faint as the horizontal branch in the halos of galaxies as distant as the Virgo cluster.

Main sequence turnoff photometry is not wholly adequate for high-precision age dating: mean metallicities and elemental abundance ratios (e.g., "alpha" elements) are also required. GSMT's large aperture and high spatial resolution (assuming a Strehl ratio of 0.3 at 0.7 µm) will enable high-resolution ($R = 40,000$), high-signal-to-noise (S/N > 25) spectroscopy of the brightest M31 red giants (with visual magnitude = 21.5) in a single night. By comparing them with synthetic spectra, lower S/N ~ 10 spectra from even shorter exposures (~1 h) would yield accurate mean heavy-element abundances ([m/H] and [α/Fe]).

Elemental abundances in metal-poor stars provide us with new means of chronometry and critical insights into nucleosynthesis from supernovae. The increasing dominance of r-process elements indicates that stars that formed before asymptotic giant branch (AGB) evolution could pollute the interstellar medium (ISM) with s-process elements. By studying the dispersion of these elements in different locations in galactic halos, researchers can learn much about the early history of heavy-element enrichment; for example, very low metallicities coupled with r-process dominance have suggested pollution by only a few supernovae or even just one.

Measurement of r-process patterns and thorium abundances are an entirely new and independent chronometer that provides crucial tests for globular cluster ages derived from stellar evolution models. Furthermore, radioactive dating provides the ages of *individual* field red giants, which have been identified throughout the halo and the disk and in the Magellanic Clouds. Most of the crucial lines, particularly thorium, are spread out in the violet part of the spectrum, where the ground-based sky is very dark. High resolving power and a high signal-to-noise ratio are essential for such work. With absolute magnitude in U approximately equal to 0.5 mag for almost all metal-poor red giants, ultraviolet seeing-limited observation with GSMT could still measure the age of a field giant as far away as the Magellanic Clouds in less than a night.

SEEING-LIMITED SCIENTIFIC APPLICATIONS

It is important to acknowledge that atmospheric conditions will preclude the AO-corrected operation of GSMT some fraction of the time and that GSMT is not likely to be diffraction-limited below $\lambda = 1 \ \mu m$. Fortunately, there are seeing-limited, wide-field applications for GSMT that will have truly dramatic impact as well. For instance, a giant aperture with a highly multiplexed, wide-field spectroscopic capability could pursue many fundamental astrophysical problems. Because it is often the case that diverse astrophysical processes contribute to general observable trends, fundamental questions must often be approached statistically, with samples large enough to allow separating these contributions. Such problems include the formation and evolution of large-scale structure, of galaxies and the Milky Way, and of stars and their planetary systems. For example,

• Densely sampled, wide-field spectroscopic surveys with GSMT will explore the evolution of large-scale structure beyond the local universe mapped by the Sloan Digital Sky Survey, i.e., at $z > 1$, allowing astronomers to distinguish among theories of structure formation. At $z \sim 1$, the 20 arcmin FOV of GSMT corresponds to a co-moving length scale of ~ 12 $h^{-1}Mpc$ ($q_0 = 0.1$), which will enable studies of the important 100 $h^{-1}Mpc$ size scale (e.g., as large as the Great Wall and Great Attractor). Redshift surveys at $z > 5$ using Lyman-alpha will map out the clustering evolution of galactic fragments and the genesis of large-scale structure. As discussed above, the absorption spectra of faint background galaxies will map the IGM in the early universe to small scales not accessible with widely separated QSOs.

• Galaxy formation is a complicated tale, with interwoven story lines of star formation, chemical enrichment, dynamical evolution and merging, and interaction with the environment. To sort out these story lines and the complex relationships between them, it is necessary to trace out galaxy age, star-formation rate, chemical abundance, and morphology as functions of mass and environment over the large redshift range $1 < z < 5$. Considering this large parameter space, the relative rarity of populating some bins (for example, very metal-poor starbursts), and the need to obtain meaningful statistics in all categories, it is likely that very large samples, including perhaps many hundreds of thousand galaxies, will be required. Obtaining spectroscopic samples this large over scales of many degrees is a feasible goal for GSMT.

• The formation and evolution of the Milky Way and other local group galaxies can be discerned from the detailed record of the age, kinematics, and chemical abundances of individual stars. To recover this record requires the study of abundant populations—dwarf stars, for example; studies of millions of stars can be used to discover the mean trends, such as age versus kinematics. Combining this information with galaxy substructure will inform such important matters as the merging history of galaxies. Its multiobject spectroscopic capability will make GSMT a powerful tool for such work. The GSMT could even investigate the merger history in the Virgo cluster, by tracing the positions and properties of intracluster red giant stars (with a typical magnitude of ~28).

THEORY CHALLENGE FOR GSMT

A theory challenge for GSMT is to develop models of star and planet formation, concentrating on the long-term dynamical coevolution of disks, infalling interstellar material, and outflowing winds and jets, as described in the report of the Panel on Theory, Computation, and Data Exploration (Chapter 6).

TECHNOLOGY BASIS

The success of the segmented Keck 10-m telescopes has provided astronomers with the technology to build increasingly large telescopes without incurring the risks associated with increasing the size of monolithic mirrors. For the first time, they have a scalable technology to apply for ever-larger telescopes.

Other key advances in computational engineering design and analysis and in the active control of optics (as exemplified by Gemini and Keck) allow predicting with confidence the performance of future large telescopes by integrated modeling of their optical, mechanical, and thermal properties. This capability has greatly reduced the risk inherent in designing and building larger telescopes.

Adaptive optics, a key component of a future very-large-aperture telescope, progressed significantly in the 1990s. AO systems now exist or are being built for all of the world's largest telescopes, and their scientific productivity is rapidly increasing. Even so, the challenges of developing AO are enormous, and existing and planned AO systems fall far short of the ideal. Therefore, the further development of AO technology for GSMT will also directly improve the performance of AO systems on 8-m-class telescopes.

The next section reviews other technology issues that affect the cost of building the GSMT.

KEY TECHNOLOGY ISSUES

The GSMT, along with the development of AO, will cost an estimated $400 million, most of it to be spent in the current decade. The panel proposes that this facility be built and operated using a combination of federal, state, and private funds, making it a collaborative effort between national and independent observatories. It also proposes that NSF be prepared to pay at least half of the total capital and operating costs, with the balance to come from private observatories and/or foreign partnerships. The costs of telescope development should be about $3 million per year for at least 3 years, ramping up to cover the construction costs as soon as the technology is ready. This should allow time to study the various fabrication trade-offs and mirror support issues, as well as to develop the basic telescope design and cost. The panel believes that the AO effort associated with the development of GSMT should be funded at $5 million per year for the next 10 years. This level of funding should be sufficient to support national, university, and industry efforts with respect to a number of key AO issues, including the following:

• Better wavefront sensors (faster, with lower read noise, more pixels, and improved IR response);

- Improved laser beacons (more affordable, powerful, and reliable; better coupling to the Na layer; ability to make multiple beacons);
- More capable deformable mirrors (more affordable and more reliable; more stroke, more degrees of freedom, and a wider choice of actuator spacing);
- Better understanding of multiconjugate adaptive optics through modeling, simulations, and experiments;
- Advanced techniques and algorithms for achieving the needed computing speed;
- Study of various hybrid systems using both natural and laser guide stars; and
- New, more efficient wavefront sensing approaches (ideas, simulations, and experiments).

The utility of a 30-m or larger aperture telescope depends crucially on its near-diffraction-limited performance, particularly in the 1- to 25-µm wavelength range. Because of the sky background faced by ground-based telescopes, truly spectacular gains are possible for unresolved sources when the effective sampling scales as the diffraction limit. Also, the general transparency and width of the atmospheric windows are strongly limited by precipitable water, arguing further for placing the GSMT at a superb site, with low water vapor included along with the usual criteria of excellent seeing and a high fraction of workable nights. Furthermore, since instrumentation for these large telescopes will be vastly simpler only if diffraction-limited performance is achieved, AO is essential to justify the large investment required. The difficulties associated with AO grow rapidly with the number of controlled degrees of freedom (dof), and the largest astronomical AO systems currently have about 350 dof. For a 30-m aperture to achieve a Strehl ratio of 0.5 at 1 µm requires control of ~7000 dof. This level of control will require significant improvements in wavefront sensors, deformable mirrors, laser beacons for artificial stars, and computational speed, as well as improved theoretical models and simulations. In addition, the problem of correcting for the atmospheric distortion of the wavefront using artificial guide stars must be generalized to the multiple-guide-star case to achieve adequate correction over significant fields of view. Fortunately, the use of multiple guide stars will provide diffraction-limited images over fields of view greater than about 1 arcmin. The AO development work will also greatly help existing large telescopes by providing superior AO systems for them, allowing them to work at the diffraction limit at shorter wavelengths and thereby greatly increasing their scientific power.

It is interesting that as the telescope grows in size, the potential for tomographic measurements of the atmosphere with natural guide stars grows as well.[6] Nevertheless, the panel believes that multiple laser beacons offer a more convenient and predictable means for tomographic measurement of the atmosphere. Both approaches need to be thoroughly explored.

Cost Issues

This chapter aims not to justify a specific design for GSMT but to point out that enough work is now under way to inspire confidence that a serious investment in technology development will bring about the necessary innovations. This is the context for the following remarks on cost issues.

To make the next-generation telescope cost-effective, significant engineering improvements are clearly required. Empirical studies have shown that the cost of a telescope scales roughly with aperture as $D^{2.6}$. The Keck II telescope cost of approximately $80 million suggests that without further innovation, a 30-m aperture would cost roughly $1.4 billion. The experience of designing and building Gemini and Keck has shown numerous opportunities for engineering innovations and development that will significantly reduce costs.

Breaking the cost curve, as the developers of the present generation of 8-m-class telescopes were able to do, is an essential part of this endeavor. In scaling up from 4-m-class telescopes, Gemini and Keck broke the cost curve by factors of 4 to 8. Clearly, then, the astronomy community has experience and success with this kind of challenge. For GSMT, costs must be reduced by roughly a factor of 4.

In comparison, to build the much more powerful ESO 100-m OWL (the acronym can be taken to stand for either Overwhelmingly Large Telescope or Observatory at a World Level) for $1.5 billion will require

---

[6]This has been an argument for building a 100-m telescope. However, although it eliminates the need for the laser itself, atmospheric tomography is just as challenging for natural guide stars, and the number of degrees of freedom needed for a given correction continues to rise as the square of the aperture, that is, by a factor of about 10 for a 100-m telescope over a 30-m telescope.

innovation that reduces costs relative to Keck by a factor of 20. Such a cost reduction is significantly more challenging and does not appear to be reasonable for a single engineering step. In fact, if OWL could be built for this price, the same technology could produce a 30-m telescope for $65 million, which would be an even more compelling next step. The panel is excited about the possibility of OWL but expects that it will probably take much more time to be developed than GSMT.

To meet the 2010 goal, development work on GSMT technologies and strategies for cost savings should begin immediately. Much of the needed human expertise is already available or will shortly be, when the current generation of 8-m-class telescopes will have been completed, so the timing is excellent. Here the panel gives a few examples of ideas that could bring about the necessary cost savings. In modern telescopes, optics and the related support and control systems account for 30 to 50 percent of the cost of the entire project. This amount includes the cost of materials, polishing, passive support systems, and active control systems. In each of these areas, the panel sees great opportunities for cost savings. For example, segment size is a key cost driver: gravity-driven deflections of mirror segments are a major difficulty, one that increases as the fourth power of the radius of the segment. Smaller segments will therefore allow much thinner segments as well as simpler passive supports. Specifically, in modest-sized pieces (~1 m), Zerodur (the material used for Keck) costs about $100 per kilogram. Accordingly, thinning the segments can produce significant cost savings. In addition, polishing costs for Keck-style segments (off-axis sections of a hyperboloid) depend on the asphericity, which scales as the square of segment size, so smaller segments allow for more economical polishing methods. The polishing of spheres using planetary polishers (with which multiple mirrors are polished simultaneously) for segments for the Hobby-Eberly telescope turned out to be extremely economical. The asphericity of GSMT's 1-m segments will be only 10 percent of that of Keck segments. A variant of stressed mirror polishing may allow GSMT segments to be warped and polished as spheres with planetary polishers, which would dramatically reduce polishing costs.[7]

---

[7]Optical fabricators are presently claiming $15,000 per square meter for polishing and testing spheres by this technique. Another option for reducing the optical costs may be to pursue a semistationary spherical primary mirror, along the lines of the Hobby-Eberly telescope. Such a design requires compromises in performance but greatly reduces cost compared with fully steerable Ritchey-Chrétien telescopes.

Active control of the telescope mirror segments will be required. Because the segments are smaller, there will be many more than the 36 in Keck. For this reason, control system innovation is essential. Fortunately, the general trend of technology with time is toward higher-performance electronic devices that are also more compact and more affordable. The Keck telescopes use interlocking-edge sensors that are relatively expensive and awkward to service. Schemes to make sensors that are basically films on the segment edges may lead to order-of-magnitude reductions in cost and greatly simplify segment servicing (e.g., recoating).

It is likely that a weatherproof covering will be needed for GSMT. A dome provides protection from both weather and wind, easing many environmental constraints. Traditionally, domes are rather expensive, but large stationary and movable geodesic structures such as those used for modern sports stadiums may suggest a technique for greatly reducing the cost of the enclosure.

CONTEXT ISSUES

It is expected that NOAO, restructured to form a strong national organization, as described in a previous section of this chapter, will play a prominent role in the development of the GSMT. This project obviously will offer a splendid opportunity for public-private partnering that could lead to a common understanding and advancement of the system. Regardless of other efforts inside or outside the United States, NOAO should initiate a solid program on behalf of (and involving) the astronomy community to explore the scientific drivers and technical hurdles associated with the GSMT. This effort will position the community for strong and effective participation. The panel can imagine a range of possible roles for NOAO in building and operating the GSMT, from leading the construction of GSMT if the program is carried out mostly with federal funds to simply forming a strong U.S. presence in a collaboration with independent observatories and/or international partners if the program is carried out with mostly nonfederal funds. For developing an O/IR ground-based system, the panel recommends one particular alternative—NOAO partnering with one or several independent observatories to build and operate GSMT—as the most desirable course. For all of the alternatives, the observing time available to the U.S. astronomy community, an amount proportional to the investment by NSF, could be distributed by NOAO (as is being done for the WIYN telescope).

Considering the difficulties in fighting gravity and wind loading, as well as the costs of enclosures to shelter a ground-based telescope, it is likely that at some point it will become more cost-effective to build a very-large-aperture space-based telescope than to build a ground-based telescope with the same aperture. From the present vantage point, GSMT is arguably on the traditional side of that comparison, according to which space-based telescopes are more expensive, but if NGST succeeds in achieving the 8-m size for its proposed budget, then ground-based telescopes with apertures larger than about twice that of GSMT may be approaching this crossover point. Experience with the GSMT and NGST technologies will help to determine whether even larger ground-based telescopes, for example, the proposed 100-m OWL telescope, will be more cost-effective than space telescopes.

ANCILLARY BENEFITS

A 30-m telescope with its superb angular resolution and sensitivity may be the preferred way to follow up extremely faint sources associated with x-ray, gamma-ray, and microjansky radio sources, particularly if more than merely a redshift is desired. Should NGST not be achieved, GSMT's existence becomes an absolute necessity for progress in the fields described above.

MAJOR INITIATIVE, PRIORITY TWO: A LARGE-APERTURE SYNOPTIC SURVEY TELESCOPE (LSST)

MISSION DESCRIPTION

The panel advocates the construction of a Large-Aperture Synoptic Survey Telescope (LSST) with the ability to map the entire accessible sky to 24th magnitude (in one optical band) over the course of three nights. (This is about a magnitude deeper than the Sun would appear at the distance of the Magellanic Clouds or the Milky Way galaxy would appear at $z = 1$.) The science objectives, which range from solar system science to cosmology, can be addressed simultaneously with the same set of images. For the first time, astronomers and the general public would have access to a motion picture of the night sky.

The requirements for the proposed system follow from the expression for signal-to-noise ratio in a sky-dominated exposure, yielding a figure of merit that scales as $A\Omega/\sigma^2$, where A is the system's effective aperture, Ω is the field of view of the focal plane, and σ is the seeing. To map out the 20,000 deg^2 of accessible sky down to 24th magnitude every few nights will require the equivalent of a 3-deg field of view on a 6.5-m-aperture telescope (one concept is shown in Figure 2.8).

The detectors of choice for the temporal monitoring task would be thinned charge-coupled devices (CCDs); the requisite extrapolation from existing systems appears to constitute only a small technological risk. An IR capability of comparably wide field would be considerably more challenging but could evolve as a second phase of the telescope's operation. Instrumentation for LSST would be an ideal way to involve independent observatories with this basically public facility.

SCIENCE WITH LSST: THE WIDE AREA VARIABILITY EXPERIMENT

The unprecedented capabilities of LSST open up the possibility of a new kind of science program, a wide area variability experiment (WAVE), whereby one or a few simple survey modes can simultaneously address a number of frontline science questions.

NEAR-EARTH OBJECTS

The orbits of many asteroids intersect the orbit of Earth. These so-called near-Earth objects (NEOs) present a threat to life on our planet, with effects ranging from the local damage inflicted by smaller members of the NEO population (for example, the blast-wave destruction of 1000 km^2 of Siberian forest at Tunguska in 1908) to global disruption of the biosphere (as occurred with the impact of a 10 km body in the Cretaceous–Tertiary event).

Extrapolations from recent surveys suggest that some 1000 NEOs are larger than 1 km in diameter (see Figure 2.9) and as many as 10^5 to 10^6 NEOs have a diameter of 100 m or more. The vast majority of these objects has yet to be discovered, but a statistical analysis indicates a 1 percent probability of impact by a 300-m body in the next 100 years. Such an object would deliver 1000 MT of energy and (assuming an average of 10 people per km^2 on Earth) result in 100,000 fatalities. The damage caused by an impact near a city or into a coastal ocean would be orders of magnitude higher. Of course, the distribution is extremely

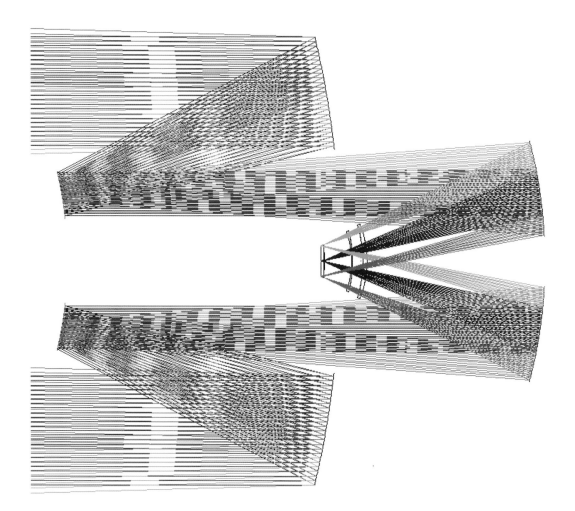

FIGURE 2.8   Optical ray trace for one concept of LSST.   This design results in a 3-deg-diameter FOV at f/1.25.   Excellent image quality is achieved at wavelengths from 0.3 to 2.2 μm, and the system can be fully baffled for either optical or IR imaging.   The diagram shows how very large secondary and tertiary optics are required to achieve such a large field, a significant increase in difficulty over present-generation large telescopes.   Courtesy of R. Angel, University of Arizona. (This figure first appeared in a paper by R. Angel et al. in *Imaging the Universe in Three Dimensions: Astrophysics with Advanced Multi-Wavelength Imaging Devices*, Proceedings from the Astronomical Society of the Pacific Conference Vol. 195, edited by W. van Breugel and J. Bland-Hawthorn, 2000, p. 81; reproduced by permission of the Astronomical Society of the Pacific.)

FIGURE 2.9 Orbits for 100 representative near-Earth objects with estimated diameters of 1 km or more; they represent only about 5 percent of the total estimated population in this size range. All orbits included have perihelion distances of 1.10 AU or less, and the orbits have been projected into the Earth-Sun (ecliptic) plane. Also shown by black dashed lines are orbits for the terrestrial planets Mercury through Mars; the positions of the planets on January 1, 1997, are also indicated by dark purple circles. The vernal equinox is to the right. Courtesy of R.P. Binzel, Massachusetts Institute of Technology.

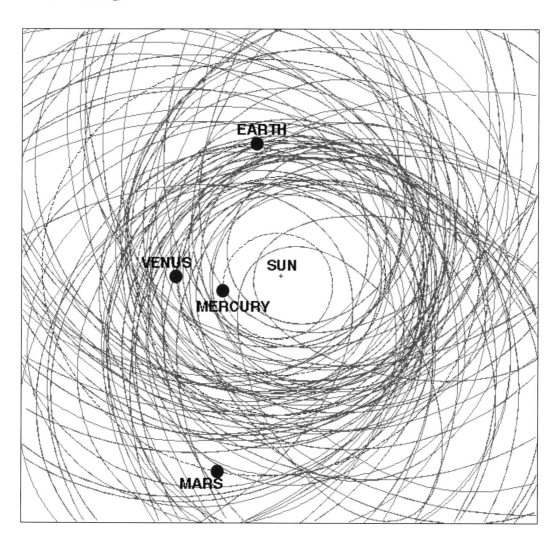

non-Gaussian: the majority of impacts would have a far smaller effect and a small fraction would have a catastrophic result. Nonetheless, such an impact clearly constitutes an extreme example of the influence of the astronomical environment on Earth.

The reality of the impact threat has been recognized by scientists only in the last 20 years. In the past few years, thanks to several reported near-miss encounters with small objects, it has become a subject of intense interest to the general public and has even been discussed in the U.S. Congress. Although it can be argued that other threats to life on Earth pose risks exceeding that of an asteroid or comet impact, the public feels clearly that any significant risk that is avoidable deserves attention. The contribution of astronomers to this task is to find these objects sufficiently far in advance (decades) that countermeasures could be taken.

A survey for NEOs demands an exacting observational strategy. To locate NEOs as small as 300 m demands a survey down to 24th magnitude, a capability five magnitudes beyond that of existing survey telescopes but well matched to LSST. NEOs spend only a small fraction of each orbit in the vicinity of Earth. Repeated observations over 10 years would be required to explore the full volume of space occupied by these objects. During this time, LSST would discover NEOs at the rate of about 100 per night and obtain astrometric information on the much larger (and growing) number of NEOs that it had already discovered. Precision astrometry is needed to determine the orbital parameters of the NEOs and to assign a hazard assessment to each object. Astrometry at weekly intervals would ensure against losing track of these fast-moving objects in the months and years after discovery.

KUIPER BELT OBJECTS

Kuiper Belt objects (KBOs) are the most primitive bodies in the solar system. Because their orbital dynamics and compositions carry an imprint of the formation of the solar system, they are arguably the most important missing piece in efforts to understand the formation process. For example, astronomers already have evidence for the injection of comets from the Kuiper Belt and for the ejection of matter to the Oort Cloud and to the interstellar medium. It would also be important to model the collisions between KBOs, which generate a dust ring around the Sun that is a local analogue of the dust rings discovered recently around nearby main-sequence stars.

KBOs are among the most challenging objects in the solar system to discover and study. A deep census with LSST is needed to establish the orbital and gross physical properties of 10,000 KBOs; such a large sample would be needed to model the complex dynamical structure of the Kuiper Belt. For example, many KBOs occupy mean-motion resonances with Neptune (as does Pluto). Dynamical models of the early solar system suggest that the relative populations of different resonances can be used to measure the rate and total distance of Neptune's migration, which is itself a measure of the mass ejected outward by Neptune toward the Oort Cloud. However, with samples as small as the one composed of the 60 KBOs with known orbits, a majority of the resonances appear empty and population ratios cannot be accurately determined.

The science requirement to accumulate 10,000 KBOs is an independent constraint for the design and operation of LSST. The luminosity function of the KBOs shows that about 10,000 such objects brighter than red 24th magnitude are to be found in the whole sky. The survey must therefore cover the whole sky in a band centered on the ecliptic and extending ± 30 deg from it to red 24th magnitude. The coverage must be repeated at intervals weekly or monthly near discovery, dropping to yearly once the orbits have been approximately determined. The panel estimates that all 10,000 KBOs could be discovered within 1 year on LSST. Subsequent astrometric observations for orbit refinement would take up a quickly diminishing fraction of the time in succeeding years. The KBO survey would operate concurrently with the NEO survey described above.

Searching for Other Planetary Systems

Two techniques will be used to search for other planetary systems: occultations and microlensing. A typical gas giant planet (a degenerate hydrogen body with a radius fixed near 10^8 m) has a cross section of about 1 percent of a solar mass main-sequence star (with a radius of about 10^9 m). In the occultation technique, stars will be monitored with subpercent photometric precision to search for the periodic dimming as a companion planet crosses the face of the star (for those orbits aligned with the line of sight). Assuming random distribution of orbital planes produces an occultation probability per planet-bearing star on the order of 10^{-3}. If, as Doppler velocity measurements suggest, 1 star in 20 possesses a close-in gas giant planet, then the probability of detecting a planet by occultation is on the order of 5×10^{-5} per star. Because accurate photometry of $\sim 10^8$ stars is completely feasible with LSST, repeat measurements at intervals shorter than the orbital period should detect some 5000 gas giant planets. Such a large sample would have great value: for example, the incidence of planets could be determined as a function of stellar spectral type (mass).

By exploiting the remarkable phenomenon of gravitational microlensing of distant stars, intensive monitoring can be used to search for planets with masses as small as that of Earth. The essential requirement is that the apparent planet-star separation be comparable to the Einstein radius. This requirement is not especially stringent. For example, if our own solar system were to be observed in this way (from a location 10 kpc distant against a lens 5 kpc away), a signal from Jupiter would be found in almost 20 percent of solar microlensing events.

The planet-search program imposes two requirements on LSST. First, LSST must generate photometry accurate to a few tenths of a percent. This is a modest requirement even for stars at 20th magnitude, given the quality of modern CCDs and the large aperture of the LSST. Second, each star must be reobserved at intervals comparable to or less than the crossing time (about 1 day for both the occultation and microlensing methods). It is probable that planet-search observations could be conducted using the same survey data obtained for the NEO and KBO surveys.

OBSERVATIONAL COSMOLOGY WITH SUPERNOVAE

Type Ia supernovae have been demonstrated to be excellent distance indicators. A program of repeated scans of the sky will detect about 100,000 supernovae per year. The nearby ones can be used to trace out departures from smooth Hubble flow caused by the nonuniform distribution of dark matter. More distant supernovae found by LSST (followed up with deeper imaging on other large ground- and space-based telescopes) will complement studies of the small-scale anisotropy of the microwave background that also measure cosmological parameters such as the density of matter and dark energy. Distant supernovae are also a powerful probe of the history of star formation over cosmic time.

STUDIES OF ASTROPHYSICAL VARIABILITY

The resulting archive of stellar variability would be a fundamental resource for studies of stellar astrophysics. The discovery of large numbers of eclipsing binaries with a broad range of masses and chemical compositions would provide fundamental data such as masses and radii for comparison with models. Temporal data for extragalactic objects, particularly AGN and lensed quasars, are powerful probes of the nature of the extreme environments surrounding massive black holes.

DETECTION OF RARE TRANSIENT OBJECTS

Mapping the entire accessible sky to 24th magnitude will open up a vast new discovery space. Optical counterparts of gamma-ray bursts are one example of unanticipated phenomena that reside in the time domain, but there are surely others. It has been suggested that some of these events become very bright (~8th magnitude) for short periods of time. If so, and if they can be detected rapidly enough, these optical transients will serve as powerful probes of the intergalactic medium, surpassing in distance and detail what can be done with quasars.

WHITE DWARFS IN THE GALACTIC HALO

Imaging of large areas at high galactic latitude repeated over a period of years using optical (VRI) colors will uncover, through their proper motions, a complete sample of halo white dwarfs. If, as the MACHO microlensing experiment suggests, an important component of the galactic halo is stellar mass objects of ~0.5 solar masses, the best candidate is a population of very old, very cool, low-luminosity white dwarfs. A deep survey at VRI colors would have the best chance of detecting both those with hydrogen-rich (and infrared H_2-dipole opacity-dominated) atmospheres and those with helium-rich atmospheres.

A DEEP DIGITAL MAP OF THE SKY

By coadding repeated scans of the sky, LSST will produce digital composite images of unprecedented depth. For example, in the course of a year, combined images with a depth 2.5 magnitudes fainter than a single image will be produced, corresponding to a limiting magnitude of 26.5. With these will be generated well-populated catalogs of rare and unusual objects for spectroscopic study by the 8-m generation of telescopes. The spatially coherent distortions of the images of faint, distant galaxies will be used to map out the structure of foreground mass concentrations, using the signature of weak gravitational lensing. Such wide-area, deep images can also be used to search for faint objects such as ultralow-surface-brightness galaxies.

A Deep Infrared Map of the Sky

With very modest integration times, a sufficiently large infrared detector array (beyond current feasibility) could be used to generate a map of the sky that is 100 times (5 magnitudes) deeper than the Two Micron All Sky Survey (2MASS) data set. This would fill the gap between the existing data and the IR limits of the current generation of large-aperture telescopes. For example, an all-sky survey to $J, H, K \sim 20$ would reach deep enough to detect what may be the majority population of field brown dwarfs in the immediate solar neighborhood. Recent simulations suggest that many local constituents are expected to have temperatures much below 1000 K; as mentioned, these could be too faint for the ongoing 2MASS and Sloan Digital Sky Survey (SDSS) surveys and too rare per unit surface area for infrared surveys covering limited regions of the sky.

THEORY CHALLENGE FOR LSST

A theory challenge for LSST is to understand the origin, relationships, and fate of small bodies in the solar system, as described in the report of the Panel on Theory, Computation, and Data Exploration (Chapter 6).

Historically, the solar system has provided us with nature's most revealing dynamics laboratory. Newton formulated his laws of gravitation largely to explain new measurements of the motions of the Moon and planets. More recently, the importance of dynamical chaos was first discovered in the solar system by astronomers studying the orbits of asteroids. The new LSST observations of vast numbers of solar system and other objects promise new material from which exciting developments in theory are to be expected.

DATA FLOW AND INFORMATION DISTRIBUTION

The LSST with WAVE is an ambitious program: it will be a pathbreaking undertaking, providing unequaled opportunities for developing real-time data-mining tools and techniques and for testing the scaling properties of database structures and algorithms.

Because the endeavor is so challenging, it is important to recognize that it builds on current successes in high-data-rate projects in optical astronomy and derives great benefit from advances in computing. The increasing availability of cost-effective computing and mass storage hardware is outstripping the increase in the rates of data produced by even the most ambitious astronomical instruments. For example, the increase by a factor of 100 in the data rate from present microlensing surveys (initiated in the late 1980s), which produce 5 to 10 gigabytes (GB) of raw image data per night, to the $\sim$1 terabyte (TB) per night rate expected for the WAVE project (starting in the middle of the current decade) will be more than offset by advances in computing and mass storage.

Hardware is only part of the solution to the data processing problem, of course. The software must (1) process the data stream in near real time, (2) detect and classify variable and moving objects, and (3) place the results in a readily accessible data repository. Considerable experience has already been gained in these tasks from microlensing surveys, and the SDSS and 2MASS are adding a wealth of new software that can be applied, both conceptually and specifically, to future projects.

An example of a step in the data reduction process is the detection of variable objects. Experience from the microlensing surveys suggests that perhaps 1 object in 1000 will exhibit statistically significant variability. Given 10 billion objects in the sky within the range of LSST, a catalog of $\sim$10 million variable astronomical sources will be generated. It seems clear that this new view will have a profound impact on astronomy and astrophysics in studies of objects ranging from quasars and AGN to gamma-ray bursters, supernovae, variable stars, planets, comets, and asteroids. General-purpose image analysis tools keyed to variability are already in the final stages of development. These routines exploit the recent progress made in supernova and microlensing search projects. By automatically scaling and transforming two images so that an accurate subtraction can be performed, the separation of variable objects from more plentiful nonvariables becomes straightforward. To date, the classification of detected variability (into solar system objects, supernovae, microlensing, and so on) has been done by humans. Several research groups are currently devising a machine-assisted way to carry this classification out. The panel notes that the development of these tools is a prime example of a cross-disciplinary activity mutually beneficial to both the computer science and astrophysics communities and one that NSF should be eager to support.

Implementing an effective database and user interface for large volumes of astronomical data is a challenge that should be addressed by the National Virtual Observatory (NVO) initiative. The NVO will add great leverage to the WAVE data set, which will contribute the temporal dimension to the aggregate data set, while taking full advantage of unified data structures and user interface developments. The WAVE database will by itself likely be the largest nonproprietary data set in the world—it will be an ideal resource for testing the scaling and efficiency of data-mining tools and techniques.

A considerable effort will be required to construct a system producing roughly a terabyte of raw data per night, with rapid data reduction of images, classification of variability, characterization of sources, and rapid distribution of the data. The panel emphasizes that it is crucial to consider LSST with WAVE as precisely that: a complex system for which the telescope and instrumentation are only the front end.

MULTIPLICATIVE ADVANTAGES AND DISCOVERY SPACE POTENTIAL

One of the most attractive aspects of LSST is that it will enable simultaneous pursuit of many of the science goals described above. The time domain is (with a few notable exceptions) largely unexplored in astronomy. WAVE will provide unprecedented access to the theater of the sky and will pay tremendous dividends for a wide variety of scientific objectives beyond the ones mentioned here. The benefit of a continuous, deep, O/IR sky survey to future space missions and ground-based radio astronomy should be considerable. The technology for the construction of the optics and instruments is well in hand, and the project could lead the way in applying state-of-the-art information technology to achieve the rapid distribution of useful data to both the scientific and lay communities.

TECHNOLOGY AND COST ISSUES

TELESCOPE AND INSTRUMENTATION

Based on the existing successful projects of this size and allowing for a more complex optical system, the panel estimates the construction cost of a 6.5-m telescope with a 3-deg FOV at $60 million. The state of the art in currently deployed mosaic CCD arrays is 10K × 12K pixels. Cameras of 18K × 18K are currently in development, and there is no fundamental impediment to achieving the roughly 30K × 30K pixel array that would be required to cover the 3 deg × 3 deg FOV of LSST. Paving the field of LSST at 0.3 arcsec/pixel with existing 2K × 4K CCDs would require 162 devices, at a detector cost of roughly $6 million. Optics, mounting, cryogenics, electronics, and system integration would add ~$10 million, for a total instrumentation cost of $16 million.

A second-generation instrument with IR detectors is an obvious next step for the system and would greatly benefit from further developments in the footprint of IR arrays.

DATA PROCESSING AND DISTRIBUTION PIPELINE

Current high-end workstations with 250 GB of disk space, a few gigabytes of memory, and multiple processors cost approximately $20,000. Real-time analysis will require roughly 10 TB of disk space and the equivalent of perhaps 50 current high-end workstations. At current prices this analysis system would cost approximately $1 million, but the cost for this level of performance continues to fall steadily with time.

The estimation of software costs is less certain. Based on experience gained with SDSS and 2MASS and the current state of the art in variability analysis, implementing the WAVE analysis system should require approximately 30 full-time-equivalent (FTE) years of programming effort, at a cost of $3 million. Developing the requisite data structures and a user interface will require a comparable effort also costing about $3 million, for a total software capitalization cost of $6 million. Maintenance and upgrades should be budgeted at an annual cost of 10 percent of the initial software investment, or 6 FTEs, for a cost of $3 million for 5 years of operation.

The data repository costs will depend on the degree of synergy with the National Virtual Observatory and the amount of user support provided. Based on experience gleaned from the microlensing projects, the panel estimates that the volume of reduced data will be roughly one-tenth the volume of the raw image data. Such a time-series database would grow at a rate of roughly 20 TB per year and would, ideally, be stored on magnetic disks, at a cost of roughly $5 million per year if implemented at current prices.

The costs of the LSST are projected to be $83 million for capital construction and $42 million for data processing and distribution for 5 years of WAVE operation, for a total cost of $125 million. Routine operating costs, including a technical and support staff of 20 people, are estimated at approximately $3 million per year.[8]

CONTEXT ISSUES

Both the construction and operation of LSST and the processing and distribution of the data present suitable opportunities for NOAO to provide a critical service to the community, in keeping with its new role.

It is clear that this type of facility and the database that it will produce represent a critical element of ground-based support for space missions. Furthermore, they would form an integral piece of the proposed National Virtual Observatory—a multiwavelength assemblage of archives from many space- and ground-based observatories with the tools to exploit the total dataset. The project will generate ~1 TB of data every night that will have to be reduced in near real time. The classification of variable sources is an interesting and challenging computational problem that is amenable to neural network and adaptive techniques. The project would produce the world's largest nonproprietary dataset, which could serve as a testbed and development platform for investigating scalability in database implementations, including access/query issues, and for exploring data mining as a tool for research.

ANCILLARY BENEFITS

The panel believes that significant educational and societal benefits will accrue from the LSST. When maps of the changing sky become readily accessible on the Web, the general public will for the first time have access to images of such phenomena as moving objects detected in the solar system and stars exploding in the distant universe. Astronomical imagery will offer time-lapse movies in addition to still photos. A motion picture of the sky should be a qualitatively new tool for K-12 education: imagine a teacher having the class track the motions of the planets on a

---

[8]The cost estimated by the Astronomy and Astrophysics Survey Committee included the $83 million capital construction cost and $57 million for all operations, including data analysis, and added 15 percent of the capital cost for instrumentation and 15 percent for grants, for a rounded total of $170 million (see Astronomy and Astrophysics Survey Committee, National Research Council. 2001. *Astronomy and Astrophysics in the New Millennium*, Washington, D.C.: National Academy Press).

weekly basis, follow the approach of an incoming comet, or comb the database for galaxies with recent supernovae or cataclysmic variable stars. In addition, the astronomy community will continue to search for objects in the solar system that pose a threat to life on Earth.

MODERATE INITIATIVE, PRIORITY ONE: TELESCOPE SYSTEM INSTRUMENTATION PROGRAM: LEVERAGING NONFEDERAL INVESTMENT AND INCREASING PUBLIC ACCESS

The U.S. ground-based astronomy inventory consists of nine 6.5- to 10-m-class telescopes (operating or under construction), nine 2.5- to 5-m telescopes, and numerous smaller instruments. Considered together, this collection of both national and independent observatory facilities represents the world's most powerful ground-based telescope arsenal, with unequaled opportunities for maintaining leadership in research. However, these facilities have traditionally not worked together as a coherent system, in contrast with astronomy facilities abroad, which are dominated by national or international observatories. The panel believes that better coordination and cooperation are essential to realizing the full potential of this system and that NSF should work to achieve such coordination and to ensure that facilities and data are made widely available to the entire astronomy community.

DEFINITION

The panel proposes a new program, the Telescope System Instrumentation Program (TSIP), modeled on the Facilities Instrumentation Program[9] of support for instrumentation and instrumentalists at the independent observatories. The TSIP would guide the evolution of the telescope system so that it becomes more powerful and more diverse; it would do this by, for example, favoring instruments with unique capabili-

---

[9]Panel on Ground-Based Optical and Infrared Astronomy, National Research Council. 1995. *A Strategy for Ground-Based Optical and Infrared Astronomy* (Washington, D.C.: National Academy Press); also known as the McCray report after the panel's chair, Richard McCray.

ties and those that would be particularly effective in reaching the scientific goals described here. The panel supports the twin goals of achieving greater public access to these facilities *and* encouraging and leveraging the contribution of institutions that contribute nonfederal funds to the U.S. astronomy enterprise, to be accomplished by equal funding for both goals.

SCIENCE DRIVERS FOR 8-M TELESCOPES WITH ADVANCED INSTRUMENTATION

With the new generation of 8-m-class, ground-based O/IR telescopes and state-of-the-art instrumentation, many of astronomy's primary science goals for the next two decades can be achieved. Some examples follow:

- Assemble very large samples of galaxy photometry and spectroscopy over the redshift range $0 < z < 2$ to measure the chemical and, to the extent possible, structural evolution of galaxies in the context of the growth of large-scale structure. This effort will require wide-field imaging and highly multiplexed moderate-resolution ($R \sim 5000$) spectroscopy. Instrumentation required: moderate-resolution, low-background spectrographs with multi-million-pixel, low-noise, near-IR array detectors.
- Identify, through low-resolution spectroscopy, distant supernovae found through large-area surveys. Use type Ia supernovae in the redshift range $0 < z < 1.5$ to measure cosmological parameters, particularly to confirm recent measurements of a nonzero cosmological constant. Compile observations of all supernovae to study the history of star formation rates over a range of galaxy types and luminosities (see Figure 2.10). Instrumentation required: high-sensitivity optical and near-IR, low- to moderate-resolution spectrographs fed by AO systems.
- Probe the stellar content of the Galactic halo to characterize its assembly and chemical enrichment history. Find thousands of blue horizontal branch stars and RR Lyraes in the Galactic halo and obtain medium-resolution spectroscopy to map the dark halo to 200 kpc. Search for extremely metal-poor stars and other subpopulations and measure their kinematics (three-dimensional motions) to reveal how the halo was assembled through early agglomeration and later accretion. Study with high-dispersion spectroscopy the r- and s-process element distributions in the most metal-poor stars to probe early chemical enrichment of the halo. Extend these studies and the studies of Galactic

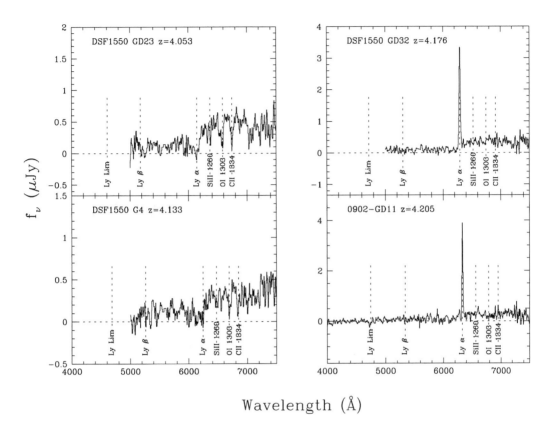

FIGURE 2.10 Low-resolution spectra obtained with the Keck telescope of $z \sim 4$ galaxies of approximately L^* luminosity. Positions of some of the prominent stellar and interstellar lines often found in the rest-frame far-UV spectra of star-forming galaxies are indicated. It is significant that despite the high rates of star formation, Lyman-alpha is seen in emission only about half of the time, as is the case here. Such spectra, which can now be obtained routinely (with a few hours' worth of integration) on 8-m-class telescopes with high-efficiency, multiobject spectrographs, have revolutionized the study of the early universe. Courtesy of C. Steidel, California Institute of Technology.

globular clusters to local group members and, to the extent possible, to more distant local supercluster galaxies. Instrumentation required: wide-field multicolor imaging to identify candidates and, for follow-up, moderate- to high-resolution spectrographs.

• Perform spectroscopic follow-up of samples of brown dwarfs in the Galaxy to study the frequency of binary systems and the evolution of their atmospheres; find evidence of chromospheres, flares, winds, and x-ray coronas. Instrumentation required: moderate-resolution, low-background, near-IR spectrographs.

• Study the infalling envelopes, accretion disks, and post-accretion disks around young stars. Use adaptive optics and ground-based interferometry in the IR to measure the structure and temperature distribution in the 10- to 100-AU region, and use high-resolution spectroscopy to probe the 1- to 10-AU region via observations of velocity-resolved molecular transitions such as those of CO and H_2O. Characterize the outflow regions of young stars and understand the transport of angular momentum and the evolution of magnetic fields. Look for structure in post-accretion disks indicative of planet formation and use coronagraphic imaging to search for high-mass planets and brown dwarf companions around neighboring stars. Instrumentation required: IR interferometers; AO-fed, near- to mid-IR, moderate-to-high-resolution spectrographs; coronographs.

• Use adaptive optics imaging to monitor weather on Mars and the Jovian planets, climatic variations on Titan, and volcanic eruptions on Io and make the first high-resolution maps of surface features of Mercury. Adaptive optics with spectroscopy will provide spatially resolved spectra of the atmospheres of Jovian planets. Measure the binary frequency of KBOs and obtain spectra of brighter objects to study composition. Instrumentation required: optical-to-mid-IR AO imaging; AO-fed, moderate-resolution, near-to-mid-IR spectrographs.

• Provide rapid-imaging follow-up of gamma-ray bursts to identify optical counterparts, and use sensitive spectroscopy to obtain host galaxy redshifts. Instrumentation required: instant-access O/IR cameras and low-resolution spectrographs.

• Use AO and IR interferometry to map the structure of AGN at very small scales in order to study the kinematics, temperature, and density structure of material close to the black hole/accretion disk. Instrumentation required: AO-fed, O/IR spectrographs; IR interferometers.

GUIDELINES FOR THE TELESCOPE SYSTEM INSTRUMENTATION PROGRAM

Because public and private resources have a history of uncoordinated development, U.S. capabilities in ground-based O/IR astronomy represent a strong—but at best loosely organized—approach to astronomical research, with agendas set by a wide range of institutions pursuing a variety of goals. Although this diversity is one of the strengths of U.S. astronomy, it is imperative that the new generation of 8-m telescopes be used as a total system in order for the nation to compete effectively. It has also become clear with the building of the 8-m telescopes that the resources necessary to instrument these telescopes properly, as well as to process, analyze, and distribute the data, are woefully inadequate. Traditionally, NSF grants have provided critical support for the processes of data gathering and reduction as well as for supporting theory and laboratory astrophysics programs essential for progress in O/IR research. However, the panel believes that NSF's already important role can become even more important in the coming decade if it enables national and independent observatories to work together, as a single system, to accomplish the scientific goals described in this chapter. Through a process of peer review, NSF can use its grants programs to focus limited federal resources in a way that will maximize the scientific return on these huge investments by supporting the development of instrumentation that provides special, as yet unavailable observing opportunities.

A key component of the nation's leadership has been the excellence of U.S. instrument builders, who have continually provided innovative, powerful, and cost-effective instrumentation for ground-based telescopes. Of particular concern is the training and support of future instrumentalists. The building of high-quality instrumentation, particularly in university environments, where students can be trained, is an essential component of a vigorous, diverse, ground-based telescope system. This instrumentation includes both traditional smaller (principal-investigator-scale) instruments as well as new, state-of-the-art facility instruments for 8-m-class telescopes. Compared with the former, the latter present special challenges: they are more expensive and more difficult to build, and they require larger groups, management structures, and longer production times. These challenges, as well as a disturbing lack of recognition in some university environments for this essential contribution to the scientific process, mean that many instrument builders see their opportunities for research hard pressed. Because instrument building at universities and observatories in the United States is crucial for the nation's continued success in astronomy, the panel believes that it is vital for NSF to focus its efforts on ensuring a healthy mix of smaller-scale instrumentation programs and large-scale facility instruments.

The facilities instruments for the new generation of 8-m telescopes are far more capable than their predecessors, but because of their scale and complexity, they are an order of magnitude more expensive, typically

$5 million to $10 million apiece (the VLT average is $11 million). Furthermore, data storage, analysis, and dissemination costs will be substantial. Fortunately, the investment required is incremental to the funds already expended at both private and public observatories for construction and continued operation of the new telescope facilities. The panel proposes a new investment beginning at $5 million per year in instrumentation for the independent observatories, concentrated on the new 8-m-class telescopes, which would leverage a scientific yield several times over. Although the panel believes that the program originally proposed in the McCray report[10] and recently reviewed and endorsed by the Committee on Astronomy and Astrophysics[11] provides the framework for the administration of this program, here it emphasizes elements of the McCray report that have not been applied so far.

The McCray report recognized that maximizing the quality and quantity of astronomical research in the United States depends on a vigorous investment from NSF; the report also acknowledged NSF's long-standing commitment to provide wide access to astronomical facilities, so that studies outlined in research proposals and rated as excellent through the peer review process could be carried out at premier facilities, public and private.

It is important to recognize that private facilities now support a large fraction of the U.S. astronomy community (~50 percent, according to the NRC report *Federal Funding of Astronomical Research*[12]), a situation very different from that existing when Kitt Peak National Observatory was founded in the 1950s. Although there are now many additional new opportunities for astronomical research, including space-based facilities, radio observatories, and data archives, it is abundantly clear that some measure of public access is vital to the health of U.S. astronomy.

The panel reaffirms the critical importance of maximizing scientific return and ensuring greater public access, and it emphasizes the crucial importance of regarding all of the U.S. O/IR facilities as a system. The

---

[10]Panel on Ground-Based Optical and Infrared Astronomy, National Research Council. 1995. *A Strategy for Ground-Based Optical and Infrared Astronomy* (Washington, D.C.: National Academy Press).

[11]Committee on Astronomy and Astrophysics, National Research Council. 1999. "On the National Science Foundation's Facility Instrumentation Program" (Washington, D.C.: National Academy Press), June 2.

[12]Committee on Astronomy and Astrophysics, National Research Council. 2000. *Federal Funding of Astronomical Research* (Washington, D.C.: National Academy Press).

NSF should administer the TSIP so as to achieve all of these objectives. By employing an approach that recognizes the important contribution of nonfederal funds for astronomy, the NSF will encourage independent observatories to participate. To encourage their participation, the TSIP must be broader than the program first implemented, when the NSF's goal of acquiring telescope time on private facilities dominated the process. Borrowing from the McCray report's recommendations, the panel strongly advocates the following guidelines:

• The TSIP should apply to facility instruments for independent observatories only, for which NSF grants at least $1 million in support of the proposed instrument. It would not replace existing Advanced Technologies and Instrumentation (ATI) or Major Research Instrumentation (MRI) programs.

• Successful proposals that include an offer of observing time would provide nights on the telescope whose value (based on amortized investment and operations) amounts to 50 percent of the granted funds. This 50/50 split properly recognizes the initiative of the independent observatory researchers in bringing nonfederal funds to astronomical research and supports their science while still attending to the important goal of providing observing time for the best peer-reviewed proposals, regardless of institutional affiliation. (The panel also notes a finding from *Federal Funding of Astronomical Research* that 50 percent of the users of ground-based O/IR facilities have access to independent observatories.) The 50/50 split should not be negotiable: to negotiate it would undermine the cooperative spirit needed to ensure the success of the overall O/IR system. The proposing institution may specify additional guidelines, for example, whether the time is available only on the proposed instrumentation or on all instrumentation, and it may include requests that specify operating modes, for example, minimum observing run duration. Such conditions are to be evaluated along with other aspects of the proposal.

• In lieu of some or all of the telescope time, proposals may be accepted that offer other comparable benefits to the astronomy community, for example, the production and dissemination of surveys and the archiving of data from this or other instruments on the telescope for which the instrument is proposed.

The panel considers the 50/50 split to be an essential part of its guidelines. It represents a fair division in which both communities benefit. It is clear that a "dollar of telescope time for a dollar of instrumentation funding" does not recognize or encourage the contribution of universities and private institutions in raising funds and does not recompense them for the talent and time of their scientists and engineers in building facilities. On the other hand, an unrestricted NSF grant of funds with no benefit provided to the broader astronomy community would frustrate the aspirations of the scientists who would like to use the unique facilities outside the national observatories. Something for both groups is the only appropriate solution; the 50/50 split has the added benefit of conveying the traditional notion of fairness. Negotiating the split, as was tried previously, promotes competition that can only undermine the goal of cooperation within the system.

Effective development of the entire suite of community facilities as a system depends on a common perception of how the parts of that system interact; a common vision of the strengths, deficiencies, and potential evolution; and an implementation plan with some level of feedback and accountability. The panel looks to NOAO for leadership in involving all segments of the community in discussions aimed at evaluating elements of the system, establishing a common vision, and devising plans for the future. Based on the results of these discussions, NOAO should develop a strategic plan for the system that includes an analysis of the benefits, costs, and risks for various prioritized implementation alternatives. Such an analysis would help the NSF decide how to invest its resources, either proactively, through solicitations for particular capabilities or negotiations for telescope time, or through TSIP, in response to instrument proposals (see Figure 2.11).

Given the wide variety of instrument capabilities and performance, scientific potential, public benefit, terms of use of telescope time, and importance to the system of O/IR facilities, the choice of successful proposals should be conducted annually by an NSF-constituted peer review committee. Any recommendations from the NOAO-led strategic planning effort should be provided to that NSF committee, but the peer review process should allow a full consideration of all factors. Every year, the NOAO-led, community-based strategic planning group should provide structured feedback to the NSF regarding the perceived efficacy of its investments in meeting the strategic goals of the community.

The panel also considered the effectiveness of direct purchase of telescope time by the NSF and concluded that in some circumstances, this could be the most efficient way for the entire community to gain access to a unique capability. However, the panel prefers the TSIP

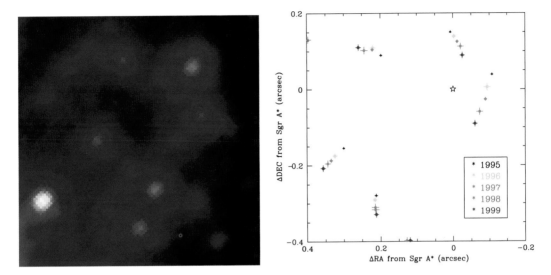

FIGURE 2.11 With the advent of diffraction-limited imaging on large ground-based telescopes, the center of the Galaxy has provided us with a unique testing ground for theories on galactic nuclei. On the left is an image of stars in the central 1 arcsec × 1 arcsec (~0.035 pc) centered on the putative black hole, obtained with the Keck adaptive optics system in a short demonstration (exposure time = 2 min). Speckle imaging at Keck over the last 4 years has been able to track the orbits of the brightest stars (right-hand panel), which reach velocities of up to 1400 km s^{-1} (0.5 percent of the speed of light). This is an example of how progress in instrumentation, adaptive optics in this case, can lead to breakthrough science. Courtesy of A. Ghez, University of California at Los Angeles.

approach because it—as opposed to the time purchase option—ensures the production of critically needed instrumentation and provides for community involvement (through peer review) in the types and capabilities of instruments that will be built and made available. For this reason, the panel recommends that the direct purchase of telescope time should be a second option, used sparingly so as not to significantly decrease the resources needed for TSIP, which would be sized according to the instrumentation needs of the independent observatories. It also urges that decisions by the NSF to buy telescope time should be guided by an understanding of the broad needs of the community, as is intended in TSIP.

The goal of building a more cohesive ground-based O/IR community—with increased incentives for private fund-raising, continued commitment to allowing public access to premier facilities, and maximized scientific creativity and output—should guide NSF policy throughout the decade. Without these investments and increased funding for research support, both provided by the NSF grants program, U.S. astronomers will not be able to continue to play their leadership role in the most basic astronomical research and will not be able to take full advantage of the powerful new space telescopes such as Chandra, SIRTF, and NGST.

TECHNOLOGY ISSUES

The NSF could enhance the system of O/IR facilities through continued investment in the development of technologies that will ultimately enable new capabilities. Development of detectors, especially large-format, near- and mid-IR arrays, is a key area. AO systems to feed near-IR spectrographs are extremely important to reduce background and work in crowded regions of the sky. Large-scale surveys require new, more efficient data-handling techniques. IR interferometry will open a new discovery space for the study of high-surface-brightness objects and may spur the evolution of ground- and space-based filled-aperture telescopes.

COST ISSUES

The investment for the first complement of VLT instruments is $91 million, with an expected continuing investment of at least $10 million per year. These figures reflect a realistic assessment of the ongoing investment needed to take full advantage of the $500 million investment in new telescopes (excluding the investment for infrastructure at Paranal). With a comparable capital investment and an even greater number of telescopes, a continuing U.S. investment for major instrumentation of at least this magnitude is required. NSF has separate commitments for instrumentation at NOAO and Gemini, and some nonfederal support for instrumentation is available, leading the panel to conclude that an additional $5 million per year for the independent observatories is critically needed.

CONTEXT ISSUES

The importance of ground-based facilities for space-based telescopes and ground-based radio astronomy is clear. Most astronomical projects, regardless of wavelength domain, have an O/IR component that is effectively addressed with ground-based O/IR facilities. The data collected should be archived and made available to the astronomy community, following the lead of the National Virtual Observatory (see the report of the Panel on Theory, Computation, and Data Exploration, Chapter 6). Support for instrumentalists is a particularly vital part of this program; the NSF can and should initiate programs that encourage and support instrument builders throughout the community. The greater participation of theoretical astrophysicists in the planning of large programs also needs encouragement and support.

OTHER ISSUES

The panel received valuable input from groups and individuals advocating programs and projects with particular scientific or technological thrusts. Described below are several that the panel found particularly meritorious:

- Spurred both by revolutionary advances in large-area detectors and data-processing capability and by the scientific promise of a new generation of large telescopes that can enable the study of large samples, surveys play an increasingly important role in astronomical research. Although the panel's main recommendations focus on the large individual effort to conduct a repetitive all-sky survey (LSST with WAVE), it is clear that the infrastructure needed for effective use of the vast amount of data collected in this and other surveys must be provided as well. Accordingly, the panel strongly endorses the National Virtual Observatory initiative, which would link many extant and future archives through common standards and protocols and would develop the tools to mine these archives effectively. The panel also calls attention to the substantial data sets from ground-based O/IR telescopes that are not integrated into archives. The development of archives for ground-based data represents a critical link between the independent observatories and the larger astronomy community. The NSF should explore (through NOAO) means for the systematic development of such archives.
- A new and exciting class of problems could be tackled using highly multiplexed multi-object spectroscopy. While much thought and effort have gone into the development of arguments for imaging facilities having a large $A\Omega$ (aperture × field of view), it is also clear that the prospect of being able to work with samples of millions of spectra opens up the possibility of other studies, for example (1) tracing the evolution of

large-scale structure (for $1 < z < 4$) and (2) understanding the dynamical history of our galaxy's halo by locating and studying the remnant streams from individual merger or infall events. These questions should be pursued by initiating preliminary studies of this type and by understanding how best to develop such a capability, possibly through incorporation into the GSMT facility or later modification of the LSST facility.

• It is clear that future very large facilities, whether they are in space or on the ground, will not be limited to filled apertures but, through interferometric imaging, will enable trade-offs in collecting area, dynamic range, and angular resolution. Although the panel endorses no specific interferometry facilities for the decade, there is a strong consensus to support development of interferometry techniques and facilities in order to understand those trade-offs.

• Another important region of parameter space that deserves increased emphasis is the time domain. Diverse science returns, such as MACHO lensing discoveries and AGN reverberation mapping, would be enabled by facilities that provide synoptic capabilities complementary to the proposed LSST with WAVE survey initiative—for example, an array of medium-size telescopes distributed around the world or dedicated photometric/imaging monitoring telescopes. Many of these projects would benefit greatly from automated telescopes, which allow routine but complete observational programs to be carried out with minimal operating costs.

• In addition to facilities that would target capabilities such as synoptic imaging, there is a need for supporting the specialized but limited projects that have been using the smaller national telescopes, for example, the spectroscopic monitoring of relatively bright stars for periodic and episodic activity, observations that are important for the theoretical modeling of stellar interiors and atmospheres. Recognizing that many important programs can be carried out with telescopes of modest aperture and that NOAO is likely to provide fewer such facilities in the future, the panel urges NSF to seek alternative means of supporting them—for example, by buying observing time at independent observatories or funding the development of specialized instrumentation.

• The panel was excited by the prospect of new ground-based sites that offer unique windows through the atmosphere. One of these is the very cold site at the South Pole, the other is the very dry site in the Atacama desert of northern Chile. The South Pole site would support wide-field imaging with a 2-m telescope operating in the 2.4- to 5-μm range, an important capability that would, for example, enable a census of brown dwarfs in the solar neighborhood.

• Although the availability of CCDs and HgCdTe arrays continues to improve, there is a pressing need for better coordinated, aggressive development of larger format arrays, particularly IR array detectors. Collaboration between NSF and NASA to develop O/IR detectors could yield benefits for both ground- and space-based applications.

• Laboratory astrophysics studies are crucial for many of the science domains discussed in this chapter. Particularly in studies of stars, where high-resolution spectroscopy enables the detection of thousands of unblended lines, it is essential to identify spectral lines of complex atoms, molecules, and their ions, together with oscillator strengths for these myriad transitions. Furthermore, quantitative models of stars are based on still-improving measurements of opacity, and nuclear physics provides the basis for understanding nucleosynthesis.

Of particular relevance for the science described here is the increasing importance of infrared observations, in the study of protostars or in the Galactic center, for example. As researchers probe the environments where stars and planets are born, an understanding of the complicated molecules and solid-state materials that dominate their observations depends on laboratory measurements. Also needed are good transition probabilities for the rare earth elements to probe the changing abundances of r-process and s-process elements, so important to the studies of chemical enrichment in the Galactic halo.

Although laboratory work of this nature is supported by a number of federal agencies, the relevant astrophysical work is dependent primarily on NASA and NSF funding. The panel urges both agencies to provide adequate support for the vital laboratory studies upon which so much astronomical work is based.

ACRONYMS AND ABBREVIATIONS

2MASS—Two Micron All Sky Survey
AGB—asymptotic giant branch
AGN—active galactic nuclei
ALMA—Atacama Large Millimeter Array
AO—adaptive optics
ATI—Advanced Technologies and Instrumentation (an NSF program)
AU—astronomical unit (150 million km)

CARMA—Combined Array for Research in Millimeter-wave Astronomy
CCD—charge-coupled device
CELT—California Extremely Large Telescope
Chandra—Chandra X-ray Observatory (NASA, launched in 1999)
dof—degrees of freedom
ELT—Extremely Large Telescope
ESO—European Southern Observatory
FOV—field of view
FTE—full-time equivalent
Gemini—NOAO/multinational Northern and Southern Hemisphere 8-m
    telescope project
GSMT—Giant Segmented Mirror Telescope
HII—ionized hydrogen
HET—Hobby-Eberly telescope
HST—Hubble Space Telescope
IGM—intergalactic medium
IMF—initial mass function
IR—infrared
ISM—interstellar medium
KBOs—Kuiper Belt objects
LSST—Large-Aperture Synoptic Survey Telescope
MACHO—massive compact halo objects
MAXAT—Maximum Aperture Telescope
MRI—Major Research Instrumentation (an NSF program)
MT—million tons of TNT, a unit of energy
NASA—National Aeronautics and Space Administration
Origins—a NASA program
NEO—near-Earth object
NGST—Next Generation Space Telescope
NICMOS—the near-infrared camera and multiobject spectrometer on
    the Hubble Space Telescope
NOAO—National Optical Astronomy Observatories
NRC—National Research Council
NSF—National Science Foundation
NVO—National Virtual Observatory
O/IR—optical/infrared
QSO—quasi-stellar object
OWL—Overwhelmingly Large Telescope or Observatory at a World
    Level, an ESO proposal for a 100-m telescope
SDSS—Sloan Digital Sky Survey

SIM—Space Interferometry Mission
SIRTF—Space Infrared Telescope Facility
SMA—Submillimeter Array
TSIP—Telescope System Implementation Program
VLT—Very Large Telescope
VRI—observations through visual, red, and infrared filters
WAVE—Wide Area Variability Experiment
WIYN—observatory run by the University of Wisconsin, Indiana
   University, Yale University, and NOAO

3

Report of the Panel on Particle, Nuclear, and Gravitational-Wave Astrophysics

SUMMARY

Particle and nuclear astrophysics and gravitational-wave astronomy offer tremendous discovery potential in the next decade and beyond. The direct measurement of gravitational waves from astrophysical sources will open new investigations in both astrophysics and the physics of strong gravitational fields. High-energy charged particles and gamma rays as well as neutrinos carry unique information about the high-energy universe that is complementary to information obtained by more traditional astronomical approaches. The quest to identify the dark matter is of the utmost importance for astrophysics and cosmology as well as for elementary particle physics.

The Panel on Particle, Nuclear, and Gravitational-Wave Astrophysics of the Astronomy and Astrophysics Survey Committee recommends that highest priority be given to the Laser Interferometer Space Antenna (LISA) because of the fundamental and novel exploration of the gravitational-wave universe it can accomplish, including the observation of massive black holes coalescing in colliding galaxies and the study of white dwarf binaries in our own galaxy. The panel's highest recommendation among ground-based projects (and second overall) is the Very Energetic Radiation Imaging Telescope Array System (VERITAS), which together with the Gamma-ray Large Area Space Telescope (GLAST) will study many rapidly variable energetic sources, including nuclei of active galaxies, and will map the gamma-ray sky with unprecedented precision. An attractive small-scale opportunity is the Advanced Cosmic-ray Composition Explorer for the Space Station (ACCESS), which will be able to measure directly the spectrum of particles to 1000 TeV and for the first time to distinguish the spectrum produced by the cosmic accelerators from energy-dependent effects of propagation in the Galaxy.

In setting priorities, the panel used three criteria: scientific importance, technological readiness, and budgetary reality. In some cases, however, where the path forward depends on results of investigations just now starting, it is not yet possible to evaluate a project even though it addresses an extremely important problem and is likely to be ready within the coming decade. The panel therefore recommends a broad program of particle astrophysics building on the important new initiatives of the past decade, including solar neutrino observatories, giant air shower detectors, neutrino telescopes, and searches for dark matter.

The scientific interest and importance of all the projects of this panel are strengthened by their multidisciplinary character. The panel therefore recommends policies to nurture such research.

SCIENCE OPPORTUNITIES

A unifying theme of many of the projects considered by this panel is the desire to study energetic processes in the cosmos not only in all wavelength ranges but with a variety of signal carriers. The idea would be to detect gravitational waves, neutrinos, hadrons, and photons from the same source and so take advantage of the complementary information they carry:

- Gravitational waves provide information on the bulk motions of matter in the most energetic events in nature, such as the coalescence of black holes.
- High-energy (nonthermal) photons trace populations of accelerated particles.
- Cosmic-ray protons and nuclei carry information about the cosmic accelerators that produced them.
- Neutrinos emerge directly from deep inside regions that are opaque to photons.

Cosmological gamma-ray bursts (GRBs) offer a good example of the potential benefits of complementary observations with more than one probe as well as at different wavelengths. The distribution of bursts is isotropic over the sky, but until 2 years ago it was not known if they were in the halo of our galaxy or at cosmological distances. Now, following a coordinated series of x-ray, optical, infrared, and radio observations, it is known that many bursts are cosmological.

Although detailed mechanisms of gamma-ray bursts are not understood and there may be substantial beaming by the sources, there is now little question that bursts represent the conversion of a significant fraction of a stellar rest mass into energy. To achieve this level of power output will probably involve ultrarelativistic motions of stellar masses drawing energy from the gravitational potential. Plausible concepts include the formation of a black hole from the coalescence of orbiting compact objects or a new class of stellar collapse resulting in a black hole. In view of the energies involved, it can be expected that models for the burst mechanism will be constrained by the measurement of (or useful upper limits on) coincident high-energy gamma rays, neutrinos, elementary particles, and gravitational waves, as well as wide-field optical and radio observations.

In what follows, the major accomplishments of the past decade and the future opportunities are grouped into four broad categories. Gravitational waves offer the potential to revolutionize our understanding of the role of massive black holes in the dynamics and evolution of galaxies. Cosmic particle acceleration (as manifested in gamma-ray, charged-particle, and high-energy neutrino astrophysics) is an essential feature of energetic processes on all scales. Neutrino astrophysics is the study of low- and medium-energy neutrinos from the Sun and energetic sources such as supernovae. Identifying the dark matter is a key goal for understanding the large-scale structure of the universe, because until this is done researchers cannot know what most of the mass of the universe is made of.

GRAVITATIONAL-WAVE ASTROPHYSICS

The role that gravitational radiation plays in the energy loss of massive, rapidly moving astrophysical systems has been established by means of the orbital period change of the binary neutron star system discovered by Hulse and Taylor. The direct measurement of gravitational waves from astrophysical sources will open up new opportunities for investigations in both astrophysics and the physics of strong gravitational fields. The gravitational waves will convey information about the large-scale motions in the dense inner regions of astrophysical systems normally not open to view in electromagnetic observations. Observation of the final inspiral of two black holes would serve as a unique probe of the strong-field limit of general relativity. The sources of gravitational waves are changing-mass quadrupole moments. Astrophysical processes can result in impulsive, periodic, and stochastic gravitational waves. Impulsive sources include the cores of supernova explosions, the metric perturbations in the formation and dynamics of black holes, and the coalescence of compact binary systems. Periodic gravitational waves may originate in the coalescence of massive black holes, in the accretion-driven excitation of normal modes in neutron stars, or in the rotation of pulsars. A stochastic background of gravitational waves would result from a collection of spectrally unresolved binary stellar systems and possibly from the metric fluctuations in the primeval universe.

The search for gravitational radiation from astrophysical sources has so far been executed primarily with acoustic bar detectors that set upper limits for gravitational-wave strains of 10^{-18} in bands several hertz wide in the spectral region around 1 kHz. To set the scale, the supernova 1987A, which gave the first evidence for neutrinos from a stellar collapse, would have produced a strain of 10^{-18} at Earth if as much as 1 percent of the rest energy of the imploding star had been converted into gravitational radiation in 10^{-2} seconds. None of the sensitive acoustic detectors was operating at the moment when the prompt neutrino signals from SN1987A arrived at Earth.

Laser interferometers currently under construction with arm lengths of 4 km will initially operate at frequencies from 40 Hz to several kilohertz with a strain sensitivity of 10^{-21}. Improvements are planned that will enhance strain sensitivity by a factor of 10 to 30 and extend the observing band to lower frequencies. Potential sources include chirps resulting from the coalescence of binary neutron star systems similar to the Hulse-Taylor system, supernovae, and formation or collisions of 1- to 1000-solar-mass black holes. By extending the search to cosmological distances, the improvements to the long-baseline detectors on Earth will make it likely that coalescing binary neutron stars will be detected.

Lowering the frequency sensitivity of gravitational-wave detectors would open the window on an important new class of sources involving the formation and interaction of $\sim 10^6$-solar-mass black holes, which are thought to lie at the centers of many galaxies. A detector with sufficient sensitivity in the frequency band between 10^{-4} and 10^{-1} Hz could expect to witness the last year in the merger of two supermassive black holes in colliding galaxies as they spiral in toward a final cataclysmic event. Observation of the characteristic orbital period as it decreases from hours to minutes would enable the gravitational-wave detector to predict the time and general location so that the final event could be observed by a variety of telescopes and detectors on the ground and in space. In addition, such a detector could observe the gravitational radiation patterns of a large number of white dwarf binaries in our galaxy.

Detecting low-frequency gravitational radiation requires going into space to escape the effects of density fluctuations in the ground and the atmosphere. Such density fluctuations cause Newtonian gravitational forces on the mirrors. The mirrors cannot be shielded nor can the forces be eliminated by vibration isolation systems. These backgrounds limit the sensitivity of terrestrial detectors to frequencies above a few hertz. A long-baseline detector in space will open up the low-frequency range and provide insight into the processes in the centers of galaxies involving dynamics in strong gravitational fields.

COSMIC PARTICLE ACCELERATION

High-energy, nonthermal particles are a prominent feature of energetic astrophysical sources ranging from supernovae and flare stars in the galaxy to accreting massive black holes in the centers of distant active galaxies, to mention just two. In this section, the study of high-energy gamma rays produced in interactions of electrons or ions in the sources is considered first. Then, the status of cosmic-ray protons and nuclei in the Galaxy is considered (their relation to specific sources and acceleration processes is still not fully understood). Next, the very highest energy cosmic ray particles, whose origin is even more puzzling, are discussed. Finally, the possibility is discussed of opening a new window on particle acceleration by detecting high-energy neutrinos produced deep inside energetic astrophysical sources.

GAMMA-RAY ASTROPHYSICS

The study of very-high-energy gamma rays is a powerful tool for understanding particle acceleration in energetic astrophysical sources in distant galaxies as well as in the Milky Way. The development of the imaging technique for Cherenkov telescopes over the past decade has revolutionized ground-based gamma-ray astronomy by dramatically lowering the diffuse background of cosmic-ray showers; this is done by rejecting events with the irregular shape characteristic of hadronic rather than electromagnetic cascades. This achievement led to the discovery of very-high-energy (VHE) gamma radiation from a variety of sources, including pulsar nebulae, shell-type supernova remnants (SNRs), and jets of active galaxies, by atmospheric Cherenkov telescopes on four continents. The Crab Nebula, the first unambiguous VHE gamma-ray source, was originally detected by the Whipple Observatory (and is now detected by a number of instruments at very high significance); it confirmed the prediction of inverse Compton radiation from synchrotron-emitting relativistic electrons. Two other pulsar nebulae, PSR1706-44 and Vela, have also been detected by the CANGAROO telescope in the Southern Hemisphere. The detection of very-high-energy emission from the shell-type SN1006 by CANGAROO (along with nonthermal x rays detected by the Japanese x-ray satellite ASCA) also indicates the presence of relativistic electrons with energies up to 100 TeV. Such source detections at high significance can now be routinely made, so that detailed studies of various emission mechanisms are possible.

The most exciting new discovery in ground-based gamma-ray astronomy in the past decade was the detection of VHE emission from active galactic nuclei (AGN), such as Mrk 421 and Mrk 501, both of which belong to the blazar class, in which the observer is in the beam of a jet of the AGN. In 1997, flares from Mrk 501 showing variability on timescales as short as 1 h were monitored by four different experiments in the Northern Hemisphere. Time variations as short as 30 min have been seen in the gamma-ray fluxes from individual AGN. These discoveries initiated large multiwavelength campaigns, using instruments at radio, optical, x-ray, and gamma-ray energies to study variability in blazars and to constrain their emission models.

These studies will be expanded to develop a detailed picture of particle acceleration in SNRs and the jets of AGN. Absorption of blazar spectra at gamma-ray energies can be used in conjunction with measurements made at infrared wavelengths to understand the radiation fields near active galaxies and the cosmic IR background. The potential for future discoveries by ground-based gamma-ray telescopes is good. So far, only a small portion of the sky has been studied at very high energies, and a major band at energies between 20 and 250 GeV has yet to be explored by any instrument. The most important instrumental innovation for the coming decade will be stereoscopic imaging Cherenkov telescopes with greatly improved sensitivity and angular resolution.

In addition to the new directions outlined above, the next-generation Cherenkov telescopes have the potential to address other exciting topics, including (1) the discovery of sources that are bright at very high energies but faint at other wavelengths, (2) the detection of evidence for proton acceleration in SNRs through measurements of energy spectra with good angular resolution and in conjunction with measurements at other wavelengths to identify the π^0 component, (3) the detailed study of the high-energy spectrum of gamma-ray bursts, (4) the detection of attenuation in the spectrum of extragalactic sources at high energy, indicating absorption by pair production on the cosmic infrared background radiation, and (5) the search for cold dark matter in the galactic center by means of gamma-ray line emission.

PARTICLE ACCELERATION IN THE GALAXY

It is generally believed that cosmic rays are accelerated by sources distributed in the Galactic disk. They subsequently diffuse through the disk and halo before they escape from the Galaxy or are lost by nuclear interactions with the interstellar medium. The only sources that seem to be capable of providing the $\sim 3 \times 10^{40}$ erg/s required to maintain this balance between acceleration and escape are supernova explosions. While there is evidence to support this idea, two key questions remain about the supernova origin of cosmic rays:

- What is the mechanism by which cosmic rays gain their enormous energies? It is widely suspected that the bulk of cosmic rays are accelerated by diffusive shock acceleration, but the evidence for this hypothesis is somewhat indirect because cosmic-ray energy spectra measured in the Galaxy ($\propto E^{-2.7}$) are apparently modified from the accelerated spectra (expected to be $\propto E^{-2.1}$) by energy-dependent diffusion through and leakage from the Galaxy. To correct for such propagation effects and obtain the source spectrum requires precise measurements of both primary accelerated species (such as H, He, C, O, Si, and Fe) and secondary species (such as Li, Be, and B) that are produced by nuclear interactions with the interstellar medium. Extending measurements of secondary/primary nuclei from the present limit of ~ 100 GeV per nucleon by at least an order of magnitude would for the first time permit an unambiguous determination of the source spectrum. Measurements of the primary nuclei are needed up to an energy approaching 10^6 GeV, where shock acceleration by supernova blast waves is expected to reach its limit, perhaps causing the spectra to steepen. Such measurements require extended exposure of a large detector in space, outside the atmosphere.
- What are the nature and source of the matter that is injected for acceleration to cosmic-ray energies? The well-known fact that cosmic rays are depleted in elements with first ionization potential more than ~ 10 eV suggests that they originate in the coronas of stars like the Sun. It was suggested recently, however, that volatility may be the relevant atomic parameter and that cosmic rays may be grain-destruction products mixed with some interstellar gas. The panel endorses the development of instruments to resolve this key issue.

At energies below a few GeV per nucleon, cosmic-ray spectra at

Earth are attenuated from those in interstellar space by an (unknown) factor of 100 or more because low-energy cosmic rays are largely excluded from the heliosphere by the solar wind. NASA's proposed Interstellar Probe mission, which would send a spacecraft beyond 200 AU, would make a broad range of measurements of matter and fields in interstellar space. As an in situ investigation, it is outside the purview of the Astronomy and Astrophysics Survey Committee, but it is mentioned here because it would include measurements of the spectra and composition of cosmic rays in the local interstellar medium (ISM). It could observe shock acceleration in situ and assess the contribution of cosmic rays to radio and gamma-ray observations and to galactic dynamics. Interstellar Probe would also study acceleration processes at the solar-wind termination shock and measure the composition of the interstellar gas.

HIGHEST-ENERGY COSMIC RAYS

As indicated above, the SNR acceleration mechanism becomes inadequate above 10^{14} to 10^{15} eV, yet the cosmic-ray spectrum is known to continue for at least five more decades in energy. New acceleration sites and mechanisms are needed. AGN may be able to accelerate particles to 10^{20} eV, and gamma-ray burst sources have also been suggested. Achieving such high energy with these sources, however, requires optimistic assumptions about the conditions and parameters of the acceleration mechanisms. Moreover, there is no clear evidence yet that singles out one particular class of astrophysical objects as the most likely source of the highest-energy events. To observe a more exotic class of sources would require that the events be produced by the decay of massive relics from the early universe, possibly topological defects in space. Detailed studies of spectral shape, particle composition, and anisotropy are required to elucidate what is going on in this energy region.

The first report of an event with energy of approximately 10^{20} eV came from the Volcano Ranch experiment in 1965. A few other such large events were gradually accumulated with large ground arrays at Haverah Park, United Kingdom; Yakutsk, Russia; and Sydney, Australia. The Fly's Eye detector has measured the profile of a shower with 3×10^{20} eV, and data from the ground array at Akeno (currently the largest) now confirm that the cosmic-ray flux continues past the predicted Greisen-Zatsepin-Kuz'min (GZK) cutoff. This cutoff in the cosmic-ray spectrum is expected to be due to the onset of inelastic interactions between 10^{20} eV protons and the 2.7 K universal blackbody radiation. If sources of such protons are at cosmological distances, the cosmic-ray spectrum should cut off near 6×10^{19} eV. A similar effect will occur for nuclei, but it will be due to photospallation. Thus, protons and nuclei with energies beyond the GZK cutoff must originate in the local super-cluster of galaxies. At these energies, charged particles should propagate nearly rectilinearly over such distances, but no obvious candidate sources such as AGN have been found in the error boxes of the seven events so far discovered. The mechanism that accelerates particles to these energies is thus completely unknown and represents the most significant departure from thermal equilibrium found in the universe. The desire to solve this mystery motivates current and planned efforts to build giant air-shower detectors with unprecedented acceptance (area × solid angle).

HIGH-ENERGY NEUTRINOS

An entirely new window into the deep interior of energetic sources could be provided by high-energy neutrinos produced in interactions of accelerated protons with gas or photons. Because neutrinos interact weakly with matter, they can escape from environments so dense that high-energy photons are absorbed or degraded in energy. For the same reason, however, neutrinos are difficult to detect, and very large detectors are needed. Moreover, neutrino detectors must be deeply buried to reduce the abundant cosmic-ray backgrounds that are present near the surface. A current example of a tracking neutrino detector is the Super-Kamiokande detector in a deep mine in Japan. At 50 kilotons the detector is big enough to detect copious solar and atmospheric neutrinos but not the high-energy neutrinos that might be tracers of acceleration processes in distant astrophysical sources. To achieve this goal, it is believed that detector volumes on a scale of at least a cubic kilometer (1000 megatons of water) will be needed. The effective volume for μ_ν-induced muons is projected detector area × muon range (>2 km for $E_\nu > 1$ TeV).

The essential characteristics of a high-energy neutrino telescope have been known for more than 20 years. All current architectures bury a sparse array of optical sensors within deep ice, deep seas, or deep lakes. The optical sensors respond to the UV-dominated Cherenkov radiation emitted by neutrino-induced muons or neutrino-induced hadronic or electromagnetic cascades. Astronomy is possible because the muon direction is aligned with the incident neutrino to within 1 deg if the energy of the neutrinos exceeds 1 TeV. The muon is detected by distributing the photon sensors (large-diameter photomultiplier tubes) over the largest possible volume of transparent medium and recording the arrival times and intensity of the Cherenkov wavefront. The detectors are deployed in string or tiered arrangements.

The path toward very large neutrino telescopes has been advanced steadily during this last decade by the commissioning of detectors at the South Pole and in Lake Baikal, in Siberia. The experience with these detectors indicates that a kilometer-scale, high-energy-neutrino telescope could be built within the decade 2001 to 2010. Such a large size is needed to have a high probability of detecting neutrinos from astrophysical sources.

NEUTRINO AND NUCLEAR ASTROPHYSICS

SOLAR NEUTRINOS

Stars emit neutrinos directly from the fusion processes that power them. Attempts to measure neutrinos from the nearest star, the Sun, began over 30 years ago. The program in solar neutrino research aims to measure the entire spectrum of solar neutrinos, but it is incomplete as the 21st century begins.

The highlights of the past decade include the first real-time, directional detection of solar neutrinos in the Kamiokande and Super-Kamiokande light-water detectors; the confirmation by these detectors that the flux of neutrinos is much lower than stellar evolution calculations predict; the study of systematic errors in the pioneering ^{37}Cl experiment, putting the experiment on a more stable foundation; and the remarkable suppression of the low-energy neutrinos observed in two calibrated ^{71}Ga detectors, GALLEX and SAGE. In addition, a series of helioseismological measurements confirms that the temperature and density profile of the Sun are essentially as predicted by stellar-structure calculations. The indications are, therefore, that the solar neutrino deficit reflects novel physical properties of neutrinos rather than some poorly understood feature of the solar model.

The discovery by Super-Kamiokande of neutrino flavor oscillations involving muon neutrinos strongly reinforces the idea that the solar neutrino problem is indeed another manifestation of the pattern of mixing and mass differences of different types of neutrinos. Claims, still unconfirmed, of yet another manifestation of neutrino oscillations by the Liquid Scintillation Neutrino Detector experiment would require the existence of a new, sterile neutrino. Complete understanding of solar neutrino observations will be an integral part of understanding fully the properties of neutrinos. This is important for astrophysics as well as for particle physics, because neutrinos play a fundamental role in many processes in the cosmos, from the Big Bang to supernovae.

Various nuclear fusion processes in the Sun's interior produce electron neutrinos. Low-energy pp neutrinos, which have a continuous spectrum up to 0.42 MeV, dominate neutrino production. The much rarer ^8B neutrinos have a continuous spectrum extending to 15 MeV, making direct detection somewhat simpler. Intermediate in flux and energy are the line sources from ^7Be neutrinos at 0.86 and 0.38 MeV and from the pep neutrinos at 1.44 MeV. A small flux from the carbon-nitrogen-oxygen processes in the Sun also contributes at energies below 2 MeV. The hep neutrinos have a continuous spectrum extending to 18 MeV; their abundance is difficult to predict but is thought to be small. The first-generation experiments mentioned above are sensitive to different components of the solar neutrino flux, but all observe fewer neutrinos than predicted by solar models. The hypothesis of neutrino mixing, or oscillation, can account for the results of all experiments so far. In this process, electron neutrinos produced in the Sun become trans-formed into another neutrino type to which the first-generation detectors are not fully sensitive. However, more may be going on, because the Super-Kamiokande experiment observes a high-energy spectral distortion that may be accounted for by a great abundance of hep neutrinos.

The second generation of solar neutrino experiments currently getting under way is designed to investigate in more detail the neutrino oscillation hypothesis. The results of these experiments will determine the future direction of solar neutrino research. Completion of the solar neutrino program will require a further series of experiments that can map the solar neutrino energy spectrum, including the low-energy pp neutrinos, separately in electron neutrinos and in all types of neutrinos. Comparison of the two series of experiments will permit the mixing of neutrinos to be measured as a function of energy. Only then will it be possible to use solar neutrinos as a precision probe of processes in the solar interior.

SUPERNOVA WATCH

SN1987A was unique among supernovae because of the information that came from detection of neutrinos emitted during the stellar collapse and from measurement of the light curve resulting from the explosion. The new generation of neutrino detectors would produce much statistical information on neutrino yields and spectra from a supernova that occurs in our galaxy. Galactic supernovae are rare, with perhaps three occurring every century. Since the neutrino signal from a supernova precedes the optical signal by hours, it could be useful to predict the onset of such a supernova to allow optical instruments to point and thereby see the early rise of the light curve. For example, the large Super-Kamiokande detector could point with an accuracy of about 5 deg within an hour to a supernova at 10 kpc. The corresponding pointing accuracy for the smaller SNO detector is about 20 deg. These detectors, supplemented by future underwater (under-ice) experiments, will form a fast alarm network to identify and point to any galactic supernova, serving to alert other detectors before the onset of the optical outburst.

As the gravitational interferometers come on line, they will join the network of neutrino detectors, adding complementary early information, possibly for supernovae as far away as the Virgo cluster if the collapse is sufficiently nonaxisymmetric.

A goal for the future is to build sufficiently sensitive detectors so that the neutrino burst from a supernova in a distant galaxy could be detected. Concepts for large detectors to see out to 10 Mpc have been discussed, but the technology is not yet in place. If this goal could be realized, researchers could hope to detect several galactic supernovae each year.

NUCLEAR ASTROPHYSICS

The cosmos is powered by gravity and by nuclear reactions. Much of what is understood about processes in the universe is learned through the study of nuclear physics and nuclear astrophysics.

It is believed that in the early moments of the Big Bang, before nuclear matter could form, the universe was a plasma of quarks and gluons. As the universe expanded and cooled, the quarks and gluons condensed into protons, neutrons, and other particles. One current thrust of research in nuclear physics is the attempt at the Relativistic

Heavy Ion Collider (RHIC) to create and study tiny volumes of space in which the quark-gluon plasma is produced by colliding relativistic heavy ions. In this way, it should be possible to recreate briefly conditions similar to those in the early universe. In addition, producing a quark-gluon plasma many times in the laboratory may make it possible to create new forms of matter. It is even possible that such new forms of matter exist in the universe but have not yet been detected.

After the quark-gluon plasma condenses into neutrons and protons, these particles in turn begin the synthesis of the light elements. Understanding this primordial nucleosynthesis requires knowledge of the neutron lifetime and the reaction rates of the light nuclei. Our present understanding of primordial nucleosynthesis explains the observed abundances of the elements through lithium and is a triumph of nuclear physics and cosmology.

As galaxies and stars evolved, the nuclear reactions responsible for stellar evolution produced elements heavier than lithium. For the most massive stars, nuclear burning continues until the nuclear fusion processes in the stars produce iron. The nuclear fuel is then quickly exhausted and the gravitational collapse of the star produces a supernova explosion. Heavy-element nucleosynthesis comes about in nuclear reactions following the explosion. The debris from the explosions feeds the interstellar medium and determines the chemical evolution of the galaxy. Much active research in nuclear physics is being carried out in this important area. Especially relevant to rapid nucleosynthesis in supernova explosions is research with radioactive beams to investigate processes far from the valley of nuclear stability. In the United States, the radioactive-beam facilities at Argonne National Laboratory, Oak Ridge National Laboratory, and Michigan State University are powerful instruments. A proposed rare-isotope accelerator would extend the capabilities of these facilities and put the United States in the forefront.

The details of the supernova mechanism are determined in part by the nuclear equation of state for the very dense matter formed in the gravitational collapse of a massive star. Further, when the remnant neutron star cools, the equation of state of the nuclear matter determines the internal structure of the neutron star. Glitches in pulsar timing may be caused by transfers of angular momentum between regions of the neutron star interior. Thus we may learn about the equation of state of cold, dense nuclear matter from these astrophysical processes in a way that is complementary to the way we learn from collisions of relativistic heavy ions.

SEARCH FOR DARK MATTER

The nature of the dark matter in the universe remains one of the central problems in astrophysics and cosmology. A number of cosmological observations favor a nonbaryonic nature. For one thing, measurements of the average density of the universe consistently yield at least four times the value of the baryon density deduced from nucleosynthesis. For another, observations of clustering of galaxies at large scales and COBE measurements of temperature fluctuations of the cosmic microwave background also favor nonbaryonic dark matter. Moreover, nonbaryonic dark matter provides the most natural explanation for the large-scale structure in terms of collapse of the initial density fluctuations inferred from COBE. Finally, a general argument comes from the implausibility of hiding a large quantity of baryons in the form of compact baryonic objects.

Several nonbaryonic candidates have been proposed, including shadow universes, condensates formed at a quark-hadron phase transition, and very massive particles produced during inflation. By selecting candidates that also solve important questions in particle physics, researchers arrive at three particularly reasonable possibilities: massive neutrinos, axions, and weakly interactive massive particles (WIMPs). Interpretation of recent data on atmospheric and solar neutrinos in terms of oscillations suggests that neutrinos have a small mass, but it is unlikely that such an interpretation provides the full explanation for dark matter because it alone cannot explain structure formation. In addition, neutrinos, as fermions, are excluded by phase space considerations from providing an explanation for the dark matter in galactic halos. Axions have been postulated to prevent dynamically the violation of charge parity in strong interactions in the otherwise extremely successful theory of quantum chromodynamics. Present limits on axion parameters are such that if they exist, they would form a significant portion of cold dark matter. Current axion searches cover only a portion of the allowed mass range. WIMPs are particles that were in thermal equilibrium in the early universe and decoupled when they were nonrelativistic. For these particles to have critical density, their annihilation rate has to be roughly the value expected for weak interactions, i.e., determined by physics at the W and Z mass scale. Conversely, in order to stabilize the mass of the W and Z particles at the 100 GeV mass scale, particle physicists are led to predict the existence of new families of undiscovered particles. The leading candidate for the class of particles created by this convergence between particle physics and cosmology is the neutralino of supersymmetry.

The most direct method to detect WIMPs is by elastic scattering on a suitable target in the laboratory. WIMPs interacting with the nuclei in the target would produce a roughly exponential distribution of the recoil energy, which could be detected by an ultralow-background detector. The challenge is to construct experiments of sufficient mass with very low radioactive backgrounds and instruments capable of recognizing WIMPs interactions. Current experiments are reaching a sensitivity that begins to explore the supersymmetric models for the WIMPs.

WIMPs can also be sought through the products of their annihilation. The most specific signature would be a gamma line at half the WIMP mass, detected from the center or halo of our galaxy. A detection possibility that covers a larger region of particle-physics parameter space and is, in addition, less model-dependent arises from the fact that WIMPs would be trapped in the center of the Sun or Earth. Their annihilation products could be detected by properly designed deep neutrino detectors. Finally, WIMP annihilation would also produce an anomalous flux of antiparticles that might be detected by particle detectors in space.

EXISTING PROGRAMS

GRAVITATIONAL WAVES

In the next 3 years, four long-baseline (0.6 to 4 km) laser interferometer detectors will start running: LIGO, with interferometers in Louisiana and Washington State; VIRGO, with an interferometer near Pisa, Italy; GEO, with an interferometer in Hannover, Germany; and TAMA, in Japan. There are also plans for an interferometer in Australia. The interferometers will operate as a network, providing multiple coincidences and correlations and thereby a means for determining the location of the sources and the polarization states of the waves. A second generation of interferometers with improved isolation from environmental perturbations, increased light power, and different optical configurations that trade bandwidth for sensitivity will be installed in 5 to 10 years as upgrades to the existing long-baseline facilities. These improvements, which were part of the initial research plan for LIGO and incorporated into the design of its long-baseline facilities, will require about $50 million over the rest of this decade.

VERY-HIGH-ENERGY GAMMA RAYS

There are numerous ground-based gamma-ray telescopes in operation around the world. The most sensitive instruments are those using the imaging atmospheric Cherenkov technique, in which large mirrors focus Cherenkov radiation created in air showers onto photodetector arrays. Imaging Cherenkov telescopes operate on clear nights at typical energies between 250 GeV and 25 TeV. The Whipple Observatory (Mt. Hopkins, Arizona) has been the leading telescope for the last decade. Other instruments include HEGRA (La Palma, Spain), CAT (Themis, France), and CANGAROO (Woomera, Australia). Outside the United States, there are three next-generation gamma-ray telescopes in development: CANGAROO-IV (Australia), HESS (Namibia), and MAGIC (La Palma, Spain). These projects have different scientific emphases and are geographically well separated in latitude and longitude, so they will complement a new telescope in the southwestern United States.

A second type of Cherenkov telescope uses large mirror arrays originally built for solar energy research to achieve energy thresholds as low as 40 GeV. These experiments, STACEE (Albuquerque, New Mexico) and CELESTE (Themis, France), observe in the largely unexplored energy region that is beyond the reach of the existing spaceborne Compton Gamma Ray Observatory and nearer than the reach of imaging Cherenkov telescopes.

Air-shower experiments detect the particles produced in the high-energy gamma-ray cascades. Operating with a wide field of view and with 100 percent duty cycle, these experiments are well suited for sky surveys and for studying transient phenomena such as gamma-ray bursts. MILAGRO (Los Alamos, New Mexico), a large, water Cherenkov detector soon to come into full-scale operation, will be the first such detector to work in the energy region between 500 GeV and 50 TeV, which is below the threshold for conventional air-shower arrays.

GALACTIC COSMIC RAYS

Direct measurements of the primary cosmic radiation made with detectors carried above the atmosphere by balloons and spacecraft now extend to energies of about 10^{12} to 10^{14} eV. Information at the top end of this range comes from balloon-borne emulsion chamber experiments and is limited to H, He, and groups of heavier nuclei rather than individual elements. Higher energies are explored only indirectly by ground-based air-shower arrays that do not measure the primaries directly but only the secondary cascades. Direct measurements with magnetic spectrometers and transition radiation detectors have measured the spectra of individual elements up to about 100 GeV/nucleon.

The Alpha Magnetic Spectrometer (AMS) is a magnetic spectrometer experiment for the International Space Station (ISS) designed to search for primordial antimatter and signatures of dark matter in the galactic cosmic rays. It will also measure the spectra of some primary species up to the TeV energy range.

At much lower energies, NASA's Advanced Composition Explorer (ACE), launched in 1997, is measuring the isotopic composition of cosmic-ray nuclei at energies below 0.5 GeV/nucleon. TREK, a joint U.S./Russian experiment to detect cosmic rays with $Z \geq 75$ that flew on Mir, has verified the existence of actinides (nuclei with $Z \geq 90$) in the cosmic radiation.

HIGHEST-ENERGY COSMIC RAYS

Japan's Akeno Giant Air Shower Array (AGASA) uses measurements of ground-level particle densities in shower cascades to infer the primary-particle energy and arrival direction. AGASA has an aperture acceptance of 200 km^{2}sr and will continue running until the newer experiments overtake it statistically.

The High-Resolution Fly's Eye (HiRes), in Dugway, Utah, was completed in 1999. The experiment images nitrogen fluorescence from shower cascades to reconstruct the longitudinal shower profile and measure the primary energy calorimetrically. The profile is also used to measure the position of the extensive air shower in the atmosphere, which is sensitive to the primary-particle composition. HiRes will have a time-averaged aperture of 1000 km^{2}sr at 10^{20} eV, implying that it will detect about 40 events above this energy in a 5-year period if the AGASA results are correct.

The Pierre Auger Observatory project is an international collaboration, which is currently building a 7000 km^{2}sr detector in Argentina. The detector consists of a large ground array of water tanks on a 1.5-km grid and atmospheric fluorescence detectors. Coincidence data taken with both the ground and fluorescence detectors will be used to calibrate the energy and type of the primary particles. The Auger South detector is schedule to be completed in 2004; based on the AGASA spectrum, it will accumulate 300 events above 10^{20} eV in a 5-year run, of which 10 percent would be in coincidence mode.

NEUTRINO ASTRONOMY

The existing underground detectors Super-Kamiokande, MACRO, and LVD currently form a Supernova Early Warning System (SNEWS), which will be joined by other neutrino detectors, including SNO and AMANDA. By comparing alerts via automated e-mail, false alarms of individual detectors can be virtually eliminated, allowing the possibility of automatic notification of the occurrence of a supernova in the Milky Way galaxy or its satellites hours before the optical outburst. It is essential to maintain and enhance this capability over the long term as old detectors are retired and new ones come on line.

There are currently two operating experiments for high-energy neutrino astronomy: Baikal (Siberia) and AMANDA (South Pole). Both experiments have succeeded in measuring upward-going muons produced by atmospheric neutrinos, demonstrating in principle that the technique works. The construction of AMANDA II, with a detection area of 30,000 m^2, has been completed; it is a prototype for a larger detector. There could also be an expansion of the Baikal detector.

Three other groups (ANTARES, NESTOR, and NEMO) have concentrated on site testing and feasibility studies at a variety of Mediterranean sites. NESTOR and ANTARES are conceived as experiments using different deployment schemes, array designs, and signal-processing technologies. ANTARES has begun construction of a detector with an effective area of up to 0.1 km^2.

SOLAR NEUTRINOS

The second generation of solar neutrino experiments includes the Super-Kamiokande light-water Cherenkov detector, the Sudbury Neutrino Observatory (SNO) heavy-water Cherenkov detector, and the Borexino and Kamland scintillation detectors. The first two detectors are sensitive to the ^8B neutrinos, while the second two are designed to study the ^7Be neutrinos. SNO will compare the total rate of interactions of neutrinos of all types with the rate of electron neutrinos, which will check directly whether electron neutrinos produced at the Sun are changing identity en route to Earth. Super-Kamiokande has been operational since 1996; SNO began operation in the summer of 1999; and Borexino and Kamland are under construction.

DARK MATTER SEARCHES

The second-generation microlensing experiments currently being started (OGLE II and Super MACHOs) should have the photometric and statistical accuracy to break the degeneracy among mass, distance, and velocity of the lensing objects. The Space Interferometry Mission (SIM) should also pin down the distance of the lenses.

The Axion experiment has published preliminary limits and will reach the required sensitivity for one generic type of axion between 10^{-6} and 10^{-5} eV/c^2. By pushing the sensitivity with SQUID amplifiers and by operating at 100 mK, an upgraded Axion experiment would cover all the present axion models in this mass range, which is one-third of the 10^{-6} to 10^{-3} eV/c^2 range allowed by astrophysical constraints.

The Cryogenic Dark Matter Search II promises to be the most sensitive WIMP experiment at the beginning of the 2000 decade. It should improve current sensitivities by two orders of magnitude and reach well into the supersymmetric region.

AMANDA II will increase the sensitivity of the search for neutrinos from WIMP annihilation and AMS will search for an excess of antiprotons and positrons in the cosmic rays.

RECOMMENDED NEW INITIATIVES

The panel was able to identify several key challenges that are ripe for progress at this time:

- To detect gravitational radiation from interacting massive objects, including massive black holes;
- To understand the origin of gamma rays of very high energy from sources such as AGN and SNRs;
- To identify positively the sources of galactic cosmic rays and to measure the output of these cosmic accelerators;
- To identify the nature and distribution of the bulk of the matter in the universe;
- To understand the origin of the highest-energy particles in nature;

- To detect neutrinos from energetic astrophysical objects and events; and
- To measure the full energy spectrum and flavor content of neutrinos from the Sun.

GRAVITATIONAL-WAVE ASTRONOMY (LISA)

Because of the fundamental and novel new phenomena that can only be studied by a long-baseline gravitational-wave detector in space, and because of its strong and well-developed science and technology plans, the Laser Interferometer Space Antenna (LISA) is the highest priority project of the panel. The mission has generated considerable interest within NASA and has become a cornerstone mission of the ESA program. It is hoped that LISA will be put forward as a joint mission for launch before the end of the first decade of this new century. LISA should observe, for the first time, the coalescence of supermassive black holes as distant galaxies merge. Although merger rates are quite uncertain, event rates are estimated to range from 1 to 100 per year. Location of events in the sky depends on frequency, ranging from several degrees at 10^{-4} Hz to 30 arcmin at 10^{-2} Hz, with an angular resolution of 1 arcmin in the last few days before coalescence. This should be sufficient for gamma-ray detectors to point and observe the final explosion. LISA will also survey the gravitational radiation from galactic white-dwarf binaries and possibly study gravitational fluctuations from the early universe. The LISA team will carry out a major theory challenge by computing the expected gravitational waveforms from black-hole mergers. This will require developing three-dimensional general relativistic codes with adaptive mesh refinements.

LISA consists of three spacecraft maintained in an equilateral triangular configuration with sides 5×10^6 km long (Figure 3.1). The system is placed in solar orbit at 1 AU with the plane of the triangle at 60 deg to the ecliptic. The orbit requires little station keeping. The three spacecraft are launched by a single Delta rocket and then deployed into the triangular configuration. The triangle enables the operation of three almost independent interferometers along adjacent pairs of sides. The interferometry is done by heterodyne detection with a single optical pass. Both the frequency range and the science of LISA are complementary to those of the ground-based interferometers (Figure 3.2).

The LISA team has identified three technical areas that would benefit from a dedicated technology mission in space: the inertial reference

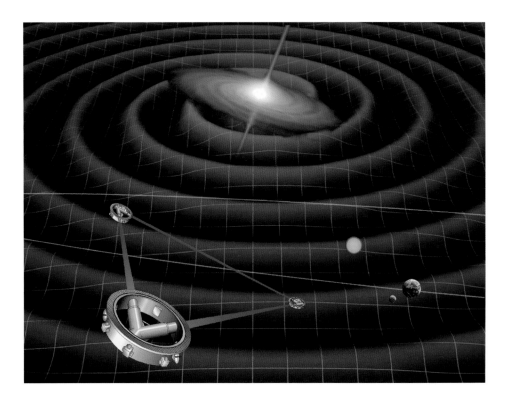

FIGURE 3.1   Artist's conception of the Laser Interferometer Space Antenna, consisting of three spacecraft in a triangular configuration separated by 5 million km. LISA is aimed primarily at studying strong-field gravity, the coalescence of massive black holes in galactic nuclei, and compact binary star systems in the Milky Way.   Courtesy of W. Folkner, Jet Propulsion Laboratory, California Institute of Technology.

mass, the precision thrusters, and the high-precision interferometry. The most critical area is the development of the sensor and control system to maintain the LISA spacecraft in pure inertial (drag-free) orbits. The technical challenge is to reduce the nongravitational disturbing forces (such as fluctuating electric fields and fluctuating radiation pressure from thermal gradients) on a reference mass used to guide the motion of the spacecraft containing the optical components. The concept of surrounding a reference mass by a spacecraft shell that follows its motion and shields it from nongravitational forces was tested in the TRIAD program almost two decades ago at acceleration levels approximately 10^7 times greater than needed for LISA. The Gravity Probe B cryogenic gyro

FIGURE 3.2 Comparison of the sensitivity of LISA with that of ground-based interferometers such as LIGO for various potential sources. Because of its lower frequency range, LISA is sensitive to the coalescence of massive black holes. It also has the potential to survey the Milky Way for binary systems involving white dwarfs, neutron stars, and stellar mass black holes. Courtesy of W. Folkner, Jet Propulsion Laboratory, California Institute of Technology.

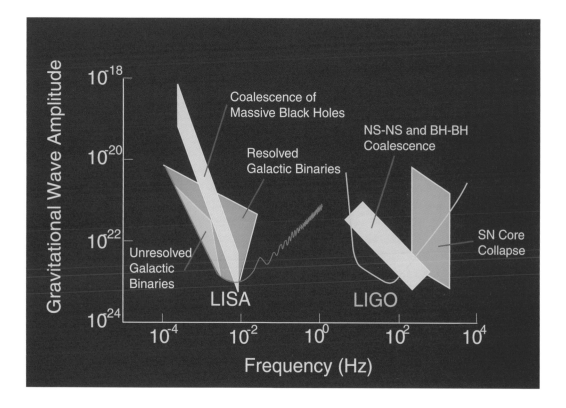

mission, to be flown in 2002, will provide a test at an acceleration level approximately 10^5 times greater.

The controllers that will be used to make the spacecraft follow the reference mass also need demonstration. Several thruster designs exist that develop proportional ion thrust control at the micronewton force level. The key technical issues are lifetime and reliability.

The laser interferometry in LISA needs to operate at displacement sensitivities of 10^{-13} m, which is less sensitive by a factor of 10^5 than the initial terrestrial interferometers. The technical challenge is to achieve the performance at low frequencies (10^{-1} to 10^{-4} Hz).

The thruster and laser interferometry can be tested on the ground.

However, to establish the low perturbation levels on the reference mass will require an integrated test of all three elements. Technology development missions are being contemplated in both the United States and Europe to test an integrated system at acceleration levels low enough to allow final design of the LISA system. The panel strongly encourages a technology development mission to reduce the risk in the LISA mission. The total cost of the LISA mission (exclusive of technology development) is estimated at close to $400 million, of which $250 million would be borne by NASA.

GROUND-BASED GAMMA-RAY ASTROPHYSICS (VERITAS)

The VERITAS project would greatly expand astronomers' understanding of the high-energy gamma-ray sky and has significant potential for achieving a breakthrough in astrophysics. It is therefore the panel's top-ranked ground-based project and second overall. VERITAS is envisioned as an array of seven 10-m-diameter reflectors, each equipped with its own imaging camera of 500-pixel elements. The reflectors can be operated either as individual telescopes or together in a stereoscopic mode. As individual telescopes, VERITAS has the greatest discovery potential and can carry out effective sky surveys. In a stereoscopic configuration, where each air shower is viewed by multiple cameras, VERITAS achieves its best sensitivity (more than an order of magnitude improvement over existing instruments; see Figure 3.3) and its optimal energy and angular resolution (better than existing instruments by a factor of between 2 and 3). For individual photons, the angular resolution of VERITAS will be as good as 3 arcmin, and the point-source location will be better than 25 arcsec. The design of VERITAS is a natural outgrowth of the successful Whipple Observatory on Mt. Hopkins. Technological advances include the development of high-speed (500 MHz) digitizers for each pixel element, which will permit the telescope to operate with a low energy threshold (50 GeV) and improved energy resolution (less than 15 percent).

VERITAS will observe TeV gamma rays from the jets of AGN and possibly also from GRBs. These highly variable energetic signals reflect violent processes occurring in the active inner regions of their distant sources. VERITAS should detect an order of magnitude more AGN than existing ground-based detectors, approximately 30 or more x-ray-selected BL Lacs and 15 radio-selected quasars. Detailed studies of AGN during

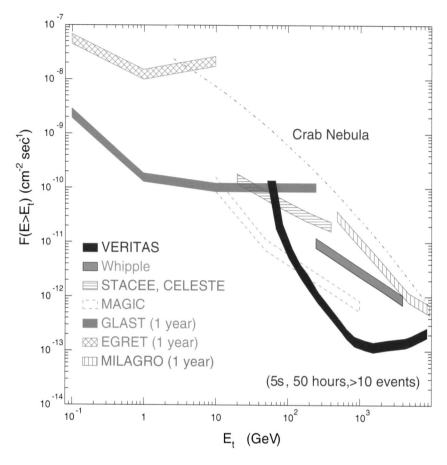

FIGURE 3.3 Comparison of VERITAS sensitivity with that of present and proposed detectors. Together, GLAST and VERITAS will study the gamma-ray sky at energies from less than 1 GeV to greater than 10 TeV. Courtesy of T. Weekes, Harvard-Smithsonian Center for Astrophysics.

states of high activity will be possible (see Figure 3.4). For distant sources, spectral cutoffs that are correlated with redshift will provide important information about the cosmic IR background radiation. An important theory challenge posed for VERITAS is to understand the origin and characteristics of the energetic signals from AGN and GRBs, including acceleration mechanisms, the relative importance of electrons and ions, and spectral shapes and cutoffs. Multiwavelength coverage, including radio, optical, x-ray, and gamma-ray bands, will be crucial in understanding these sources.

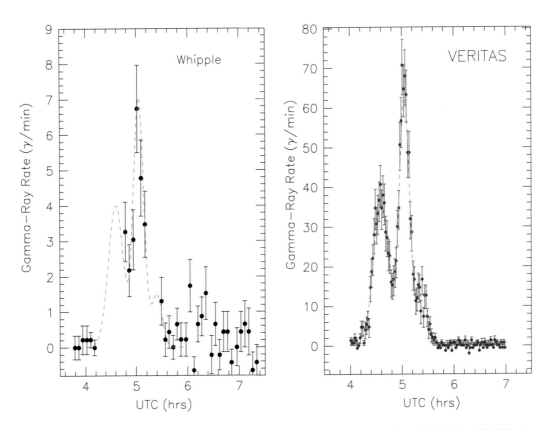

FIGURE 3.4 Example of potential offered by the increased sensitivity of VERITAS. *Left:* Whipple observations of a rapid flare from Mrk 421. Dashed curve is a possible intrinsic flux variation. *Right:* Simulated response of VERITAS to such a flare showing the much better resolution that would be possible with its greater sensitivity. Courtesy of R. Ong, University of Chicago.

VERITAS will also study the emission from shell-type SNRs. TeV energies are well suited for detection of this emission, owing to the lower Galactic diffuse background and superior angular resolution of VERITAS relative to satellite-borne instruments. VERITAS should enlarge the sample of VHE pulsar nebulae by a factor of 3 or 4 and should map the emission from strong sources such as the Crab; it would, as well, provide a unique test of pulsar wind models.

VERITAS will perform the first sky survey at very high energies with good sensitivity. It should be possible to complete a survey of the Galactic plane within a year for sources as weak as 2 percent of the Crab above 300 GeV. Spectral measurements made with GLAST and VERITAS at energies from 1 GeV to 1 TeV should be able to identify uniquely a significant fraction of the 170 EGRET unidentified sources, 40 or so of which are in the Galactic plane, where the good angular resolution of VERITAS is particularly advantageous. VERITAS will also be an effective survey tool for potential sources of VHE neutrinos and UHE cosmic rays.

VERITAS has significant potential for new discoveries in several areas. By operating at higher energies, it will complement GLAST in the search for dark matter in the form of supersymmetric particles near the Galactic center, whose annihilation products would include photons, or in the form of cold molecular clouds, which would be revealed by concentrations of cosmic-ray-produced photons. With its larger sky coverage relative to existing instruments, VERITAS will carry out better searches for VHE sources not seen at other wavelengths, such as primordial black holes.

The construction cost for VERITAS is estimated at $20 million, including instrumentation.[1] It is expected to be fully operational by 2003 and will operate until the end of the decade, thus completely overlapping in time with GLAST. The performance of VERITAS could be significantly enhanced by technological improvements to the photodetectors (high quantum efficiency) or the optics (wider field of view). The appropriate design for an instrument beyond VERITAS is not obvious at the present time, but technology development for future telescopes should be encouraged.

PROGRAM IN PARTICLE ASTROPHYSICS

To take advantage of several important scientific opportunities, the panel recommends a balanced and coherent program in particle and nuclear astrophysics. Such a program will include, in addition to the gamma-ray observations carried out by GLAST, VERITAS, and other ground-based detectors, opening up the neutrino astronomy window and understanding the origins of the high-energy cosmic radiation. Further studies of solar neutrinos will be necessary to understand fully the properties of neutrinos and their role in astrophysical processes. More

---

[1] The estimated cost for VERITAS that appears in the survey committee report includes grants and operations in addition to instrumentation, as described in this report's preface.

sensitive searches are needed to identify and study the nature of the dark matter. This section discusses the projects the panel expects to need funding in the decade 2001 to 2010. In several cases, the design of the new detectors will depend on the outcome of current experiments. For this reason, the panel did not assign a higher priority to one or another of these projects. In addition, the panel highlights one small space experiment (ACCESS) because of its readiness and potential scientific payoff.

A SMALL SPACE MISSION FOR GALACTIC COSMIC RADIATION

The Advanced Cosmic Composition Explorer on Space Station (ACCESS) is a small NASA experiment with an estimated cost of $100 million designed to make direct measurements of individual cosmic-ray nuclei to the highest possible energy allowed by extended exposure of a calorimeter on the ISS (Figure 3.5). The experiment is motivated by suggestions that the standard picture of cosmic-ray acceleration in supernova remnants followed by energy-dependent diffusion out of the galaxy has difficulties accounting for the relative fractions of various elements in the cosmic radiation. ACCESS addresses this problem in two ways. First, by measuring individual elements to energies approaching 1000 TeV, the experiment will show definitively whether different species have the same momentum spectra, as expected in the simplest supernova-diffusive escape model. A possible outcome is that structure in the energy spectra may indicate that some acceleration sources have reached a maximum energy within the ACCESS energy range. Second, the lever arm in energy is large enough so that the contributions of diffusion and acceleration to the observed energy spectrum can be distinguished. This will be accomplished by measuring the ratio of secondary to primary cosmic-ray nuclei. Thus, ACCESS will for the first time allow a measurement of the energy spectrum of galactic cosmic particle accelerators rather than the convolution of acceleration and diffusion. A theory challenge posed by ACCESS is to identify signatures that discriminate among models of the origin of the most energetic galactic cosmic rays. In the supernova picture, for example, this would require relating a realistic and detailed distribution of various supernova types, including characteristic spectra of particles that they accelerate, to elemental composition at PeV energies as observed locally after propagation in the galaxy.

The ISS is a good platform for ACCESS because the detector does

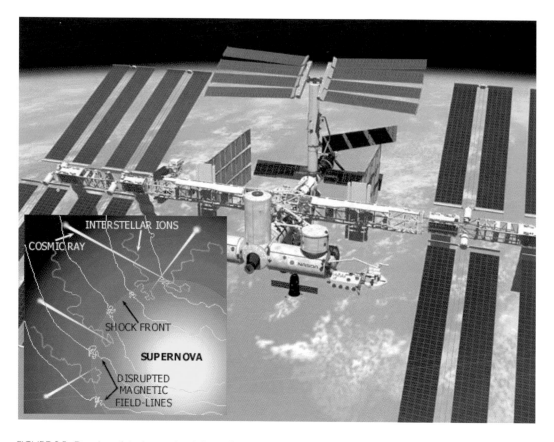

FIGURE 3.5 Drawing of the International Space Station, proposed site of the Advanced Cosmic Composition Explorer on Space Station (ACCESS). The inset illustrates the process of particle acceleration at a blast wave driven by a supernova explosion. With the aid of turbulent magnetic fields, large-scale kinetic energy is believed to be transferred to some interstellar ions, accelerating them to relativistic energies. ACCESS is designed to probe the limits of this widely accepted but still hypothetical picture of the origin of Galactic cosmic rays. Courtesy of R. Mewaldt, California Institute of Technology.

not require accurate pointing. An accommodation study shows that the project is feasible. It is technologically ready and constitutes an obvious next step in the exploration of high-energy cosmic rays. Moreover, it has the potential to change the paradigm for the origin of Galactic cosmic rays if the source spectrum is significantly steeper than expected in the model of first-order diffusive shock acceleration by supernova blast waves.

QUEST FOR HIGHEST-ENERGY COSMIC PARTICLES

HiRes and the Auger South experiment will clearly be able to confirm the continuation of the UHE cosmic ray spectrum past the GZK cutoff (Figure 3.6). If the sources of these particles are AGN or other pointlike objects, then anisotropy or clumping on the scale of 5 deg is expected in the direction of the sources. A Northern Hemisphere experiment with

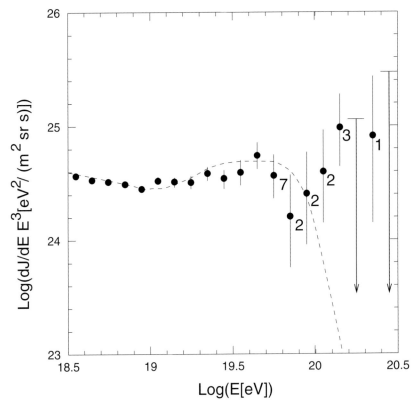

FIGURE 3.6 The spectrum of the highest-energy cosmic rays as reported by the AGASA experiment. The dotted line shows the spectrum that would be observed for a uniform distribution of sources in the universe. The few events above 10^{20} eV must have sources that are relatively nearby on a cosmological scale. The numbers beside the high-energy data points indicate the number of events on which each datum is based. The HiRes group Fly's Eye Experiment has reported a similar number of events above 10^{20} eV, confirming by a different technique that the spectrum indeed extends beyond the expected cutoff from photopion production during propagation from cosmological distances. Source: M. Takeda et al., "Extension of the Cosmic-Ray Energy Spectrum Beyond the Predicted Greisen-Zatsepin-Kuz'min Cutoff," *Physical Review Letters* 81 (1998): 1163.

sensitivity comparable to that of Auger South will then be necessary to study and catalog these sources over the whole sky and to establish correlations with visible, radio, x-ray, and gamma-ray observations. A site in Millard County, Utah, near the HiRes experiment has already been identified by the Auger collaboration, and a proposal for a northern site is expected around 2002.

A good understanding of the composition of cosmic rays is critical to unraveling the acceleration mechanism. For example, in a top-down process like the decay of topological defects, a large gamma/hadron ratio is expected, in contrast to the proton flux expected from AGN. Fluorescence detection allows measuring the longitudinal profile of each shower and hence makes possible the most direct calorimetric measurement of the primary energy. It also gives a measure of the identity of the primary particles different from and complementary to the Auger ground array. An international collaboration is proposing to build a chain of HiRes-type detector stations around the proposed Auger North site with a time-averaged aperture of near 8000 km²sr. In an enhanced northern detector, half the events will undergo determination of the atmospheric profile (by Telescope Array) and half will have ground array information gathered (from Auger North). This will allow a comparison of the spectrum and composition results at the highest energies based on two independent techniques with equal statistical strength. Ten percent of the data will be observed in coincidence and will be usable for intercalibrating the energy scales.

If the cosmic ray flux extends beyond 10^{21} eV, as predicted in many top-down models, its study requires detectors with apertures an order of magnitude larger. Since detectors with such a large area are impractical on the ground, the possibility of observing atmospheric fluorescence of giant showers from a detector in space is being considered. Several downward-looking detectors at 500- to 700-km orbits can achieve time-averaged apertures of 50,000 km²sr. Such a detector would look for the characteristic rebound in the UHE cosmic-ray spectrum near 10^{21} eV due to particles produced with energies well beyond the GZK cutoff. It would also be sensitive to UHE neutrino fluxes, predicted in some models.

There are many technical challenges to observing fluorescence from space, including the development of very-wide-angle optics, million-pixel photon detectors with high quantum efficiency, data acquisition electronics with very low power, and accurate determination of obscuring cloud cover and atmospheric transmission. Groups in the United States (the

Orbiting Wide-angle Light Collector (OWL)) and Italy (AIRWATCH) are cooperating on research and development to address these issues, and both NASA and the Italian Space Agency have provided preliminary funding. A balloon flight is planned to measure the ambient UV background looking down at Earth.

High-energy Neutrinos

Observation of high-energy neutrinos from astrophysical sources would open a new window on the cosmos because neutrinos can reach the observer over cosmological distances from deep inside the sources. The discovery potential is therefore great. Although predicted event rates vary substantially, detection of high-energy neutrinos would provide unambiguous proof of proton acceleration and would likely produce a better understanding of the dynamical role of hadrons in the astrophysical milieu. Plausible models of neutrino production in GRBs and active galaxies, which predict tens of events or more per year per square kilometer, can be tested with a kilometer-scale neutrino telescope. A few dramatic neutrino events correlated with GRB observations would provide strong evidence that these spectacular objects are the sources of the highest-energy cosmic rays.

The IceCube proposal for a kilometer-scale neutrino detector is designed to expand significantly the reach of high-energy neutrino telescopes. IceCube capitalizes on the success of AMANDA and on the laboratory and logistical infrastructure of the Amundsen-Scott South Pole Station, which is currently being expanded and modernized.

In addition to searching for high-energy neutrinos from distant sources, IceCube will detect about 5000 atmospheric neutrino events per year, which will be used for calibration. It may also be possible to see manifestations in the atmospheric neutrinos of the neutrino oscillation phenomena found in Super-Kamiokande. There will also be searches for neutrinos from Galactic compact objects and for emission from the Galactic center region. IceCube will search for bursts of low-energy (10 to 20 MeV) neutrinos coming from supernova collapse by monitoring the photomultiplier tube (PMT) dark counting rates for short-term increases. It would be sensitive to neutrinos from WIMP annihilation in the Sun, with good sensitivity for WIMP masses >200 GeV. If extragalactic sources are found, neutrino mass can be investigated over baselines of unprecedented length. Because hadronic processes in the source produce ν_e and ν_μ, the appearance of ν_τ from a potential cosmological source can probe neutrino mass with a sensitivity of $\delta m^2 \geq 10^{-17}$ eV^2. Tau neutrinos with PeV energies can be identified by the characteristic pair of separated showers associated with the production of a tau lepton followed by its decay.

There has already been substantial development of the IceCube concept, including aspects of string installation, detector operation, and event reconstruction. Strings of sensors have been deployed in vertical holes to a depth of nearly 2500 m, sufficient for IceCube. The reference design (Figure 3.7), which is based on the transmission of analog signals over optical fiber, requires no new technology development. Scaling from the experience with AMANDA II, IceCube construction will take 6 years. The operational lifetime will be a minimum of 4 years. An initial proposal for IceCube was submitted in November 1999. AMANDA II, with an acceptance about an order of magnitude larger than the current AMANDA, was completed in early 2000 and can be considered as a prototype for the full IceCube.

SOLAR NEUTRINOS

A third generation of solar neutrino experiments will be required to complete the program for measuring the energy spectrum and flux of solar neutrinos. The goal is to make a complete set of precise measurements of neutrinos from an astrophysical source. The new experiments will be directed at measuring both the electron-neutrino component of the dominant low-energy *pp* flux and the total *pp* flux. It is highly likely that one or more of these experiments will be proposed for construction early in the decade 2001 to 2010.

It is important to compare the measured flux and spectrum of neutrinos with predictions based on stellar evolution calculations and laboratory data. As both the neutrino data and the laboratory data improve, the astrophysical calculations are put to more stringent tests.

Specific examples of detectors under development are HERON, utilizing a 10-ton superfluid ^{4}He target; LENS, employing ^{176}Yb or ^{160}Gd dissolved in liquid scintillator; and HELLAZ, using high-pressure gaseous ^{4}He in a time-projection chamber. The helium experiments, which are sensitive to the total flux and spectrum of low-energy neutrinos through elastic scattering of neutrinos by electrons, are complementary to the LENS experiment, which measures the flux and spectrum of low-energy electron neutrinos directly. Very preliminary cost estimates for these

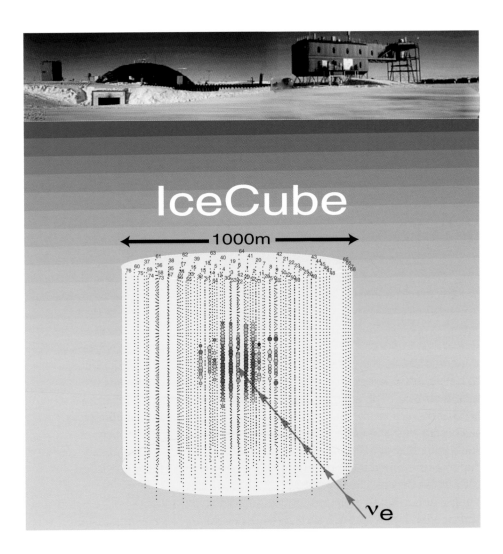

FIGURE 3.7 Artist's drawing of the proposed IceCube high-energy neutrino telescope at the South Pole. The composite photograph at the top shows the dome over the present Amundsen-Scott South Pole Station together with the Martin A. Pomerantz Observatory, which houses astrophysical facilities, including the data acquisition system for the present AMANDA experiment. Each dot in the diagram represents one of the approximately 5000 optical modules that will make up the detector at depths from 1.5 to 2.5 km in the clear Antarctic ice. The display here depicts a large electromagnetic cascade initiated by an electron neutrino interacting near the center of the detector. A high-energy muon neutrino would appear as an elongated series of hits along the path of a high-energy muon produced when the neutrino interacts, either inside the detector or in the surrounding ice. Courtesy of S. Barwick, University of California, Irvine.

projects put them in the same range as moderate-cost, ground-based projects.

The scientific motivation for these experiments is to make the most precise measurements possible of the entire solar neutrino spectrum and of the physics parameters involved in neutrino mixing. The solar neutrino experiments will lead to an understanding of the emission spectrum of neutrinos by the Sun and the fundamental physics of neutrinos. Without this knowledge, it is not possible to make progress in other areas of astrophysics where neutrinos play an important role. For example, neutrino oscillations will affect the dynamics of core-collapse supernova explosions and nucleosynthesis through the n-process and the r-process. This in turn has implications for the chemical evolution of our galaxy.

DARK MATTER

Deciphering the nature of dark matter remains one of the most important goals of astrophysics. With the Axion and CDMS II experiments in the United States, the nation is currently engaged in searches that are probing cosmologically important regions. It is very likely that at the end of these experiments (around 2005), there will be a need to start at least one second-generation dark matter experiment. Although such an experiment could be motivated by the need for greater sensitivity, it will become compelling if a discovery is made in the current nonbaryonic dark matter searches or if a new feature of particle physics pertinent to the dark matter problem is uncovered at accelerators. If, for instance, supersymmetry is discovered, the next question will be, Is it responsible for the dark matter in the universe? If the new, direct searches now getting under way find a signal, this will determine the nature of detectors needed for more sensitive studies. If and when particle dark matter is discovered in direct searches, dark matter detectors would be able to map the local velocity distribution of the Galactic halo. This information would revolutionize the study of the distribution of dark matter in the Galaxy as well as theories of galaxy formation. It would allow us to overcome the current fundamental limit to galaxy-formation theories that arises from the fact that researchers do not know the spatial or velocity distribution of the dominant mass component.

TECHNOLOGY FOR THE FUTURE

Technology development is critical for the future of particle, nuclear, and gravitational astrophysics. The LISA technology program was described in the section on gravitational-wave astronomy. Dark matter searches and solar neutrino experiments both need the kinds of advances that will bring about low backgrounds and ultrasensitive detectors with low thresholds: advanced analysis methods to select the purest materials; hardware that reduces inert mass; and manufacture of critical parts underground. Particularly important for WIMP searches would be directional detection of nuclear recoil, using very large (10^4 m^3), low-pressure-gas, time-projection chambers or detection of athermal phonons in isotopically pure crystals.

Sensors and amplifiers need to approach the quantum limit, especially for gravitational-wave detection and axion searches. For example, SQUID amplifiers in the gigahertz range that are being developed for axion searches are nearly quantum limited and may also have important applications in radio astronomy. Photolithography, micromachining, low-temperature techniques, and optimal filtering are essential elements of this development direction.

Affordable optical photon detectors with higher quantum efficiency are essential for future Cherenkov telescopes and shower detectors, including OWL, which also needs low-cost, large-aperture optics.

POLICY ISSUES

One theme of this panel report is the need for multimessenger as well as multiwavelength astronomy to unravel the mysteries of the cosmos. To understand fully the most violent events in the universe will require the detection of gravitational waves and neutrinos as well as observations throughout the electromagnetic spectrum and the information provided by the flux and composition of cosmic rays (Figure 3.8).

For the astrophysics community, the emerging field of particle and nuclear astrophysics provides new approaches, a new population of enthusiastic scientists, new scientific cultures, and new funding sources, all of which if properly assimilated will offer wonderful scientific opportunities. However, the cross-cutting nature of the tools employed and the problems addressed do not fit neatly into the traditional categories of wavelength or ground- versus space-based observations. For the field to

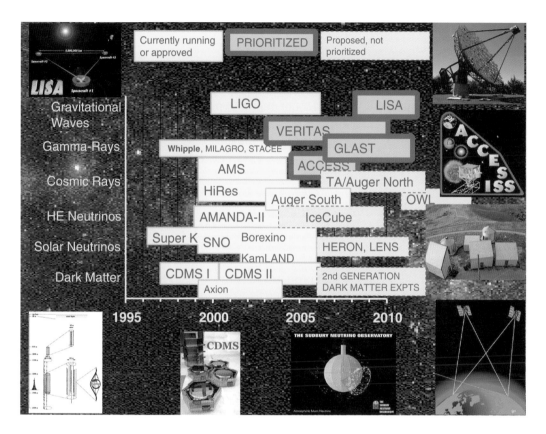

FIGURE 3.8 Time line for studies of gravitational waves, gamma rays, cosmic rays, high-energy neutrino astronomy, solar neutrinos, and dark matter. Currently running or approved projects are shown in yellow boxes; prioritized projects are in pink; and proposed projects, not yet prioritized, are in blue boxes. (GLAST was prioritized by the Panel on High-Energy Astrophysics from Space.) Courtesy of T. Gaisser, University of Delaware, and A. Harding, NASA's Goddard Space Flight Center.

thrive, the astronomy and astrophysics and physics communities and the funding agencies must work to overcome these boundaries and focus on using multiple approaches to solve diverse but interconnected scientific problems. The frontiers of particle and nuclear astrophysics, whether in space, on the ground, or underground, should be viewed as essential tools for answering fundamental questions of physics as well as astrophysics.

FACILITIES

Several existing and proposed experiments are carried out at special sites or observatories that require substantial investments in infrastructure and, in some cases, development. Generally, the diversity of needs means that establishment of one or two central laboratories for the whole field is not feasible. Rather, sites appropriate for each experiment are identified. Some may be in existing facilities (as in the use of the Soudan underground lab for CDMSII); others need to be developed (as in the case of the site in Argentina for the Auger South Observatory). Other examples are Dugway, Utah, for the HiRes experiment and the proposed Telescope Array, the Whipple Observatory on Mt. Hopkins in Arizona for atmospheric Cherenkov telescopes, and the Amundsen-Scott South Pole Station for the proposed IceCube detector.

NASA is currently developing an Ultralong Duration Ballooning (ULDB) program that promises to provide round-the-world flights of up to 100 days, for payloads of up to 1 or 2 tons. The first 100-day demonstration flight in this program is currently planned for 2001. The ability to fly a large payload above the atmosphere for a fraction of a year, at a fraction of the cost of a satellite mission, offers a great opportunity for high-energy astrophysics investigations as well as for investigations using solar and infrared instruments. The panel strongly recommends that NASA support technology developments in balloon, telemetry, and fine-pointing systems so that the ULDB program can develop a reliable science platform of broad utility to the community.

The International Space Station (ISS) can provide a useful platform for certain classes of heavy and/or large-area payloads (up to ~5 ton) that do not require fine pointing. Examples include high-energy cosmic-ray instruments and all-sky gamma-ray or x-ray monitors. The AMS experiment serves as a pathfinder for future ISS payloads. The proposed ACCESS mission is well suited to ISS. The panel endorses the use of the ISS for appropriate astrophysical experiments.

RECOMMENDATIONS FOR THE FUNDING AGENCIES

Particle astrophysics has developed into a coherent field and should be recognized as one: the funding agencies should institute robust mechanisms to support exciting new projects that cut across traditional funding categories. This requires cooperative funding and project coordination both within and across agency borders.

- The Department of Energy should officially recognize that particle and nuclear astrophysics fall within its charter. Investigations in these disciplines probe the fundamental forces and the nature of matter in ways that directly complement accelerator-based experiments. Indeed, much of the evidence for physics beyond the standard model comes from particle astrophysics and cosmology, including evidence for neutrino masses from solar and atmospheric neutrinos; evidence for baryon-number violation and CP violation from the baryon asymmetry of the universe; cosmological evidence for a nonzero cosmological constant; and indications for a new physics at an ultrahigh-energy scale associated with inflation in the very early universe. Investigations such as the search for dark matter and solar neutrino experiments address fundamental physics problems, from supersymmetry to the origin of mass. Giant-air-shower experiments investigate the role of high-energy particles and their interactions in astrophysical settings. The panel expects that the DOE-supported community of high-energy and nuclear physicists will continue to make important contributions in these areas, from advanced instrumentation and detection techniques to data acquisition and analysis. An example where laboratory experiments contribute directly to astrophysics is the study of quark-gluon plasma at the Relativistic Heavy Ion Collider (RHIC), which is relevant to early universe physics and to neutron star interiors. The national laboratories will play an important role in particle astrophysics experiments, which are increasing in size: They can provide technical support for the deployment of large numbers of highly sophisticated components, and they are uniquely equipped to manage such large projects. It is not surprising, therefore, that most of DOE's laboratories—Brookhaven National Laboratory, Fermi National Accelerator Laboratory, Lawrence Berkeley National Laboratory, Los Alamos National Laboratory, Lawrence Livermore National Laboratory, and the Stanford Linear Accelerator Center—are deeply involved in such programs.

- At the National Science Foundation, this field spans two divisions, Astronomical Sciences (AST) and Physics (PHY), with different cultures and customs. The Division of Mathematics and Physical Sciences (MPS) recently initiated a program activity in Nuclear and Particle Astrophysics within PHY. This initiative cuts across several units of NSF, including programs in high-energy physics, nuclear physics, gravitational physics, and theory, as well as AST and the Office of Polar Programs. The panel strongly endorses this effort as an important step toward ensuring that this exciting interdisciplinary science is coherently supported by NSF.

• The panel strongly supports the increasing collaboration between the various agencies in this field, such as the collaboration of DOE and NASA on AMS and GLAST. Interest in two NASA projects, GLAST and LISA, has been expressed by NSF-supported groups and by NSF officials. Such cooperation would, for example, allow groups with relevant expertise developed under the aegis of one agency to participate readily in a project supported primarily by another. The panel expects that as more projects are cross-funded, there will be efforts to rationalize the procedures and customs at the different agencies.

• To help agencies in this endeavor and to serve the needs of this growing community, there is a clear need for an interdisciplinary advisory structure to review proposed projects and to help set long-term priorities. The panel strongly supports the continuation of the Scientific Assessment Group for Experiments in Non-Accelerator Physics (SAGENAP) by DOE and NSF, with NASA participation as an observer, to assess cross-cutting projects. It is important for all relevant divisions and agencies to participate fully in this coordinated process, so that projects are not required to pass through multiple review committees. The agencies should also regularly seek long-range, coordinated strategic advice on the main scientific priorities, leading to a strategic plan for projects involving astrophysics. This is particularly important as the proposed projects become larger.

• The funding mechanisms should also be adapted to the field. The NASA concept of missions is probably well suited to the large experiments being considered in particle and nuclear astrophysics: experiments should be of fixed duration, and their costing should include operations and the extraction of science. Various approaches to an important science theme could be coordinated by an organization set up for a fixed time.

• A strong theoretical effort is essential for the field. In addition to individual researchers, relatively large theory groups that maximize interactions between postdoctoral fellows have made important contributions by exploring the interface between particle and nuclear physics and astrophysics. Such theoretical work lays the foundation for experiments over the next decade. Funding for theoretical astrophysics has gradually eroded, and all three agencies are urged to strengthen their support for theory in particle astrophysics, including the support of large, multidisciplinary groups.

- International collaboration will become the norm in this field, because of the growing scale of particle, nuclear, and gravitational-wave astrophysics experiments. This will ultimately reduce costs and take advantage of the global diversity of expertise; it should help the community avoid a proliferation of competing, often subcritical projects. On the other hand, long-range planning and coordination of large international projects presents a challenge, because of differences of culture, procedures, and budgetary timescales. In this vein, the panel supports the International Union of Physics and Applied Physics in its creation of the Particle and Nuclear Astrophysics and Gravitation International Committee (PANAGIC). Its mission is to increase the circulation of information in the field and promote convergence on large international projects. The panel urges the agencies to coordinate with the community in this endeavor.

ACRONYMS AND ABBREVIATIONS

ACCESS—Advanced Cosmic-ray Composition Experiment for the Space Station, a cosmic-ray experiment on the ISS

ACE—Advanced Composition Explorer (NASA)

AGASA—Akeno Giant Air Shower Array (Japan)

AGN—active galactic nuclei

AMANDA—Antarctic Muon and Neutrino Detector Array

AMS—Alpha Magnetic Spectrometer

ANL—Argonne National Laboratory (DOE)

ANTARES—Astronomy with a Neutrino Telescope and Abyss Environmental Research

ASCA—Advanced Satellite for Cosmology and Astrophysics (Japan)

AST—Division of Astronomical Sciences (National Science Foundation)

AU—astronomical unit: basic unit of distance equal to the separation between Earth and the Sun, about 150 million km

Baikal—an underwater neutrino telescope in Lake Baikal, Russian Federation

BL Lacs—BL Lacertae objects; galaxies with an extremely bright active galactic nucleus, sometimes referred to as a blazar

BNL—Brookhaven National Laboratory (DOE)

CANGAROO—Collaboration of Australia and Nippon (Japan) for a Gamma Ray Observatory in the Outback

CAT—Cherenkov Array in Themis, France

CDMS—cryogenic dark matter search

CELESTE—Cherenkov Low Energy Sampling and Timing Experiment at Themis, France

COBE—Cosmic Background Explorer; a NASA mission launched in 1989 to study the cosmic background radiation from the Big Bang

CP violation—charge-parity violation

DOE—Department of Energy

EGRET—The Energetic Gamma Ray Experiment Aboard the Compton Gamma Ray Observatory

ESA—European Space Agency, the European equivalent of NASA

FNAL—Fermi National Accelerator Laboratory (DOE)

GALLEX—an international solar neutrino research project that measured the solar neutrino flux produced inside the Sun by proton-proton fusion

GEO—a laser-interferometric gravitational-wave observatory (Hannover, Germany)

GLAST—Gamma-ray Large Area Space Telescope, a NASA-DOE mission

GRBs—gamma-ray bursts

GZK cutoff—upper limit to the cosmic-ray energy spectrum of around 10^{19} eV as specified by the theory of Greisen, Zatsepin, and Kuz'min

HEGRA—High-energy Gamma Ray Astronomy experiment, a project that features a gamma-ray telescope in La Palma, Spain

HELLAZ—proposed French solar neutrino detector

HERON—a solar neutrino detector using superfluid helium

HESS—High-Energy Stereoscopic System; gamma-ray telescope in Namibia

HiRes—High-Resolution Fly's Eye

IR—infrared

ISM—interstellar medium

ISS—International Space Station

LANL—Los Alamos National Laboratory (DOE)

LBNL—Lawrence Berkeley National Laboratory (DOE)

LENS—international Laboratory for Nonlinear Spectroscopy; also known as Solar Neutrino Interactions through Real-time Excitation of Nuclei (SIREN)

LIGO—Laser Interferometer Gravitational-Wave Observatory

LISA—Laser Interferometer Space Antenna

LLNL—Lawrence Livermore National Laboratory (DOE)

LSND—Liquid Scintillation Neutrino Detector experiment; searches for neutrino oscillations and explores other aspects of particle and nuclear physics

LVD—Large Volume Detector (Gran Sasso, Italy)

MACHO—massive compact halo object; MACHOs are dark stars or planets that may make up the Milky Way's dark halo

MACRO—Monopole, Astrophysics, and Cosmic Ray Observatory (Gran Sasso, Italy), a detector for atmospheric neutrinos and magnetic monopoles

MAGIC—gamma-ray telescope in La Palma, Spain

MILAGRO—large, water Cherenkov detector at Los Alamos, New Mexico

Mir—space station of the Russian Federation

MPS—Division of Mathematics and Physical Sciences (National Science Foundation)

MSU—Michigan State University

NASA—National Aeronautics and Space Administration

NEMO—Neutrino Mediterranean Observatory, an international collaboration to study double-beta decay without the emission of neutrinos

NESTOR—Neutrino Experimental Submarine Telescope with Oceanographic Research; a deep-sea neutrino-detector in the Mediterranean

NSF—National Science Foundation

OGLE—Optical Gravitational Lensing Experiment, a program to search for dark, unseen matter using the microlensing phenomena

ORNL—Oak Ridge National Laboratory (DOE)

OWL—Orbiting Wide-angle Light collectors

PANAGIC—Particle and Nuclear Astrophysics, and Gravitational International Committee; created by the International Union of Physics and Applied Physics

PHY—Division of Physics (National Science Foundation)

PMT—photomultiplier tube

RHIC—Relativistic Heavy Ion Collider at Brookhaven National Laboratory (DOE)

SAGENAP—Scientific Assessment Group for Experiments in Non-Accelerator Physics (DOE and NSF)

SAGE—Soviet-American Gallium Experiment

SIM—Space Interferometry Mission

SLAC—Stanford Linear Accelerator Center (DOE)

SNEWS—Supernova Early Warning System

SNO—Sudbury Neutrino Observatory, a heavy-water Cherenkov detector

SNRs—supernova remnants

SQUID—Superconducting Quantum Interference Device

STACEE—Solar Tower Atmospheric Cherenkov Effect Experiment, a Cherenkov telescope at Albuquerque, New Mexico

Super MACHO project—Massive Compact Halo Object project (United States/Australia)

Super-Kamiokande—large, water Cherenkov detector for cosmic particles based at the University of Tokyo (Japan/United States)

TAMA—a 300-m laser-interferometer gravitational-wave antenna (Japan)

TREK—detector aboard Mir that probes the composition of the galactic cosmic rays

TRIAD—Tucson Revised Index of Asteroid Data

UHE—ultrahigh-energy

ULDB—ultralong-duration ballooning

VERITAS—Very Energetic Radiation Imaging Telescope Array System

VHE—very high energy

VIRGO—French-Italian gravitational-wave interferometry project

WIMP—weakly interactive massive particles

4

Report of the Panel on Radio and Submillimeter-Wave Astronomy

SUMMARY

Radio astronomy covers five orders of magnitude in wavelength (300 mm to 30 m) and provides unique as well as complementary windows on the origins of the universe, galaxies, stars, and planets. Radio astronomers sample milliarcsecond scales and millisecond periods. Radio astronomers alone can view the early universe directly through the cosmic microwave background (CMB) and probe large-scale structure independent of redshift using the Sunyaev-Zel'dovich (SZ) effect. Radio waves offer the only clear view of the earliest stages of star and planet formation, both locally and in distant galaxies, by directly probing the dust, magnetic fields, gas dynamics, and rich molecular complexity in the highly obscured environments where galaxies, stars, and planets form.

Not surprisingly, astronomy in the radio and submillimeter wavelength range is driven by technology advances. The last decade has seen the success of the Cosmic Background Explorer (COBE); the completion of the Very Long Baseline Array (VLBA); the launch of the Japanese Very Long Baseline Interferometry (VLBI) Space Observatory Program (VSOP) mission, which pioneered the technique of very-long-baseline interferometry from space; the upgrade of the Arecibo telescope; and the development of millimeter-wave interferometry and submillimeter capabilities. The Green Bank Telescope (GBT), a unique and powerfully flexible instrument exploiting new technology for radio-wave active optics, was dedicated in August 2000. The Atacama Large Millimeter Array (ALMA), the first-ranked major project of the Astronomy and Astrophysics Survey Committee's Panel on Radio Astronomy a decade ago (it was known then as the Millimeter Array), is currently approaching its construction phase.[1] The ALMA project is far more exciting and capable than originally envisaged and will provide the means to explore the dusty sites of planet and star formation and the hearts of the earliest galaxies. The Panel on Radio and Submillimeter-Wave Astronomy reaffirms the high priority given to ALMA by the 1991 Astronomy and Astrophysics Survey Committee and emphasizes that its construction schedule should be maintained.

The panel recommends as its highest priority for major new funding

---

[1]Astronomy and Astrophysics Survey Committee, National Research Council. 1991. *The Decade of Discovery in Astronomy and Astrophysics* (Washington, D.C.: National Academy Press).

the Expanded Very Large Array (EVLA). While the Very Large Array (VLA) remains the most powerful and productive centimeter-wave telescope in the world, advances in technology make possible an order-of-magnitude improvement in both sensitivity and angular resolution, combined with a more than 1000-fold improvement in spectroscopic capability. The new EVLA will be an essential tool for astronomers investigating a wide range of scientific problems. For example, submillimeter studies have shown that a substantial part of the energy release at high redshifts occurs in regions obscured by dust, but the origin of the energy is in question. The EVLA will uniquely distinguish between massive star formation and accretion onto a supermassive black hole as the underlying energy source, allowing researchers to decode the history of star and galaxy formation as well as the role of supermassive black holes in galaxy evolution.

The panel recommends the following moderate projects[2] in order of priority:

- A well-orchestrated technology development program leading to the future construction of the Square Kilometer Array (SKA), an international next-generation radio telescope;
- Immediate construction of the Combined Array for Research in Millimeter-wave Astronomy (CARMA), to precede and ultimately to complement ALMA;
- Development of the Advanced Radio Interferometry between Space and Earth (ARISE) mission for space VLBI to achieve the highest spatial resolution at centimeter and millimeter wavelengths; and
- Construction of a large, single-aperture telescope at the South Pole equipped for wide-area surveying at submillimeter wavelengths, the 10-m South Pole Submillimeter Telescope (SPST).

The panel emphasizes that continued investment in the pursuit of a complete understanding of the CMB, particularly its polarization properties, is critical. A host of experiments—on the ground, balloon-borne, and in space, including the Microwave Anisotropy Probe (MAP) and Planck missions—will characterize the CMB anisotropy within the next few years. Detection of the signature of gravitational waves on the CMB

---

[2]Capital costs in the range $5 million to $50 million.

polarization would provide a unique measurement of the energy scale of the inflationary potential, allowing the origin of the Big Bang to be explored. Radio astronomers will probe energy scales of 10^{16} GeV, well above the reach of particle accelerators, with a sensitivity to the gravity-wave background well beyond that possible with direct gravity-wave detectors. Discoveries by the MAP and Planck missions as well as ground-based investigations will suggest the direction for follow-on NASA missions. NASA should take the necessary steps to initiate those missions rapidly when the optimal strategy becomes clear.

The panel endorses construction of the Low Frequency Array (LOFAR), an international and interagency project. This unique instrument will provide the first real capability to image the low-frequency (meter- and decameter-wave) sky and, along with the One Hectare Telescope (1HT), will constitute the first of the stepping-stones to the SKA.

The panel further recommends an aggressive and vigorous program of technology development. SKA development activities should focus on low-cost, high-performance electronics and processors, techniques and technologies for radio frequency interference (RFI) identification and compensation, array optimization, and radio-wave adaptive optics. For future space missions, development should emphasize inexpensive large apertures, large-format arrays, receivers with the lowest possible noise, high-capacity, space-qualified refrigerators, and enhanced telemetry bandwidth.

The panel supports the recommendation of the Panel on Ultraviolet, Optical, and Infrared Astronomy from Space that NASA pursue technology development leading toward a far-infrared (FIR)/submillimeter interferometer in space (see Chapter 7). The Single-Aperture Far Infrared (SAFIR) Observatory is the logical first step toward the long-term goal of space FIR/submillimeter interferometry, which could provide high-resolution imaging of star formation sites both locally and at high redshift. Such a capability will provide an excellent FIR/submillimeter complement to ALMA, EVLA, and SKA.

The panel emphasizes that the National Science Foundation (NSF) has a special responsibility for radio astronomy, because it is primarily a ground-based activity. This responsibility is reflected in NSF's current support for the university radio observatories as well as the national centers. The panel strongly endorses continuation of this support and recommends enhanced efforts to support full utilization of these essential facilities by their general users.

The radio astronomy community is proud of the national radio astronomy centers. At centimeter wavelengths, the National Radio Astronomy Observatory (NRAO) and the National Astronomy and Ionosphere Center (NAIC) lead the world, and U.S. astronomers rely almost entirely on them for telescope access. The NSF needs to provide increased support to operate, maintain, and continually upgrade the radio facilities to keep them at the cutting edge, and it should seize the opportunity to develop subarcsecond imaging capabilities, complementing those of the VLA and the Next Generation Space Telescope (NGST) at shorter wavelengths (with ALMA) and at longer wavelengths (with LOFAR).

The panel is concerned that NSF funding for critical activities such as data analysis, theory, correlative studies, and student support is not commensurate with its investment in facilities. The NSF should provide sufficient funds for individual investigators to maximize the scientific output of both national and university facilities. It should consider innovative ways to support astronomers in obtaining, analyzing, and publishing data, along with training a new generation of astronomers. This new support should complement the traditional grants program, for which the fraction of proposals that can be funded within the budget has fallen to dangerously low levels.

The panel emphasizes that preservation of portions of the spectrum for future radio astronomical research is vital. The NSF plays a critical role in setting spectrum management policy and in increasing public awareness of its importance. Continued vigilance is required at both the national and international levels to ensure that spectrum allocation balances commercial and research interests. At the same time, investments must be made in the development of hardware and signal-processing techniques to mitigate the effects of human-generated radio interference, which will otherwise drown out the much weaker cosmic radio signals.

In summary, technological advances in telescope hardware spanning the entire wavelength range of interest to radio and submillimeter astronomy now permit the construction of a new generation of powerful instruments. These powerful instruments will provide crucial information on the leading astronomical questions of the decade, especially on how the universe formed—from its superclusters, clusters, and galaxies to their constituent stars and planets. Full exploitation of these instruments will require adequate support for facilities and investigators and preservation of portions of the radio spectrum.

SCIENCE OPPORTUNITIES

Radio astronomy in the last decade has made fundamental contributions to the most important issues in astrophysics. By directly observing CMB radiation, radio astronomers have shown that its spectrum is described to remarkable accuracy by a thermal Planck spectrum, verifying the Big Bang model and illustrating beautifully the simple physics needed to describe the universe at only 2×10^{-5} times its present age. The present imaging of the weak anisotropy in the CMB indicates that the universe may be flat, a strong prediction of the inflationary theory. Ongoing observations of the CMB promise to constrain to high precision a host of parameters that describe our universe, including its curvature and, therefore, the energy density; the Hubble constant; the baryonic and dark matter content; and Einstein's cosmological constant. High-resolution radio imaging allows the detailed measurement of the kinematics in galactic microquasars and their more powerful extragalactic counterparts and measurement of the speed of expansion of gamma-ray bursters. Such observations provide compelling evidence of a supermassive black hole in the heart of the nearby galaxy NGC 4258 and suggest the presence of many more massive black holes in galactic cores. In the past decade, radio astronomers have shown that most forming stars are surrounded by disks, and they have watched the expansion of material ejected from dying stars. Radar probes of Mercury have shown the existence of water ice in the polar craters of the planet closest to the Sun. Quite unexpectedly, it was the exquisitely precise timing of a pulsar spinning on its axis 161 times per second that led to the discovery by radio astronomers of the first extrasolar planetary system.

In areas where radio astronomy is not the only channel of information, it provides a crucial complementary view. In nearly all forefront areas of astronomical research, from the solar neighborhood (star and planet formation) to the furthest reaches of the universe in space (high-redshift galaxies), time (the CMB), and energy (gamma-ray bursts), radio astronomy complements optical, infrared, ultraviolet, x-ray, and gamma-ray observations, delivering ever more detailed views of the cosmos. The imaging capabilities of the new arrays will produce fantastic and fascinating pictures of the otherwise invisible internal workings of molecular clouds producing planetary systems, of the active engines buried within the hearts of galaxies, and of the intricate wisps and filaments lacing the interstellar medium in the core of our galaxy, in the lobes of radio galaxy halos, and in the outermost reaches of distant clusters of galaxies. Furthermore, radio telescopes have already proven to be effective hunters of the faint ephemeral signatures of the titanic explosions signaling star deaths in the distant universe, as heralded by flashes of gamma rays. It is ever more clear that in the coming decades the diverse views provided by the whole of the electromagnetic spectrum will be synthesized into a coherent physics picture of the clockwork of the universe. The radio and submillimeter programs and facilities described in this chapter will be a vital part of this intellectual adventure.

THE LARGE-SCALE STRUCTURE OF THE UNIVERSE

Studies of the relic radiation of the Big Bang, the CMB, lie entirely within the province of radio astronomy. The CMB contains the fossil record of conditions existing in the very early universe before the time of recombination. The mapping of fluctuations in the brightness of the CMB—its anisotropy—provides a snapshot of conditions in the universe at a redshift $z \sim 1000$, equivalent to an age of 300,000 years. The CMB contains the imprints of structures that later grew to produce the large-scale structure we see today. Studies of the CMB radiation and how it has propagated from the time of its generation to the current time of observation thus allow us to take inventory of the matter and energy content of our universe.

A comprehensive program is under way to characterize the CMB anisotropy using ground-based (e.g., CBI, DASI, VSA, Viper, POLAR, Polatron), balloon-based (e.g., BOOMERANG, TopHat, BEAST, MAXIMA), and space-based (MAP, Planck) missions. The majority of these will be operational in the first few years of this decade, and their results will determine the specifications for the suite of missions to follow. As illustrated in Figure 4.1, observations of the details of the CMB will allow radio astronomers to directly test models of the emergence and evolution of large-scale structure, leading to the formation of the super-clusters, clusters, and galaxies we see today. Before the end of the decade, radio astronomers will provide high-precision determinations of the primary cosmological parameters.

Measurements of the polarization of the CMB will determine the contribution to CMB anisotropy from gravitational waves excited by the decaying inflationary potential in the early universe ($t = 10^{-30}$ s). Detection of the unique signature of gravitational waves on the CMB polarization will provide a measurement of the energy scale of the inflationary potential, on the order of 10^{16} GeV, well beyond the reach of particle

174

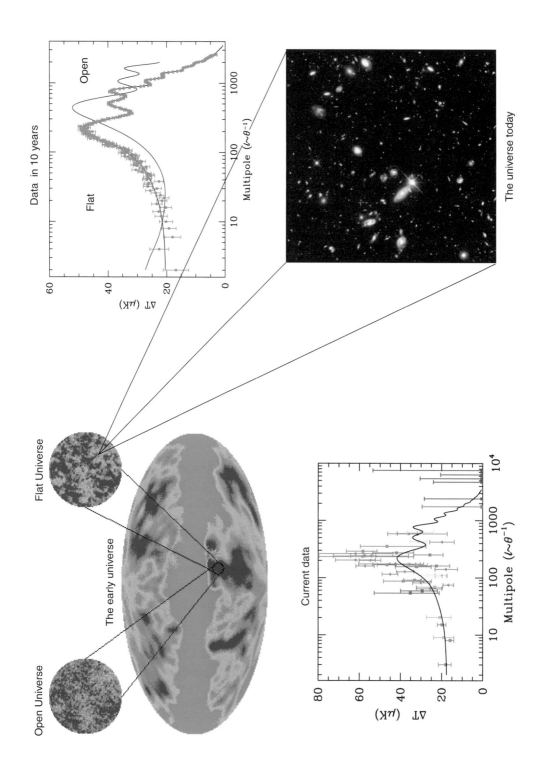

Data in 10 years

Open

Flat

ΔT (μK)

Multipole ($l \sim \theta^{-1}$)

The universe today

Flat Universe

Open Universe

The early universe

Current data

ΔT (μK)

Multipole ($l \sim \theta^{-1}$)

accelerators and with a sensitivity to the gravity-wave background well beyond that possible with direct gravity-wave detectors. Such scales are of great interest not only to cosmologists but also to particle physicists working on constructing a fundamental "theory of everything." Polarization measurements will be difficult and will require, depending on just what the universe is up to, a combination of ground- and space-based instruments toward the end of this decade and beyond, but the importance of these measurements is tremendous. In the end, the observations made by these instruments will tell us whether the inflationary model for the origin of the Big Bang is correct or whether our understanding of the origin of the universe must undergo another major revision.

The small distortion of the background radiation induced when CMB photons scatter off electrons in the hot gas contained within clusters of galaxies is known as the Sunyaev-Zel'dovich (SZ) effect. On its own, the SZ effect is an effective probe of the baryon density in the cluster intergalactic medium. In concert with x-ray observations of free-free emission from the same gas, the SZ effect provides an independent determination of the Hubble constant from local as well as high-redshift clusters. Figure 4.2 illustrates the promise of SZ mapping for tracking the census of galaxy

FIGURE 4.1 Cosmology with the cosmic microwave background (CMB) radiation as promised by the planned ground-based (e.g., CBI, DASI, VSA, Viper, POLAR, and Polatron), balloon-based (e.g., BOOMERANG, TopHat, BEAST, and MAXIMA), and space-based (MAP, Planck) missions. The enlarged regions of the COBE CMB map of the whole sky show the vastly increased resolution expected from future experiments. The simulated high-resolution observations are shown for both a low-density "open" universe and a "flat" universe, which is favored by the inflationary theory for the origin of the universe. The slight variations in the intensity of the CMB measure the inhomogeneities in the universe 300,000 years after the Big Bang. These inhomogeneities eventually grew into the rich structure observed today, as seen, for example, in the HST image shown bottom right. The most efficient way of extracting the cosmological information from CMB maps is to look at the amplitude of temperature fluctuations at different angular scales. This power spectrum is indicated bottom left for today's experiments and top right for the level of accuracy expected within the next 10 years. The two curves shown are for open and flat cosmological models—whether current models are correct and whether the universe is open or flat will be easy to tell. The combination of results promised for the next decade will determine the correct cosmological model for our universe and measure its parameters with high precision. Courtesy of J. Carlstrom, University of Chicago, D. Scott, University of British Columbia, and M. White, Harvard-Smithsonian Center for Astrophysics.

clusters through time, thus strongly constraining models of structure formation. The second-order SZ effect, due to the CMB dipole that the cluster itself sees, can also provide a measurement of the peculiar velocity in its own frame of rest (i.e., the kinematic SZ effect). The smallest antennas proposed for CARMA will be critical for conducting large-scale SZ surveys of the high-redshift universe, while submillimeter observations with the SPST will measure the spectral dependence of the SZ effect. The large collecting area and high spatial resolution provided by ALMA at 7- to 10-mm wavelengths will allow sensitive detailed imaging of the SZ effect from distant clusters.

Radio astronomy also provides its own direct probes of the expansion history of the universe and the allowed values of the cosmological

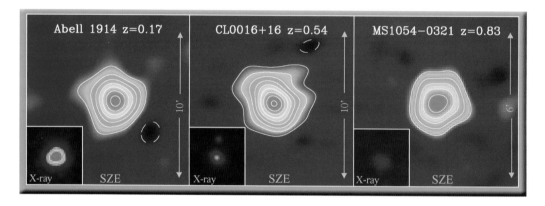

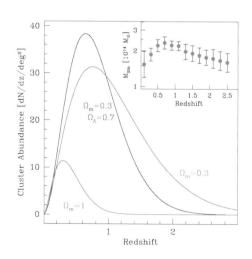

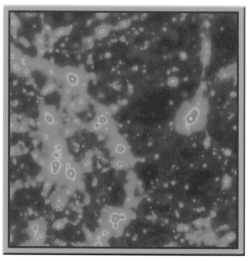

constant. Gravitational lensing, in particular that due to individual galaxies, which causes image splitting on scales of arcseconds, gives measurements of the Hubble constant through time-delay observations. Additionally, the gravitational mass, luminous plus dark matter, in the lensing galaxy is measured. Ongoing and future large surveys of lensing in samples of compact radio galaxies and quasars will determine parameters such as the cosmological constant, matter density, and curvature. In addition to providing large lens samples, the EVLA will offer great improvements in the ability to monitor lens systems for time-delay measurements. The order-of-magnitude increase in sensitivity promised by the future SKA will likewise offer a leap forward in lens statistics.

Radio-source surveys provide the bread-and-butter source count statistics, object identifications, and parent samples that enable high-profile cosmological studies. In particular, gravitational-lens surveys are possible only after a catalog of target sources is available. Furthermore, an understanding of lens statistics requires knowledge of the radio-luminosity relationship as a function of redshift. Other radio-source-based cosmological tests are important: angular size versus redshift, superluminal jet speed versus redshift, and geometric distance measurements made by observing maser hot spots and supernova expansion. Additionally, CMB studies must be carefully corrected for foreground

FIGURE 4.2 Montage of observations and simulations of the Sunyaev-Zel'dovich (SZ) effect. The top montage shows SZ effect images, obtained using dedicated centimeter-wave receivers on the OVRO and BIMA millimeter-wave arrays, of three galaxy clusters at redshifts $z = 0.17$, $z = 0.54$, and $z = 0.83$ (at $z = 0.83$, the universe is approximately one-third of its present age). The same contour level is used for each SZ image. Insets show x-ray images of each cluster, with the same intensity scale used for each cluster. The signature of the SZ effect is comparable at all three distances, whereas the detected x-ray emission (in fact, any emission) decreases rapidly with distance. Hence SZ effect observations allow detection of clusters throughout the observable universe. As illustrated in the bottom left panel, observations of the abundance of clusters at different redshifts, corresponding to different epochs, will allow discriminating between cosmological models. The inset shows the limiting cluster mass M_{lim} as a function of redshift for a realistic SZ effect survey instrument. The bottom right panel illustrates the expected image of the SZ effect for a simulated supercluster (by D. Bond, Canadian Institute of Theoretical Astrophysics). Future very sensitive SZ observations promise to image the intercluster gas (green and blue in the image), a diffuse cosmic web that is likely to be the largest reservoir of baryons or normal matter in the universe. CARMA, ALMA, and SPST will provide the range of sensitivity, resolution, stability, and sky coverage needed to exploit all aspects of the SZ effect for cosmology. Courtesy of J. Carlstrom and J. Mohr, University of Chicago.

(Galactic and extragalactic) radio contamination. The two recent large-scale, centimeter-wave continuum surveys, the FIRST survey and NVSS, have been important products of the current VLA, providing views of the radio sky comparable to those from surveys produced at other wavelengths. Data mining of these surveys continues to engage various segments of the astronomy community at large. An important task for the EVLA and other future arrays will be to produce the next generation of such multiuse surveys, including surveys of gravitational lenses and compact galactic objects.

Models of large-scale structure and galaxy formation predict that between redshifts of about 20 and 5—the dark age—detectable signatures of the first structures should exist in the form of inhomogeneities in the primordial hydrogen heated by infalling gas or by the first generation of stars or quasars. More recently than $z \sim 5$, the development of the cosmic web of neutral hydrogen and the formation and evolution of galaxy halos should be observable. Calculations of the sensitivity required for these measurements led to the need for a radio telescope with 1 square kilometer of collecting area, giving the SKA its name. Depending on the redshift of this epoch, studies of highly redshifted atomic hydrogen (HI) at low frequencies might be undertaken by LOFAR and/or the SKA, allowing researchers to follow the evolution of the baryons quite possibly even before the formation of the first stars and quasars. Figure 4.3 shows what the primordial HI might look like, either before or after the first generation of galaxies/quasars. The first generation of luminous objects will heat the primordial hydrogen, giving observable signatures of their properties and the epoch of their formation, potentially revealing the source of the radiation that reionized the universe. The correlation properties of the hydrogen structures should reflect the initial conditions provided by the quantum fluctuations of the early universe, in effect probing the same linear density field as the CMB, although at a much later stage, and thus will provide an independent check of cosmic background radiation studies and of the effects of reionization on structure formation.

THE FORMATION AND EVOLUTION OF GALAXIES

One of the important observations carried out during the 1990s was the imaging of submillimeter-luminous galaxies at intermediate and high redshift. Much like the discovery of ultraluminous infrared galaxies by IRAS in earlier decades, the newly discovered objects give researchers a

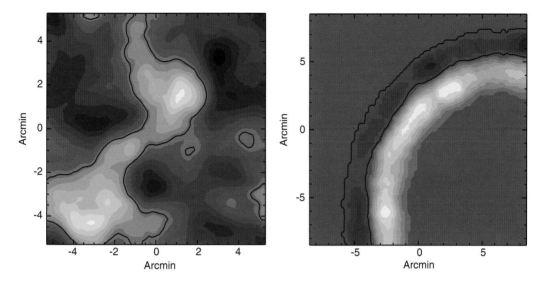

FIGURE 4.3 Simulations of observations of highly redshifted atomic hydrogen during the dark age. *Left:* A map of the 21-cm radio HI line at redshift $z = 8.5$ in an open universe dominated by cold dark matter. The surface brightness fluctuations are due to inhomogeneities in the density of neutral hydrogen that were seeded by the primordial density fluctuations. *Right:* The outwardly propagating front of absorption and emission caused by the interaction of the neutral gas with the Lyman-alpha photons emitted by a newly formed quasar. It is reasonable to expect that a centimeter-wave radio telescope with a very large collecting area would be capable of detecting these two very different signatures of the first structure formation in the universe after recombination. The possibility of carrying out such measurements is a prime driver for the SKA. With permission of P. Tozzi, Johns Hopkins University. Courtesy of J. Hewitt, Massachusetts Institute of Technology.

window onto an otherwise invisible critical phase in the life cycle of galaxies. These galaxies include some of the most distant objects in the universe, and their radio-to-submillimeter spectral index serves as a promising indicator of redshift even in the absence of detection at optical wavelengths. It is too soon to tell if the submillimeter-luminous galaxies are signposts for the all-important principal star-forming phase, but indications are that they represent a population comparable in significance to the Lyman-break galaxies discovered in the near infrared (IR) and most likely represent an early ultraluminous star-formation phase in a forming or merging galaxy. As illustrated in Figure 4.4, high-resolution observations made with the current millimeter arrays demonstrate the likely link between low-redshift, gas-rich, ultraluminous infrared galaxies and the high-redshift submillimeter sources; however, the co-moving number density of the distant objects is 100 times higher and their far-infrared luminosities are greater than those of the nearer population. Many of the ultraluminous galaxies appear as faint radio continuum sources, invisible to ground-based optical images and to the Hubble Space Telescope (HST). During this decade, observations of these objects at all wavelengths will be critical to our understanding of them.

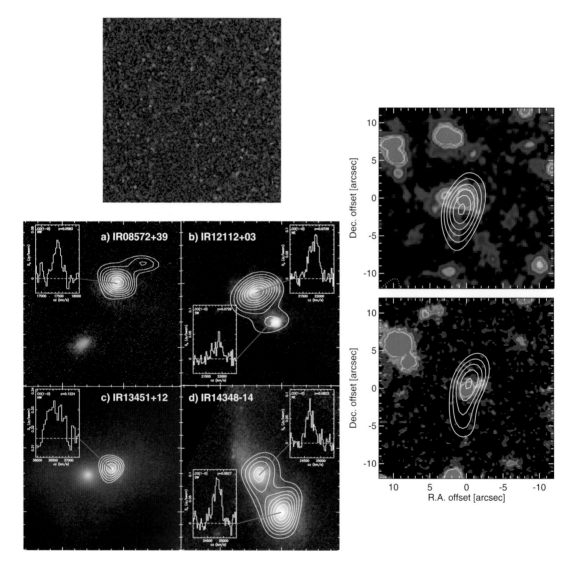

The increased resolution offered by EVLA and CARMA will be followed by the superb capabilities of ALMA, which will detect high-redshift galaxies with great ease. Sensitive surveys using multiband bolometer arrays on the SPST will provide a complete characterization (source flux densities and spectral characteristics) of the source population and target lists for the ALMA programs.

Deep-field radio continuum observations obtained with the EVLA and SKA will revolutionize the study of radio galaxies and their environments. Figure 4.5 offers a glimpse of such a deep field observed after Phase I of the VLA expansion. High-redshift radio galaxies can be found by low-frequency continuum spectral selection (i.e., compact steep-spectrum sources) and, if prevalent, may indicate vigorous structure formation at early epochs. Relic radio haloes, low-energy cosmic-ray electrons, and low-energy radiation due to shocks in large-scale structures in galaxy-cluster environments are a relatively unexplored realm and are accessible only through long-wavelength radio astronomy. The pilot studies using the VLA at $\lambda = 4$ m are extremely promising, proving that the effects of the ionosphere can be removed. It is well known that radio continuum luminosity correlates tightly with the star formation rate and that radio continuum observations, unlike their optical counterparts, do not suffer from dust extinction. Because the EVLA will detect synchrotron radiation from galaxies undergoing massive star formation to $z = 5$, it will serve as an excellent tracer of the history of star formation

FIGURE 4.4 Future studies with ALMA, CARMA, and the EVLA will identify galaxies at high redshift and the processes responsible for their high luminosities. The upper left panel shows a simulation of a 2.56-arcmin^2 deep field observed at 350 GHz (a wavelength of 870 mm) with ALMA. It illustrates the ease with which ALMA will find galaxies at redshifts greater than 3 (red symbols). Observations made with the OVRO millimeter array already show the power of spectroscopic studies of the ultraluminous infrared galaxies found by IRAS and SCUBA. The lower left panel shows contours of the CO (2-1) emission superposed on HST/NICMOS images of four double-nucleus ULIRGs; insets show the CO spectra at the emission peaks. The right panel shows the CO (3-2) emission in two high-redshift ($z \sim 2.6$ to 2.8) submillimeter sources detected by SCUBA. Nearly all of the local ultraluminous merging systems contain massive central molecular gas disks. They appear to be analogues of the high-redshift ultraluminous dusty galaxies, although the co-moving number density of the submillimeter population is more than 100 times higher than that of the low-redshift objects. The suite of high-resolution studies enabled by the future arrays will allow the construction of detailed models of merging systems and an understanding of their importance through cosmic history. Courtesy of NRAO (top left) and Caltech (lower left and right panels).

through the ages. Following Phase II of the VLA expansion, the resolution will be improved to only a few hundredths of an arcsecond, comparable to the spatial resolution of ALMA and NGST, and distant galaxies will be imaged with sufficient resolution to distinguish star-forming phenomena from active galactic nuclei (AGN) related to a massive central engine.

Molecular gas, as traced particularly by carbon monoxide but also by other species, represents the star-forming component of the interstellar medium in galaxies. The dynamics of the gas in bars, in spiral density

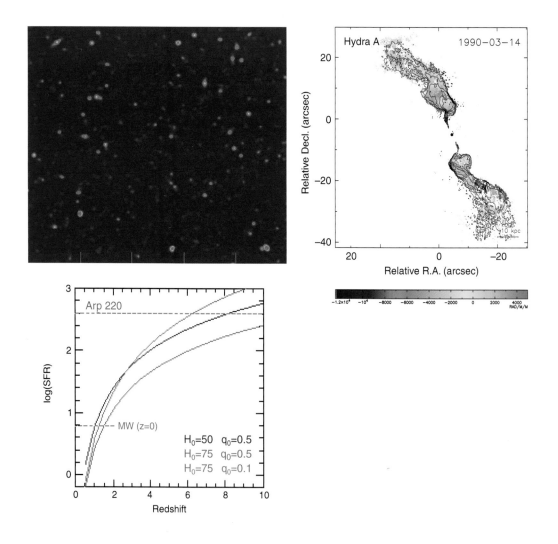

waves, in regions of starbursts, and in the vicinity of AGNs is often gleaned only from molecular studies, yielding clues to the radial redistribution of gas that must drive at least in part the evolution of galaxy disks. The all-important rate of star formation and its evolution through cosmic history is constrained by studies of redshifted molecular lines and atomic recombination lines. The study of the neutral hydrogen 21-cm line emission has been a staple of observational cosmology, providing diverse clues to the structure of galaxy disks, the circumstances of star formation within them, and the history of galaxy interactions. The molecular, atomic, and continuum studies of galaxies provide the foundation on which detailed galaxy formation models will be built, while the information extracted from the velocity fields via spectroscopic imaging of atomic and molecular species reveals the history of encounter dynamics. Figure 4.6 illustrates the potential for future studies with EVLA, ALMA, and CARMA to provide rich details on the processes of tidal disruption and mergers tracing the kinematic disturbances on both large and small scales.

Late-type spiral and irregular galaxies are rich in neutral hydrogen, which often extends far beyond their optical boundaries. In gas-rich dwarf galaxies, the HI line provides the most efficient, if not the only, means of measuring the redshift. The HI line width, in combination with

FIGURE 4.5 Future centimeter-wavelength deep-field observations will revolutionize studies of faint radio sources. *Top left*: Simulation of a deep integration comparable in area to the Hubble Deep Field (160 arcsec × 160 arcsec) made with the EVLA at 1.4 GHz for 500 h, yielding about 500 radio sources stronger than 500 nJy. The full field imaged by the EVLA will be much larger (45 arcmin × 45 arcmin but with reduced sensitivity in the outer parts) and will contain 40,000 sources with a resolution of a few arcseconds. Studies of these sources will be used to trace the star formation history with redshift, irrespective of dust extinction. Additionally, it will yield 1000 rotation measures that probe the structure of the large-scale magnetic field. In the future, the SKA will achieve comparable sensitivity in only 1 hour. *Top right*: The rotation measure structure of Hydra A at 0.3-arcsec resolution as seen with the current VLA; the EVLA will allow similar mapping of sources that are an order of magnitude fainter. *Bottom left*: The 5σ detectability of star-forming galaxies in the deep EVLA radio continuum field in terms of the star formation rate (vertical axis, in logarithmic units of solar masses per year), assuming the local relationship between star formation rate and the radio continuum flux density holds. At this level of sensitivity, it will be possible to detect galaxies similar to our own Milky Way (with a star formation rate of 6 solar masses per year) out to $z = 1$ and star-forming galaxies such as Arp 220 out to $z = 10$. Courtesy of NRAO.

optical imaging, gives an easily measured, redshift-independent distance estimate via the Tully-Fisher relation. Extension of such studies to the more distant universe awaits improvements in spectral resolution and coverage through the VLA correlator enhancement and in low-frequency-array sensitivity through LOFAR and, later, the SKA. With these advances, the otherwise invisible network of lower-density, large-scale structure bridging the clusters, superclusters, and galaxy groups will be delineated by HI line emission and/or absorption. Because the HI can

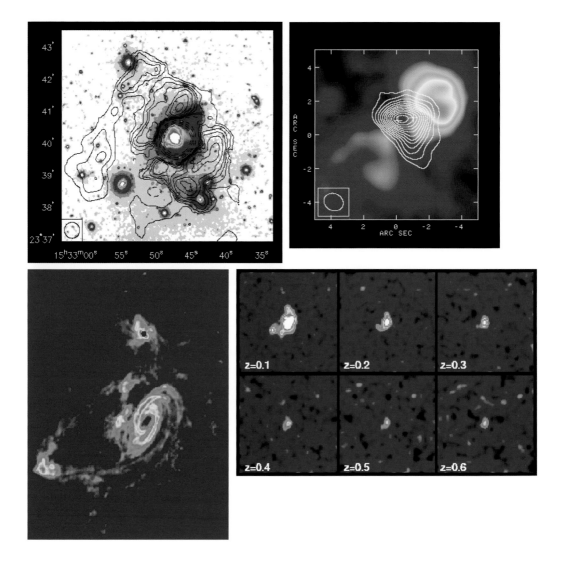

be traced in many objects far beyond the optical edge, the HI rotation curve strongly constrains the dark matter halo parameters. It is recognized, but not understood, that in many galaxies the radial surface densities of gas closely follow the inferred dark matter surface densities. Improvements in VLA spectral resolution and coverage promise advances in understanding the shapes and structures of haloes as well as constraints on the nature of the dark matter within them.

Magnetic fields are suspected or known to be important in virtually all astrophysical contexts. Synchrotron and maser emissions are both intimately connected to magnetic phenomena and provide direct probes of magnetic field distributions, orientation, and strength. High-fidelity polarization imaging and Faraday rotation measurements trace out the magnetic field structure in normal galaxies and distant clusters and locate distant ($z > 2$) clusters of galaxies.

About 10 percent of bright elliptical galaxies and quasars are strong radio sources, often traced by radio jets and lobes stretching as far as a million light-years from the central compact core. Determination of the origin and evolution of this radio emission remains among the most challenging problems in contemporary astrophysics. The radio emission almost surely arises from the synchrotron radiation of highly relativistic electrons moving in weak magnetic fields, but the energy required in the form of relativistic particles and magnetic fields is enormous—up to 10^{61} ergs. It is widely speculated that this energy originates in a supermassive black hole accreting material from its surroundings.

FIGURE 4.6 Spectroscopic studies of disturbed gas disks bear the spatial and kinematic signatures of the dynamics of galaxy encounters. HI traces the interaction history of the merger on large scales, while studies of the CO and other species elucidate the effects of the merger on small scales. *Top left:* The HI emission obtained from VLA mapping of Arp 220 superposed on a deep R-band image. The large-scale HI distribution and corresponding velocity field reveal that two gas-rich galaxies have merged, producing widespread tidal tails. *Top right:* A high-resolution image of CO (2-1) emission obtained with the OVRO millimeter array overlaid on a deep H_α image of the central pixel of the HI map. The details suggest that the bulk of the molecular gas has settled into the center faster than have the two individual galaxy nuclei. *Bottom left:* VLA observations of the HI in the M81/M82/NGC3077 system reveal the tidal disruption. *Bottom right:* Simulation of what the EVLA would see in a similar system at redshifts out to 0.6. Future spectroscopic observations with the EVLA, ALMA, and CARMA will reveal the details of merger histories as well as the evolution of quiescent gas disks. The combination of atomic, molecular, and continuum imaging permits comparison of the interaction and starburst time scales. Courtesy of NRAO.

Figure 4.7 illustrates the complex structure of the radio continuum emission and its wavelength dependence, seen over many orders of magnitude in scale in the giant elliptical galaxy M87. How is the energy of accretion converted to relativistic plasma, and what collimates this plasma into narrow beams and relativistic jets that extend far beyond the optical galaxy? VLBA observations of circular polarization in some of these relativistic jets have provided compelling arguments that the plasma is composed primarily of electrons and positrons rather than electrons and protons. Future VLBA observations of the polarized morphology and kinematics of quasars and AGN with angular resolution better than 0.001 arcsec (a few parsecs at cosmological distances) will give new insight into the energy and mass content of relativistic jets and help us to understand how they are created. Indeed, VLBA observations at a resolution of 100 μarcsec (about 30 Schwarzschild radii, R_s) have already shown that the jet is not fully collimated at a radius of about 100 R_s. Measuring and understanding the collimation mechanism of the jet is crucial to understanding its origin. It can be expected that new images obtained with ARISE will be able to settle the controversy about whether jets are powered by the accretion disk or the spin of the black hole. Space VLBI observations with VSOP2, the planned Japanese follow-on mission to VSOP, and, later, ARISE, will improve the linear resolution to better than a parsec for distant quasars and AGN and to within 3 R_s for nearby AGN.

As illustrated in Figure 4.8, high-precision observations with the VLBA of the positions, velocities, and accelerations of the water vapor masers reveal in detail the dynamical structure of the central molecular disk in the nearby Seyfert galaxy NGC 4258. Although the rotation period of the disk is about 800 years, the high angular (200 μarcsec) and velocity (0.1 km s^{-1}) resolutions of the VLBA measurements allow tracking the rotation of the water vapor masers within the disk over a time frame of only a few years (at 32 μarcsec/yr). The molecular material that traces a very thin disk is in near-perfect Keplerian motion over a wide range of radii. The central mass contained within 0.1 parsec exceeds 3×10^7 solar masses, with a mass density exceeding 10^{10} solar masses per cubic parsec; such a high density provides compelling evidence that the central object is a supermassive black hole. In the new decade, similar observations will be extended to much higher sensitivity with the large, ground-based radio telescopes and arrays and to unprecedented angular resolution with ARISE.

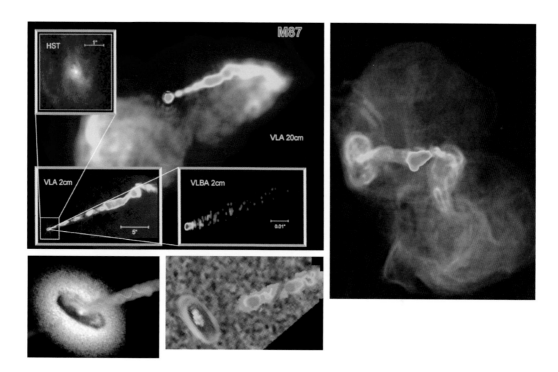

FIGURE 4.7 Current and future observations of the radio emission from the giant elliptical M87, one of the first known radio galaxies. *Right panel:* Recent observations with the VLA at 90 cm tracing the intricate structure of the cloud of relativistic plasma on the largest scales. The radio lobes extend over 100 kpc, nearly 10 times larger than the optical extent of the galaxy. This radio halo is embedded in an even larger x-ray-emitting gas. The bright plumes emanating from the inner lobes suggest that the halo is still "alive" and is being supplied by energy from the central black hole. *Top left panel:* Progressively higher-resolution observations, made with the VLA and VLBA, of the inner set of radio lobes and the relativistic jet. The inset is an HST Wide Field Planetary Camera 2 (WFPC2) image of the nucleus of M87. The scale of the VLBA image is represented by one pixel in the HST image. The radio jet appears one-sided because of the relativistic boosting of the radio brightness along the direction of motion; the bright side is the side moving toward us. The VLA image at 2 cm shows the jet with a resolution of about 0.2 arcsec. The 2-cm VLBA image has an angular resolution of better than 0.001 arcsec and shows the inner few parsecs. The jet appears to flow at a velocity of only about one-tenth the speed of light, whereas the VLA and HST observations further outward indicate motions much closer to the speed of light. This increase in velocity implies that acceleration is taking place along the jet rather than in the black hole. The VLBA image shows evidence of a corkscrew motion or periodic oscillation, characteristic of a precessing nozzle or a plasma instability. At the highest resolution of the VLBA, 100 μarcsec, or 30 Schwarzschild radii (image not shown here), the jet appears to narrow toward its base. *Bottom left panel:* A cartoon of a supermassive black hole, surrounding accretion disk, and jet like the one in the center of M87. The cartoon consists of an actual HST image of the inner accretion disk in a nearby galaxy [NGC4261] with a simulated jet superimposed to conceptually show greater detail. *Bottom center panel:* Simulation of the image of the inner accretion disk and jet that ARISE would obtain at a wavelength of 3 mm. Courtesy of NRAO (VLA and VLBA images), the Space Telescope Science Institute and NASA (HST images), and NASA's Jet Propulsion Laboratory (simulation and cartoon).

0.5 ly

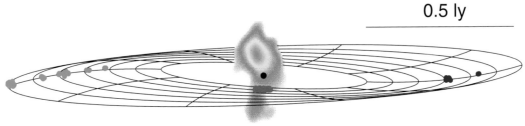

FIGURE 4.8 The H_2O maser and continuum emission from the nearby Seyfert galaxy NGC 4258, also known as Messier 106. *Top panel:* A cartoon drawing of the accretion disk surrounding the black hole in NGC 4258 and the synchrotron jet emerging along its spin axis. The spectrum of the H_2O maser at 1.3-cm wavelength is shown below. *Bottom panel:* The image of the H_2O maser. The wire grid diagram depicts the accretion disk determined from the positions, velocities, and accelerations of the maser spots. The continuum emission near the dynamical center of the system appears to be the innermost portion of a jet. Proper motions of the systemic group of masers have been measured at about 32 μarcsec per year, providing direct evidence of the rotation of the disk as well as a method for determining the distance to the galaxy to an accuracy of 4 percent. Future kinematical studies of masers around supermassive holes will be greatly enhanced by use of the VLBA in conjunction both with the GBT and with orbiting antennas such as VSOP-2, RADIOASTRON, and ARISE. Top image prepared for M. Inoue, National Astronomical Observatory of Japan, by John Kagaya, Hoshi No Techou. Both images courtesy of J. Moran and L. Greenhill, Harvard-Smithsonian Center for Astrophysics.

Precision astrometry achieved through VLA and VLBA observations has enabled accurate determinations of Milky Way structure, including direct estimates of the distance to the Galactic center and the angular rotation rate. The latter compares well with recent estimates, based on results obtained with the Hipparcos satellite, of the shear and vorticity (Oort's constants A and B) in the solar neighborhood. The apparent proper motion of Sgr A*, about 6.0 milliarcsec/yr relative to extragalactic sources, can be attributed entirely to the Sun's orbital motion about the Galactic center. The lack of a measurable peculiar motion of Sgr A* itself indicates that the subastronomical-unit-scale radio source is more massive than 10^3 solar masses and that it anchors the dynamical center of the Galaxy. Coupled with proper motions of stars in the central cluster, the case for a black hole of a few million solar masses in the center of the Galaxy is very strong.

Observations of the position of Sgr A* relative to extragalactic sources may soon be able to achieve an accuracy of 30 marcsec, allowing a determination of the distance to Sgr A* via a trigonometric parallax with an uncertainty of about 5 percent. The angular rotation of the Sun around the Galactic center will then be determined to much higher accuracy (<1 percent), and thus the Sun's circular velocity will be known with an accuracy limited only by knowledge of the Galactic center distance. The new antennas of the EVLA, when used in conjunction with the VLBA, will provide the increased astrometric accuracy. Ultimately, the SKA, if sited in the Southern Hemisphere, operated at high frequencies (≥ 20 GHz), and capable of multibeaming (simultaneous observations toward different sources), might yield a trigonometric parallax to the Galactic center accurate to 1 percent.

Gamma-ray bursts (GRBs) have been one of the long-standing mysteries vexing astronomers over the past decades. Radio observations delivered the identification of the host galaxies of GRBs and vindicated the cosmological fireball model for the origin of GRBs. One gamma-ray burst was found to expand at an apparent superluminal speed and to be the product of a very massive supernova, given the epithet "hypernova." Highly accurate radio positions and light curves obtained with the EVLA will complete the view of these transient high-energy phenomena when combined with the automated gamma-ray alert network, x-ray localization, and optical redshifts (and, lately, early flashes).

THE FORMATION AND EVOLUTION OF STARS

Star formation is a pivotal process about which both galaxy formation and planet formation turn: the formation of galaxies is ultimately controlled by how they form stars, and planets form out of disks that are intimately connected to the star-formation process. Post-main-sequence stars provide the heavy-element enrichment of the general interstellar medium and the kinetic energy that stirs it, affecting subsequent generations of star formation as well as the likelihood of habitable planets and the building blocks of life. Yet all these phenomena are poorly understood.

Star formation occurs in dense cores within molecular clouds, where dust hides the process from observations at optical wavelengths. In the earliest stages, only radio/submillimeter and far-infrared observations can peer deep within the hearts of star-forming cores. Dust emission in the Rayleigh-Jeans limit increases rapidly with frequency, and a rich spectrum of molecular transitions arises at millimeter wavelengths. As a result, radio observations of star formation have been most intensely pursued at millimeter and submillimeter wavelengths, revealing the rich complexity of molecular clouds. Such observations have only just begun to unveil the details of the accretion disks that feed the stars and eventually form planets.

Stars of low mass like the Sun are of particular interest. They can be studied in detail in nearby clouds, where some form in relative isolation rather than in rich clusters. A paradigm for the formation of isolated low-mass stars exists, but is it correct? Observations of dust continuum emission across the radio/submillimeter band, combining fine spatial resolution with sensitivity to extended structures, can trace the flow of matter from the outer envelope through the disk to the forming star. Bolometer arrays on the Caltech Submillimeter Observatory (CSO) 10.4-m antenna, the Large Millimeter Telescope (LMT), the GBT, and the SPST will map the largest scales, while the Submillimeter Array (SMA), CARMA, and ultimately ALMA will enable study of smaller scales. Studies of spectral lines with excellent spectral resolution with the same telescopes will reveal the kinematics, physical conditions, and chemical complexity.

Massive stars, and probably most stars, form in embedded clusters within massive dense cores. We lack a clear paradigm for clustered star formation. Can our understanding of low-mass star formation be extended to massive stars? The most massive stars, being rarer, are found principally in more distant regions, and observations with better angular resolution are needed to provide a firm comparison to the formation of low-mass stars. In addition to enabling submillimeter studies, the EVLA will be able to image, simultaneously and with sufficient angular resolution, multiple transitions of molecules like ammonia in the dense, high-temperature material close to the star. The new centimeter-wave continuum observations of nascent ionized hydrogen (HII) regions and stellar jets will have the high spatial resolution necessary to directly image the essential structures in these objects, providing crucial clues to the evolutionary state of different objects.

Studies of star formation in clustered environments will also address questions about the origin of the stellar initial mass function (IMF). The mass function of molecular clouds or of clumps within clouds is not the same as the IMF. New results suggest that interactions in dense clumpy cores may cause the mass function of clumps to evolve toward the IMF. Imaging a variety of regions with large detector arrays on single dishes and with arrays of telescopes like the SMA, CARMA, and ALMA, combined with new theoretical insights, will ultimately yield an understanding of the IMF and how it depends on initial conditions. Such an understanding will be crucial to unraveling the formation of stars at high redshift and the origin of the first generation of stars in the universe.

One of the most striking facts about stars is that about half of them exist in multiple systems, mostly binaries. Why is it so close to a 50:50 split, rather than almost all single or almost all binary systems? Some crucial parameter must be poised close to a critical point so that half the time a single star forms. Rotation is the obvious candidate, but rotation is not dynamically important on the scales of individual star-forming cores. Spectroscopic observations of Doppler shifts with CARMA, ALMA, and the EVLA can trace rotational motions to smaller scales, where rotation does become important, identifying the stages at which multiplicity should develop. ALMA, observing dust continuum emission with exquisite sensitivity to mass, will identify binaries at the earliest stages and determine the effects of binaries on disks, resolving long-standing issues about the method of binary formation and the possibilities for planets in binary systems.

The wild card in star-formation studies is the magnetic field. Astronomers know little about the strength, orientation, and organization of magnetic fields in regions forming low-mass stars. What they know about regions forming more massive stars suggests that magnetic fields play an important role. Zeeman measurements of molecular lines probe magnetic field strengths along the line of sight, while polarization of dust continuum and line emission will provide the field direction projected on the plane of the sky. Recent progress in detecting the Zeeman effect in lines tracing denser gas indicates that the future, more sensitive arrays on higher-spatial-resolution instruments will allow study of the magnetic field strength in more condensed regions closer to the forming stars. The polarization of continuum and line emission has also been detected with current arrays in a few cases, although the interesting structures in these objects are somewhat blurred by the limited spatial resolution of the arrays; CARMA and ALMA will allow systematic study of the role of magnetic fields in star formation.

From the red-giant phase to their final state, stars eject dust grains and molecules back to the interstellar medium (ISM). Future studies of evolved stars will focus on imaging the expulsion of matter at various stages. Figure 4.9 shows sequences of images obtained for two dramatic examples: the extragalactic supernova SN1993J (left) and the x-ray nova CI Cam (right). In both cases, radio images trace the expansion of the material into the surrounding medium. Measurement of the expansion of the supernova, in combination with the observed Doppler broadening of the optical H_α line, also provides a simple geometric determination of the distance to the host galaxy, M81. While such ejection of material from old stars is commonplace, many puzzles about the processes of ejection and their consequences remain. How, in detail, is dust made in the outflows from late-type stars? How do the anisotropic winds observed in certain late-type stars form prior to, and during the transition to, the planetary-nebula stage, and why is the large-scale morphology of these objects often bipolar? Studies of obscured supernovae are the domain of long-wavelength radio astronomy. The EVLA, LOFAR, and the SKA will be important tools for the study of supernovae and, along with the Arecibo Observatory and the GBT, the pulsars they leave behind. Studies of pulsar Doppler shifts promise further exciting discoveries, including more pulsar planets and, with any luck, binary systems containing a pulsar and a black hole. The tremendous increase in sensitivity offered by the SKA will make possible the submicroarcsecond timing of several hundred millisecond pulsars, the ultimate pulsar timing array, not only to serve as a time standard for referencing the solar system barycenter but also to be capable of sensing the passage of long-period gravitational waves and to probe the gravitational-wave background.

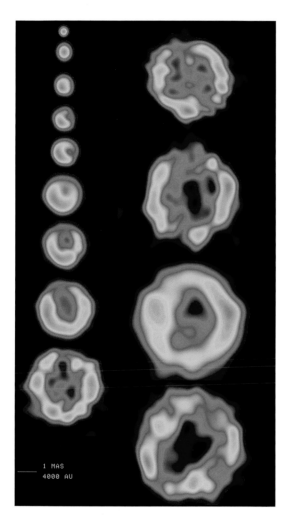

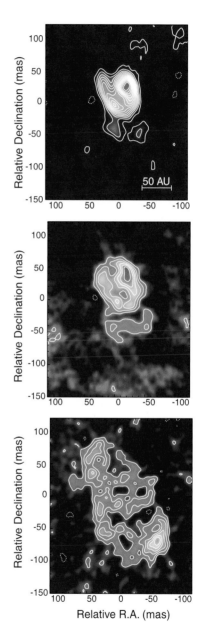

FIGURE 4.9 Time-lapse imaging of exploding stars. *Left:* A sequence of VLBA images of the expansion of the supernova SN1993J observed in the nearby galaxy M81. The images were taken from May 16, 1993 (top left), 6 weeks after the explosion, to November 14, 1997 (bottom right). Courtesy of M. Rupen (NRAO). *Right:* A sequence of VLBA images of the evolving x-ray nova CI Cam taken 75 (top), 93 (middle), and 163 (bottom) days after the x-ray outburst. Some of the fade-out of structural detail in the last image is an artifact of the current imaging capability, as the source becomes larger than ideal for the VLBA but is still too small to be resolved by the existing VLA. The EVLA will allow tracking of such expansion without degradation of the image quality. Courtesy of A. Mioduszewski, NRAO/Joint Institute for VLBI in Europe.

THE FORMATION AND EVOLUTION OF PLANETS

It is now well established that most young solar-mass stars are surrounded by circumstellar accretion disks with masses of $\sim 10^{-3}$ to $10^{-1} M_\odot$ and sizes of ~ 100 to 400 AU. In addition to serving as a conduit for mass accretion onto the young star, the disks also serve as a reservoir of gas and dust for the formation of planetary systems. Figure 4.10 illustrates the promise that ALMA offers for imaging protoplanetary systems, including the ability to detect giant planets in formation. The canonical theory for the formation of our own solar system holds that the Jovian planets formed in the solar nebula at or near their present location, while the terrestrial planets were assembled largely in gas-poor conditions long after the formation of the outer gas giants. In this model, the delivery of volatiles from primitive planetesimals such as chondrites and comets may have played an important role in the surface chemistry of Earth, Venus, and Mars. The highly reduced organic contents of these chondrites and comets are difficult to generate in young planetary atmospheres and may thereby have contributed substantially to the prebiotic evolution of terrestrial planets.

The recent discovery of massive extrasolar planets in orbits well inside 1 AU—the so-called "hot Jupiters"—casts considerable doubt on our present understanding of the formation and evolution of (habitable) planetary systems. All the theoretical mechanisms for inducing the migration of large planets during star formation require the presence of a massive gas and dust disk that can exchange angular momentum with the Jovian protoplanet. The gas densities required of such massive disks imply large optical depths in the optical and infrared. Gaps cleared by

FIGURE 4.10 Montage of future observations of protoplanetary disks around young stellar objects (YSOs) embedded within complex molecular clouds. The large-scale molecular distribution will be studied with the LMT, the GBT, and the Arecibo telescope, while higher-resolution studies of localized regions with ALMA and CARMA will focus on individual star-forming cores. *Top:* ^{13}CO (1-0) map of the Taurus region made with the BIMA millimeter array. *Bottom left:* HST-NICMOS image of the accretion disk around the YSO IRAS 04302+2247, which lies to the north and west of the BIMA field. The contours present ^{13}CO (1-0) emission mapped with the OVRO millimeter array. *Bottom right:* Simulations of potential ALMA images of disks around embedded young stars showing the formation of giant planets. The model disk has a central hole 3 AU in radius and protoplanet-driven overdensities at 9 AU, 22 AU, and 37 AU. The word ALMA in the model images has a line stroke 4 AU wide and 35 AU high and is used as a gauge of the image fidelity. Simulations courtesy of L. Mundy, University of Maryland, and G. Blake, Caltech.

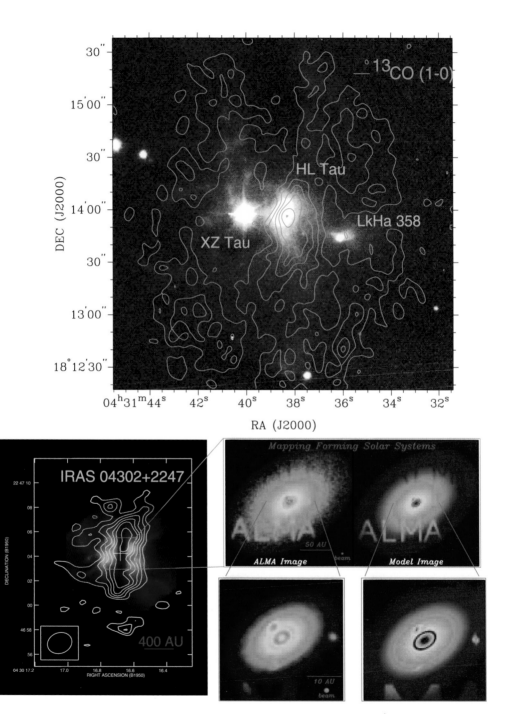

the planet as it migrates will certainly be observable at short wavelengths, but to probe the earliest stages of the process and to access the nebular mid-plane, where the bulk of the planetesimal growth occurs, observations at radio and submillimeter wavelengths are essential. Initial forays early in this decade will utilize the SMA, CARMA, and the EVLA; toward the end of the decade ALMA will examine planetary migration mechanisms directly in the nearest star-forming clouds. Observations by these arrays, and in the 2010s by the SKA, will extend our understanding of planet formation and migration to the most deeply embedded, protostellar stages. Large bolometer arrays on the CSO, the LMT, the GBT, and the SPST will examine the coldest components of debris disks, complementing studies in the infrared. Follow-up studies with CARMA and ALMA will reveal the structure of these debris disks.

ALMA and the SKA will be able to detect the photospheres of stars out to distances of at least 30 pc. Provided that the centroids of the photospheres are sufficiently stable, astrometric searches for planets will be feasible. The superb long-term precision and accuracy of radio astrometry will allow the study of long orbital periods and will complement the measurements from spaceborne instruments, which are necessarily of limited duration. Subarcsecond imaging with ALMA and the EVLA, in tracers of both the gas and dust, will be crucial for answering questions such as the following: Do habitable planets form around stars more massive than the Sun? What kinds of binary and multiple stars have planetary systems capable of supporting life? Does the migration process in such systems differ dramatically from that which operates in the disks around single low-mass stars?

Radio and submillimeter instruments will also make possible the first detailed studies of the chemical composition of circumstellar accretion disk analogues of the solar nebula. In particular, by combining images from the SMA, CARMA, and ALMA with spectra of various gas-phase species and solid-state features from the CSO, GBT, LMT, the Stratospheric Observatory for Infrared Astronomy (SOFIA), and the Space Infrared Telescope Facility (SIRTF), astronomers will be able to probe, for the first time, the radial chemical variations so clearly observable in the composition of small bodies in the solar system. The detailed comparison of these results with what is known about our own solar system will be critical in unraveling the complex suite of chemical and physical processes that lead to the formation of potentially habitable worlds.

An important component of this investigation will be the new studies of our solar system made possible by advances in radio and submillimeter instruments. In the outer solar system, the combination of optical/IR surveys and millimeter-wave flux measurements can directly constrain the size-albedo relationship for Kuiper Belt objects (KBOs). High-resolution images from the SMA, CARMA, and ALMA can be combined with images from heterodyne focal-plane arrays at the CSO, GBT, LMT, and SPST to examine the comae of comets, including, for the first time, those with short periods. Direct imaging of the nuclei of comets and investigations of large particles in their comae will be possible with the Arecibo radar system. Observations of the continuum emission from comets and KBOs will become possible for objects 10 times smaller than those currently observable. The inventory of such objects reflects the properties of the solar nebula at the time of their formation. In addition, dust grains from small bodies may have provided a substantial fraction of the organic compounds on the early Earth, and so it is important to understand their composition in detail. Such studies will include precision measurements of isotopic ratios such as the deuterium-to-hydrogen ratio in various tracers, studies that can now be done for only the brightest objects.

Studies of the planets themselves and of smaller, inner solar system bodies such as asteroids will also be revolutionized during this decade. The great boost in sensitivity given by the recently completed upgrade will be exploited for ever-more-detailed Arecibo radar studies of the surfaces of the terrestrial planets as well as the satellites of Jupiter and Saturn. The Arecibo radar system now allows imaging with 20-m scales of a large number of near-Earth asteroids, giving clues to their surface and dynamical properties and their collision histories. The subarcsecond imaging capability envisioned for CARMA, ALMA, and the EVLA will enable observations of solar system objects at radio and submillimeter wavelengths with resolution similar to that of HST and NGST. In continuum mode, such observations can be used to examine the properties of the surface and subsurface layers of terrestrial bodies, while spectroscopic tracers at radio wavelengths directly probe the atmospheric circulation and composition. The former observations can detect, among other things, the presence of ice versus that of rock, while the latter probe both the vertical and the horizontal profiles of the atmosphere. Such observations, which are nearly impossible in any other spectral region, provide essential comprehensive adjuncts to spacecraft missions, in addition to being of great interest in their own right.

THE ORIGIN AND EVOLUTION OF LIFE

In addition to the evolution of the Universe at the macroscopic level, complexity also evolves at the microscopic level, eventually leading to life, possibly sentient. From a sea of quarks, baryons emerge and organize themselves into progressively more complex nuclei, first in the era of Big Bang nucleosynthesis and later in stars. With more-complex nuclei, molecules and dust can appear when conditions are suitable. What is the history of nucleosynthesis and chemical complexity in the universe? Preliminary evidence suggests that molecules and dust are common out to a redshift of at least 4, but the detailed history remains to be unraveled by observations of atoms (primarily carbon, nitrogen, and oxygen), molecules (carbon monoxide and others), and dust at high redshift. These will be accessible to ALMA, CARMA, and the EVLA. In nearby molecular clouds, these same instruments will trace the evolution of chemical complexity in the dense cores that will form stars. It is already clear that molecules accrete onto dust grains in cold regions, are sublimated when those grains are heated by forming stars, and trigger further increases in complexity when ejected back into the gas phase. The study of these processes in collapsing molecular cores will connect to studies of the chemical evolution in disks and in our solar system to provide a more complete picture of how the building blocks of life are delivered to planetary surfaces.

Recent developments have led to the emergence of astrobiology as a new, interdisciplinary field. The past decade has witnessed the addition of more complex organic molecules to the list of interstellar ingredients, the detection of extrasolar planetary systems, and the identification of more potential sites for liquid water in our solar system. On Earth, life has been found in extreme and unexpected environments, organisms of unexpectedly small size have been claimed, and, most controversially, such organisms have been suggested to exist in Martian meteorites. Within the next few decades, we may acquire a crude chemical assay of the atmospheres of nearby extrasolar terrestrial planets. Indirect evidence for the existence of extraterrestrial life may follow.

Are we alone? This question, more than any other, fascinates the general public. The detection of atmospheric signatures of life around other stars may become possible in the next few decades. The most exciting development would be unambiguous evidence that life elsewhere has evolved intelligence and a technological civilization. Evidence for another technological civilization would have a profound impact on humanity. Observations currently under way at a number of radio telescopes may succeed in detecting radio signals from other civilizations, but past searches have barely stirred the surface of the cosmic haystack. It is important to support observational programs that attempt to detect signals at increasingly large distances. Modest re-sources are needed to develop new signal-processing techniques and new hardware and to establish innovative search programs. Such programs should be part of a broad strategy to search for life elsewhere in the universe.

The 1HT will be the first telescope built to search for extraterrestrial signals and will pioneer new radio techniques. Later, the SKA will add a powerful capability: it will be capable of detecting signals—comparable to our own planet's television emission—from planets around nearby stars.

EXISTING PROGRAMS

Because radio astronomy is almost exclusively a ground-based activity, primary funding comes from the National Science Foundation.

NATIONAL CENTERS

Through cooperative agreements, the NSF supports two national centers for radio astronomy: the National Radio Astronomy Observatory (NRAO), operated by Associated Universities, Inc., and the National Astronomy and Ionosphere Center (NAIC), operated by Cornell Univer-sity. At centimeter wavelengths, NRAO and NAIC facilities are the best in the world, and unlike the situation in optical astronomy, there are no competing private facilities; university astronomers working at centimeter wavelengths rely almost entirely on NRAO and NAIC for their research.

NRAO operates the 27-element Very Large Array (VLA) in New Mexico; the 10-element VLBA, spread throughout the United States; and the newly completed 100-m GBT in West Virginia. The NRAO also operates a ground station for space VLBI and a VLBI antenna for the U.S. Naval Observatory at Green Bank. A cutting-edge technology develop-ment program is conducted at the NRAO central headquarters in Charlottesville, Virginia. The NRAO spearheads U.S. involvement in ALMA and has aggressively and successfully pursued its design and development.

Activities at the NAIC revolve around the 305-m spherical antenna in northwestern Puerto Rico but span the fields of atmospheric and ionospheric modification and planetary remote sensing as well as passive radio astronomy. The recent installation of the Gregorian optical system has provided unprecedented sensitivity and new frequency coverage and flexibility.

The panel endorses the statements about NRAO and NAIC made in the Executive Summary of the survey committee report. The radio astronomy community is justifiably proud of both its national centers, NRAO and NAIC, for their expertise, leadership, and dedication to providing the most advanced instrumentation to the nation's scientists. In addition to supporting their respective scientific programs, both NRAO and NAIC maintain effective education and outreach programs.

UNIVERSITY RADIO FACILITIES

Until ALMA begins operation, the university radio facilities will provide the only U.S. high-resolution capabilities at millimeter and submillimeter wavelengths. Furthermore, even when ALMA is fully operational, they will continue to play a crucial role. These facilities include the following:

• *Caltech Submillimeter Observatory.* In operation since 1988, the CSO, operated by Caltech as an open-access facility, consists of a 10.4-m-diameter antenna in a dome near the summit of Mauna Kea. It provides the only regular access to the crucial submillimeter region for U.S. astronomers, and the hands-on operation and state-of-the art instrument development provide vital experience for young astronomers. A full suite of heterodyne receivers covers the atmospheric windows from 350 to 1300 μm, and large-format detector arrays are being developed.

• *Five College Radio Astronomy Observatory.* FCRAO currently operates a 14-m antenna at New Salem, Massachusetts, working at millimeter wavelengths. It offers a heterodyne array and an advanced spectrometer for large-scale mapping. Like the CSO, it is a hands-on facility used by many students, and it supports instrumentation development by young astronomers. In the decade 2001 to 2010, operations will be shifted to the LMT, a joint U.S.-Mexican 50-m active optics antenna at a high (4600-m) site in Mexico. The telescope will greatly enhance flexibility and sensitivity in the 0.85- to 3.4-mm window.

• *Berkeley-Illinois-Maryland Association array.* The BIMA array consists of ten 6-m dishes operating at 1 and 3 mm and located at the Hat Creek Radio Observatory in northern California. The antennas can be configured in various patterns to give baselines from 7 m to 2 km.

• *Owens Valley Radio Observatory millimeter array.* The present OVRO array consists of six 10.4-m dishes in a reconfigurable pattern that provides baselines out to 440 m at an altitude of 1200 m. SIS hetero-dyne receivers cover the 1- and 3-mm spectral regions, with analog and digital correlators serving as the continuum and spectral line back ends. A flexible, high-bandwidth digital correlator based on field-program-mable gate array technology will soon be completed.

• *Combined Array for Millimeter Astronomy.* The BIMA and OVRO arrays will be merged, moved to a higher site, and enhanced to produce the CARMA facility, one of the moderate initiatives recommended by the panel in the next section.

• *Submillimeter Array.* Now under construction near the summit of Mauna Kea, the SMA will consist of eight 6-m elements in a reconfigurable array that will provide baselines from 8 to 508 m and will cover all bands from 1.6 mm to 300 μm. A collaborative project of the Smithsonian Astrophysical Observatory and the Academia Sinica Institute of Astronomy and Astrophysics of Taiwan, the SMA is an exploratory instrument designed to provide high-resolution imaging capability in the wavelength regime intermediate to the wavelengths covered by the existing millimeter arrays and the far-infrared band accessible from air- or spaceborne instruments. Although the SMA is not strictly a university facility, U.S. astronomers will have access to it.

• *Coordinated Millimeter VLBI Array.* The CMVA facilitates 1- to 3-mm VLBI observations of galactic and extragalactic sources using a worldwide array of millimeter-wave radio telescopes. It offers submilliarcsecond resolution and provides correlated VLBI data to its users.

Support for university radio facilities provides the astronomy commu-nity access to unique observational capabilities and maintains a strong university involvement in scientific and technological development. The NSF-supported university radio facilities all provide substantial commu-nity access, typically 30 to 50 percent, through their guest observer programs. Student involvement in cutting-edge research is crucial for the education of the nation's next generation of scientists. Growing numbers of graduate students incorporate multiwavelength data into their research. It is therefore crucial to increase their knowledge of the capabilities of instruments in different wavelength regimes. Enhanced support for research at the university radio facilities is a vital part of the nation's investment in research.

RECOMMENDED NEW INITIATIVES

The panel recommends the following new major- and moderate-category projects[3] for the decade, in order of priority:

Major Project	*Estimated Cost (millions of dollars)*
EVLA	140 (NSF) less ~50 (cost sharing)

Moderate Projects	
SKA program	22 (NSF)
CARMA	11 (NSF plus external)
ARISE	300 to 400 (NASA)
SPST	50 (NSF)

As illustrated in Figure 4.11, the proposed facilities will contribute wide-ranging new capabilities, among them increased angular resolution and wavelength coverage, thus opening up large areas of discovery space. The objective is to construct unique new facilities and design future ones that provide maximal observing capability, often exploiting facilities partly or largely funded through nonfederal sources. For example, the millimeter/submillimeter portfolio will include the ALMA, CARMA, CSO, SMA, LMT, and SPST, each contributing uniquely to the overall effort. International collaboration is already incorporated into ALMA, the EVLA, LOFAR, the LMT, the SMA, and the SKA. The NSF Office of Polar Programs may fund the SPST. The U.S. Department of Defense and the Netherlands Foundation for Research in Astronomy (NFRA) will provide the major funding for LOFAR. All proposed instruments assume provision of significant, if not total, community access, even when capital funding comes from other sources.

---

[3]Capital costs above $50 million for major and between $5 million and $50 million for moderate. Cost estimates include technology development plus funds for operation, new instrumentation, and facility grants for 5 years.

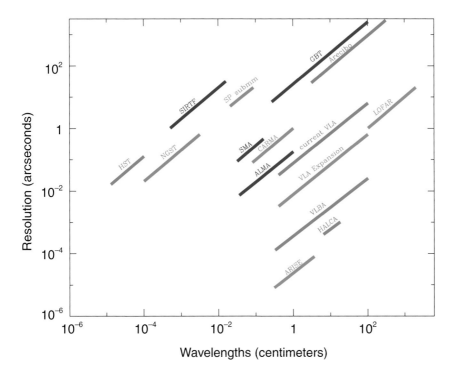

FIGURE 4.11 Angular resolution as a function of wavelength for some individual radio, infrared, and optical telescopes. Colors indicate existing facilities (green), facilities under construction (blue), and proposed projects (red). The radio and submillimeter programs recommended in this chapter will add wide-ranging new capabilities, among them increased angular resolution and wavelength coverage, thus opening up large areas of discovery space. Note that the SKA is not shown; it will cover the 1.3-cm to 1-m band with a resolution comparable to that of the expanded VLA, but its primary objective is to increase sensitivity by orders of magnitude. If the SKA is located in the Southern Hemisphere, it will provide uniquely high resolution for southern centimeter-wave targets.

EXPANSION OF THE VLA

By most measures, the VLA is by far the most powerful and productive centimeter-wave telescope in the world. As a general-purpose facility, the current VLA serves 600 scientific programs per year. However, much of the instrumentation uses 25- to 30-year-old technology because refurbishment has been inadequate. The modernization represented by the EVLA will provide dramatic new capabilities for the world's most powerful radio telescope.

The overdue leap forward in technology associated with the EVLA will provide an order-of-magnitude improvement in sensitivity and angular resolution combined with a greater than 1000-fold improvement in spectroscopic capability. The resolution of this new VLA will be comparable to that of ALMA and NGST, facilitating complementary multiwavelength studies. Of particular interest will be the study of protogalaxies as well as protostars and protostellar disks that are often optically thick or obscured by dust at shorter wavelengths. With the greatly increased sensitivity and resolution of the VLA at centimeter wavelengths, it will be possible to study star formation at the earliest epochs, to distinguish starburst from AGN phenomena even for $z > 5$, and within our Galaxy, to see into the regions corresponding to planetary formation on scales of 10 AU and less.

The EVLA project will proceed in two phases: Phase I includes the construction of an advanced correlator/spectrometer, installation of the wideband fiber-optic data links required for improved sensitivity, the replacement of the 15-year-old unsupported monitor-control computer, enhancement of the archive system, installation of sensitive new receiver-feed systems to replace the aging receivers and to cover new frequency bands, the development of new software tools, the enhancement of antenna performance, and the development of a means of extending continuous frequency coverage below 1 GHz. In Phase II, as many as eight new antennas will be sited within 250 km of the VLA and connected to the new VLA correlator by fiber-optic links. The new antennas will be used together with both the VLA (to increase the angular resolution by an order of magnitude, with high sensitivity and image quality) and the VLBA (to increase its field of view and sensitivity to low-surface-brightness structures). These antennas are important because they will improve image fidelity and angular resolution, as can be seen in the simulation of observations of a supernova remnant in the Andromeda galaxy displayed in Figure 4.12.

Image quality is of critical importance in tracing the time evolution of cosmic sources, as evidenced in Figure 4.9. Transient radio sources associated with x-ray and gamma-ray bursts, relativistic jets from microquasars, and early phases of nova outbursts will be followed in much greater detail than previously possible. To image the regions of planet formation may require wavelengths as long as 1 cm and, for nearby systems, angular resolution spanning spatial scales provided by the new antennas. A fiber-optic system is now in place connecting the VLA to the innermost VLBA antenna. This single connection provides a twofold improvement in resolution over the stand-alone VLA and

FIGURE 4.12 Simulations of observations of a supernova remnant like the famous Cassiopeia A remnant if it were located in the Andromeda galaxy, M31. *Left panel:* The model source, based on the actual VLA image of Cas A. *Center panel:* The image that would be obtained with the current VLA. *Right panel:* The image that would be obtained with the expanded VLA, illustrating the great improvement in image fidelity that would be provided by eight additional antennas spread throughout New Mexico. Courtesy of NRAO.

demonstrates the feasibility of extending the real-time operation to the Los Alamos VLBA antenna, the proposed eight new antennas, and eventually to the entire VLBA.

Although the EVLA will provide dramatic new observational capabilities, it does not involve the creation and operation of new facilities and hence will require only a small increment in the current operating costs. While the SKA will ultimately provide even more sensitivity than the EVLA at the longer wavelengths, nothing proposed replaces the VLA at the shorter wavelengths. Furthermore, the SKA may be placed in the Southern Hemisphere, leaving the VLA preeminent for northern skies. The need for Phase I is urgent and should be undertaken immediately. Phase II should generate worldwide interest, leading to considerable cost-sharing.

SQUARE KILOMETER ARRAY

The SKA is a proposed centimeter-wave radio telescope with 10^6 m^2 of collecting area. The SKA is the next-generation radio telescope, an instrument motivated by the desire to observe, as illustrated in Figure 4.3, the formation of the first structures and the first luminous objects during the dark age of the universe. Considerations of antenna element cost and improvements in computation technology steer the design of the SKA toward an array composed of a large number of elements, offering dramatic new imaging capabilities. The SKA is being pursued as an international collaboration with an expected total cost of $600 million, the U.S. share of which would be about one-third. The United States must play a major role in the conceptual design and development of the SKA. The panel recommends that our country establish an SKA development program during this decade that will, in collaboration with the international radio astronomy community, develop the technology and techniques for SKA construction in the next decade.

An increase of two orders of magnitude in sensitivity, with vastly improved imaging capabilities, will revolutionize fields newly accessible to study by centimeter-wave astronomy. It will be possible to image the HI jets in forming stars and the inner 1 AU of optically thick disks forming around stars. Molecular Zeeman studies will trace how magnetic fields control cloud collapse. The effects of weak gravitational lensing will map the dark matter on the largest scales. Thousands of new pulsars will be discovered, including, possibly, the first black-hole–pulsar binary or pulsar–pulsar binary. Extensive observations of complex organic molecules at various stages of preprotostellar evolution will be possible. With the improvements in sensitivity and imaging, the ability to form images with many beams at once, and the tremendous dynamic range in intensity and scale will come unanticipated discoveries. A dramatic example would be the detection of radio signals from nearby civilizations as astronomers eavesdrop on the local neighborhood.

The SKA has been an international project from the outset. An international steering committee, with representatives from the participating nations, has been formed. The U.S. activities are coordinated by the U.S.-SKA Consortium, which is university-led; NRAO and NAIC are both involved by virtue of their membership on the consortium board. The prominent participation of the academic community provides an excellent opportunity for innovative development in the competitive university environment and will train the next generation of instrument builders and radio telescope users. It is anticipated that some aspects of the development will be carried out in collaboration with industrial partners, providing important cross-fertilization between the industrial and academic communities. The goals of the SKA are ambitious, and not all are achievable with current technology. However, it is reasonable to expect that a coherent development program over the next 5 to 10 years will lead to technology advances and techniques that will allow the current science objectives to be met over a longer time frame.

COMBINED ARRAY FOR MILLIMETER ASTRONOMY

CARMA is the planned combination of the BIMA and OVRO millimeter arrays at a new high site (2700 m) in the Inyo Mountains of southeastern California. In addition to the current ten 6-m BIMA antennas and six 10.4-m OVRO dishes, an array of ten 2.5-m antennas, critical for SZ mapping, will be added. The resultant array will have much-improved sensitivity and will be able to work at shorter wavelengths than the current separate arrays. Furthermore, because of its mixture of antenna sizes, CARMA will offer unique imaging capabilities for the study of structures on all scales, with particular sensitivity to low-surface-brightness, extended emission. Operation down to $\lambda = 860$ μm should be possible.

The hybrid CARMA array will uniquely address wide-field millimetric imaging, in particular that of the SZ effect and star formation at all epochs. Complementary to ALMA with respect to both its hybrid characteristics and northern location, CARMA has an educational mission and a role in technique development, both of which are likewise critical. The project is highly leveraged and cost-effective, and the panel recommends that it begin as soon as possible.

ADVANCED RADIO INTERFEROMETRY BETWEEN SPACE AND EARTH

ARISE, a proposed NASA mission, will place a sensitive ~25-m radio antenna into space that can operate at wavelengths as short as 3 mm in an elliptical orbit reaching as high as 50,000 km and giving baselines as long as ~5 Earth diameters. Together with Earth-based radio telescopes, ARISE will allow radio imaging with a sixfold improvement in angular resolution compared with the VLBA. In combination with the large ground-based telescopes such as the VLBA, GBT, and LMT, the 25-m space antenna, equipped with state-of-the-art receivers, will be more than an order of magnitude more sensitive than the 8-m Japanese HALCA antenna.

The main goal of ARISE is to study jet regions in AGN as close to the supermassive black hole as possible, as illustrated in Figure 4.7, and to map the spatial-velocity structure of masers, as seen in Figure 4.8, to determine the mass of the black hole and the physical conditions in the accretion disk. The highest angular resolution of ARISE corresponds to 150 AU for a nearby active galactic nucleus like M87 or about 3 R_s for the $3 \times 10^9 M_\odot$ black hole estimated to power that source. Even for a quasar of typical redshift ($z \sim 0.5$ to 1), the resolution of ARISE is very fine—a few light-months.

The panel recommends that NASA continue to support the development and operation of large apertures in Earth orbit for VLBI in order to achieve angular resolution of ~10 μarcsec. An instrument such as that on the ARISE mission will allow mapping of the emission from the inner accretion disks and the masers in the outer accretion disks in AGN as well as the x-ray- and gamma-ray-emitting regions from the jets of blazars. The panel recommends that the ARISE mission be funded so that it can (1) be launched before the end of the decade, (2) operate at wavelengths down to and including 3 mm, and (3) use either an inflatable or a solid structure with sufficiently wide bandwidth and sufficiently sensitive detectors so that its sensitivity remains close to the proposed level.

The panel further recommends that NASA continue to support foreign missions such as VSOP-2 and the Russian mission RADIOASTRON by making available to them ground tracking stations and other logistical support.

SOUTH POLE SUBMILLIMETER TELESCOPE

The panel recommends the construction of a large, single-aperture telescope operating at submillimeter wavelengths that would take advantage of the superb atmospheric transmission conditions at the South Pole. The South Pole is the best known site in the world for its combination of low opacity and stable seeing conditions at submillimeter wavelengths. The SPST should be equipped with wide-field array detectors to survey the dusty universe, study arcminute anisotropy of the CMB, determine the spectral dependence of the SZ effect, and identify sources such as primordial galaxies. The South Pole station is undergoing modernization, and new infrastructure is being built that will enable the deployment of the large telescope. To optimize the scientific output of the SPST and to provide community access to the South Pole astronomical facilities, a scientifically driven umbrella organization is needed to continue the efforts of the Center for Astrophysical Research in Antarctica, an NSF Science and Technology Center with a grant that ends in

February 2002. A high-speed data link to the South Pole is also a requirement.

OTHER HIGH-PRIORITY PROJECTS

Projects of smaller cost and scope will also be critical to the continued study of the universe. Among the several promising areas identified
by the panel for smaller-scale investment are the following, which are
neither exhaustive nor ranked: construction of LOFAR, a ground-based
array for long-wavelength (2- to 20-m) astronomy; studies of CMB
polarization and the SZ effect; development of far-infrared/submillimeter-
wave interferometry in space; laboratory astrophysics; and solar radio
astronomy.

LOW-FREQUENCY ARRAY

The LOFAR will provide the first real capability to image the low-
frequency sky; important applications across all areas of astrophysics will
mean an exciting potential for new discoveries in this relatively unexplored part of the electromagnetic spectrum. As currently conceived,
LOFAR will be an electronically controlled, broadband, ground-based
array operating in the wavelength range from 20 to 2 m and, with
reduced sensitivity, at 1 m. Digital signal processing will be used to
suppress interference adaptively. LOFAR will have multiple simultaneous beams that can be steered instantaneously over the sky. LOFAR
might use the VLA site as its primary location, but additional remote sites
would give an overall dimension of a few hundred kilometers. LOFAR
will achieve a two- to three-orders-of-magnitude improvement in both
sensitivity and resolution over existing instruments and may provide the
opportunity to detect HI structures at the epoch of reionization ($z > 6$) at
these long wavelengths.

LOFAR is a joint project between the Netherlands Foundation for
Research in Astronomy (NFRA) and the U.S. Naval Research Laboratory
(NRL), with the cooperation of the NRAO. Initial development has
begun. NFRA and NRL plan to contribute equally toward a combined
investment of $20 million. The panel recommends that NSF contribute
the additional $8 million required if the facility is to offer access to a
broader segment of the science community. LOFAR will employ many of
the design concepts intended for the SKA, including electronic beam
switching, ionospheric compensation, real-time interference excision,

and remote operation. At the longest wavelengths, LOFAR will have about a million square meters of collecting area—equivalent to that of the SKA.

COSMIC MICROWAVE BACKGROUND EXPERIMENTS

Because of the large number of imminent observations and the likelihood of surprises, the panel elects not to specify in detail future CMB missions and projects. Clearly, however, further study of the CMB radiation will produce unique and fundamental clues to the processes and evolution of the earliest epochs. In particular, the panel recommends the following:

• Continued support of the ambitious suite of ground-based and balloon-borne experiments, satellite missions, and extensive theoretical studies of the CMB radiation that will allow direct determination of the fundamental parameters that govern the cosmology describing the origin and evolution of our universe. The program will culminate with the completion of the MAP and Planck space missions during this decade.
• The construction of dedicated instruments to carry out large-scale imaging surveys and high-sensitivity spectral observations of the SZ effect in order to obtain a redshift-independent inventory of the cosmic web of large-scale structure, to determine the expansion history of the universe, and to measure the velocities of massive galaxy clusters and superclusters relative to the Hubble flow.
• The establishment of small exploratory programs to investigate and develop the techniques and technology for extracting the detailed polarization structure of the CMB that will directly test the ingredients of the inflationary model for cosmology. The results of these efforts will guide planning for the Planck mission and also set the stage for post-Planck mission planning in the CMB field.

FAR-INFRARED/SUBMILLIMETER INTERFEROMETER IN SPACE

The far-infrared/submillimeter spectral region contains about half the luminosity of the universe and carries vital diagnostics for galaxy, star, and planet formation. An integral part of the radio/submillimeter vision is thus the development of a capability for FIR/submillimeter interferometry in space. As can be seen plainly in Figure 4.11, a FIR space interferometer will fill a crucial gap between NGST and ALMA in the wavelength-resolution plane. The science addressed by such an instrument requires superb sensitivity in continuum and high-spectral-resolution modes ($\lambda/\Delta\lambda \geq 10^5$). During this decade, technology development and precursor missions are needed.

LABORATORY ASTROPHYSICS

Laboratory studies underpin the interpretation of astronomical observations at all wavelengths. This is particularly true as new spectral windows such as the submillimeter and far infrared are exploited. Support for laboratory astrophysics is becoming increasingly difficult to obtain and must be enhanced to meet the scientific objectives of the existing and planned radio and submillimeter facilities. Relevant laboratory measurements are an integral component of the analysis of astronomical data, and funds for such research should be included in both NASA- and NSF-supported astronomy programs.

SOLAR RADIO ASTRONOMY

Observations of the Sun at radio wavelengths contribute significantly to the study of transient energetic phenomena, the nature and evolution of coronal magnetic fields, and the solar atmosphere. A suitable dedicated instrument such as the proposed Frequency Agile Solar Radio telescope (FASR) discussed by the Panel on Solar Astronomy appears to be feasible and would be ideal for the continued investigation of these areas. While it is possible that some of the objectives of FASR, particularly studies of transient phenomena, might be met by modifying current or planned instruments such as the 1HT, the EVLA, and LOFAR, the panel recognizes the importance of constructing a facility dedicated to the study of the Sun.

TECHNOLOGY FOR THE FUTURE

The coming decades promise truly remarkable developments in radio and submillimeter astronomy. Projects under way or just beginning (ALMA, SMA, SOFIA, and FIRST/Planck) will improve the sensitivity resolution product of submillimeter observations by one to two orders of magnitude. Projects on the 2010 horizon (SKA, a FIR/submillimeter interferometer in space, and space-based CMB polarization observations) will push the sensitivity × resolution product orders of magnitude further. To achieve these ambitious goals, technology development must be pursued aggressively throughout the chain, from telescope to detector to processing electronics and data reduction/image processing. As the pace of technological innovation quickens, it is essential to pursue a technology development strategy that maintains the health of a range of different institutions and approaches, rewards innovation, and maintains a base of talented researchers well connected to the science of the field. To this end, the panel recommends emphasizing a number of areas, which have been classified as ground-based or space-based.

GROUND-BASED NEEDS AND OPPORTUNITIES

The large arrays that are under development rely on very wide receiver bandwidths to improve their sensitivity and on a large number of elements to generate high-fidelity images. The resulting downstream processing must deal with bandwidths greater than 1 Tbps, which places severe demands on correlator technology. While the rapid pace of semiconductor innovation will help, it is likely that novel approaches to megacorrelator design will be needed. The wide bandwidths also lead to dramatically increased sensitivity to radio-frequency interference (RFI). With the advent of high-redshift astronomy, astronomical research must extend beyond the small, protected frequency windows established for radio astronomy into spectral ranges (necessarily) allocated for commercial and other purposes. Radio astronomers, accordingly, must develop ways to characterize and excise man-made signals in order to recognize and study cosmic ones. It will thus be imperative to characterize and excise interference where and when it occurs. Eventually, it should be possible to construct the entire receiver/intermediate-frequency (IF) chain after the mixer in digital circuitry at radio frequencies, opening up a wealth of new possibilities. At their largest extents, ground-based arrays suffer from atmospheric or ionospheric phase distortion, and phase-correction schemes (the radio equivalent of adaptive optics) must be further developed. Finally, large-scale surveys for a wide variety of objects would most profitably be carried out by large detector arrays at single-dish instruments, as would spectral line imaging with large-format, heterodyne receiver (SIS) arrays.

SPACE-BASED NEEDS AND OPPORTUNITIES

The extraordinarily low natural backgrounds at radio and submillimeter wavelengths lead to superb sensitivities for space-based telescopes. To take advantage of this sensitivity, a new generation of large-format, incoherent arrays is needed. Bolometer development must be continued, as must the design of superconducting detectors that can also be used as energy-resolved photon counters in the optical and UV. A major stumbling block at present is the design of multiplexers for large-format arrays. The telescopes as well as the detectors must be kept at very low temperatures, so the development of high-capacity, space-qualified refrigerators must be accelerated. In certain cases (e.g., spectral line work), heterodyne receivers may provide alternatives to direct detectors provided they operate at or near their quantum noise limit. As detectors and processing electronics approach their fundamental sensitivity limits, the only means left for improving sensitivity is to increase the collecting area. Large, inexpensive apertures, including inflatable ones, are therefore another high-priority item for space-based radio and submillimeter astronomy. Finally, instruments with large numbers of detectors or imaging elements produce data streams that cannot be dealt with using existing downlinks, and new means must be found to increase the telemetry bandwidth for future missions (true for all wavelengths).

POLICY ISSUES

OPEN SKIES POLICY

The panel reaffirms that the open skies policy—allocating telescope time based purely on scientific merit—is the policy that enables the best science to be undertaken. However, the panel is concerned about how future cost-sharing arrangements may affect this policy. Traditionally, some foreign radio facilities have provided comparable open access. The panel encourages continued dialogue to enable U.S. astronomers to have open access to all facilities, particularly those operating at millimeter wavelengths.

RADIO SPECTRUM MANAGEMENT

Radio-frequency interference is a worldwide problem that transcends national boundaries and policies. For decades, U.S. radio astronomers have been active in the spectrum management activities of the International Astronomical Union (IAU), the International Union of Radio Science (URSI), and the International Telecommunication Union (ITU). Radio astronomers from around the world have collaborated closely to preserve their common interests in the face of powerful commercial, government, and military interests. The recent award-winning film produced by NSF on the need for preservation of the radio spectrum is an outstanding example of NSF's proactive effort in this arena. The panel recommends continued U.S. participation and vigorous involvement in spectrum management issues.

THE NATIONAL RADIO ASTRONOMY OBSERVATORY AND THE ATACAMA LARGE MILLIMETER ARRAY

It is imperative that the United States maintain its leadership and critical involvement in ALMA. Given the demonstrated effectiveness of the NRAO organization, the panel urges that to operate ALMA, no new institution be created as an entity separate from NRAO. At the same time, NRAO's other unique and vital facilities must be run in synergy with ALMA, not in competition with it.

AGENCY FUNDING AND MANAGEMENT POLICIES

The NSF needs to provide adequate support for operating, maintaining, and continually upgrading federally funded radio facilities (both the national centers and the university radio facilities) to keep them at the cutting edge. Increased and continuing investment is needed.

The NSF's funding for ancillary activities such as observing preparation, data analysis, theory, and correlative studies is not commensurate with its investment in facilities.

The NSF should provide sufficient funds to allow individual investigators to maximize the scientific output of the facilities it supports. In particular, NSF should plan to make available sufficient funds for data reduction and analysis as well as for the maintenance and operation of the new facilities. The panel endorses the recommendations laid out in the survey committee report with regard to the funding of new programs and facilities.

The NSF should also establish a national postdoctoral program similar to the Hubble Fellowship program that includes support for outstanding young scientists pursuing research associated with the NSF-supported radio facilities.

Making the provision of leveraged or matching funds a criterion for grant support may in some cases compromise the opportunity to pursue individual initiatives. The NSF should examine the circumstances under which such support is required and should ensure that review panels do not give undue weight to the availability of matching funds in programs that do not require them.

ACKNOWLEDGMENTS

The panel benefited from written comments received from many individuals through the American Astronomical Society (AAS) discussion forum and oral comments at the public forums conducted during the AAS meetings in January and June 1999 and at the URSI meeting in January 1999. Presentations to the panel were made by J. Baars (University of Massachusetts), D. Backer (University of California at Berkeley), F. Bash (University of Texas), T. Bastian (NRAO), L. Blitz (University of California at Berkeley), R. Brown (NRAO), M. Davis (NAIC), R. Dickman (NSF), D. Gary (New Jersey Institute of Technology), R. Ekers (Australia Telescope National Facility), P. Goldsmith (NAIC), N. Kassim (Naval Research Laboratory), T.J. Lazio (Naval Research Laboratory), D. Leisawitz (NASA), J. Mather (NASA), H. Moseley (NASA), R. Perley (NRAO), J. Peterson (Carnegie Mellon University), P. Schloerb (University of Massachusetts), A. Stark (Harvard-Smithsonian Center for Astrophysics), R. Taylor (University of Calgary), H. Thronson (NASA), M. Turner (University of Chicago), J. Ulvestad (NRAO), P. Vanden Bout (NRAO), K. Weiler (Naval Research Laboratory), and D. Woody (California Institute of Technology).

ACRONYMS AND ABBREVIATIONS

1HT—One Hectare Telescope
AAS—American Astronomical Society

AGN—active galactic nuclei

ALMA—Atacama Large Millimeter Array

ARISE—Advanced Radio Interferometry between Space and Earth, an orbiting antenna that will be used in concert with the ground-based VLBA

AU—astronomical unit. A basic unit of distance equal to the separation between Earth and the Sun, about 150 million km

BEAST—Background-Emission-Anisotropy Scanning Telescope; a long-duration, balloon-borne cosmic microwave background experiment

BIMA Array—Berkeley-Illinois-Maryland Association Array

BOOMERANG—Balloon Observations of Millimetric Extragalactic Radiation and Geophysics; a balloon-borne telescope that circumnavigated Antarctica

CARMA—Combined Array for Research in Millimeter-wave Astronomy, a millimeter-wave array in the Northern Hemisphere

CBI—Cosmic Background Imager, a 13-element interferometer located in northern Chile

CMB—cosmic microwave background

CMVA—Coordinated Millimeter VLBI Array

COBE—Cosmic Background Explorer, a NASA mission launched in 1989 to study the cosmic background radiation from the Big Bang

CSO—Caltech Submillimeter Observatory, a 10-m telescope operating on Mauna Kea, Hawaii, used for observations of millimeter and submillimeter wavelength radiation

D/H—deuterium/hydrogen ratio

DASI—Degree-Angular-Scale Interferometer for imaging anisotropy in the cosmic microwave background

EVLA—Expanded Very Large Array

FASR—Frequency-Agile Solar Radio telescope

FCRAO—Five College Radio Astronomy Observatory

FIR—far infrared

FIRST—European Far Infrared Space Telescope

GBT—Green Bank Telescope

GRB—gamma-ray burst

HALCA—Highly Advanced Laboratory for Communications and Astronomy, the Japanese VSOP satellite launched in February of 1997

HI—atomic hydrogen

HII—ionized hydrogen

HST—Hubble Space Telescope, a 2.4-m-diameter space telescope designed to study visible, ultraviolet, and infrared radiation, and the first of NASA's Great Observatories

IAU—International Astronomical Union

IMF—initial mass function

IR—infrared

IRAS—Infrared Astronomical Satellite, a NASA Explorer satellite launched in 1983 that surveyed the entire sky in four infrared wavelength bands using a helium-cooled telescope

ISM—interstellar medium

ITU—International Telecommunication Union

JPL—Jet Propulsion Laboratory (NASA)

KBO—Kuiper Belt object

LMT—Large Millimeter Telescope

LOFAR—Low Frequency Array, a joint Dutch-U.S. initiative to make observations at radio wavelengths longer than 2 m

MAP—Microwave Anisotropy Probe mission

MAXIMA—Millimeter Anisotropy Experiment Imaging Array; a balloon-borne millimeter-wave telescope designed to measure the cosmic microwave background

NAIC—National Astronomy and Ionosphere Center, Arecibo, Puerto Rico

NASA—National Aeronautics and Space Administration

NFRA—Netherlands Foundation for Research in Astronomy

NGST—Next Generation Space Telescope, an 8-m infrared space telescope

NICMOS—Near Infrared Camera and Multi-Object Spectrometer, an instrument on the Hubble Space Telescope

NRAO—National Radio Astronomy Observatory

NRL—Naval Research Laboratory

NSF—National Science Foundation

NVSS—NRAO VLA Sky Survey

OVRO—Owens Valley Radio Observatory

POLAR—Polarization Observations of Large Angular Regions, an instrument designed to measure the polarization of the cosmic microwave background

Planck Surveyor—A European-led space mission to image anisotropies in the CMB.

Polatron—a bolometric receiver with polarization capability designed for use at the Owens Valley 5.5-m radio telescope

RADIOASTRON—A Russian satellite designed to conduct VLBI observations of radio sources in conjunction with the global ground VLBI network

RFI—radio-frequency interference

SAFIR—Single Aperture Far Infrared Observatory

SCUBA—Submillimeter Common-User Bolometer Array, a British-French-Canadian ground-based telescope located in Hawaii; it operates at wavelengths between 350 and 2000 μm.

SIRTF—Space Infrared Telescope Facility, NASA's fourth Great Observatory, will study infrared radiation

SIS heterodyne receivers—devices that use a SIS superconducting junction, a junction consisting of two layers of superconducting metal (niobium) separated by a few nanometers of insulator (aluminum oxide)

SKA—Square Kilometer Array, an international centimeter-wave radio telescope

SMA—Submillimeter Array

SOFIA—Stratospheric Observatory for Infrared Astronomy, a 2.5-m telescope flown above most of the Earth's water vapor in a modified Boeing 747 aircraft to study infrared and submillimeter radiation

SPST—South Pole Submillimeter Telescope

STScI—Space Telescope Science Institute

SZ effect—Sunyaev-Zel'dovich effect, the small distortion of the CMB radiation induced when CMB photons scatter off electrons in the hot gas contained within clusters of galaxies

TopHat—A NASA-sponsored experiment in which a telescope was placed on top of a balloon to measure cosmic microwave background radiation anisotropy

ULIRG—ultraluminous infrared galaxy

URSI—International Union of Radio Science

Viper—A 2-m telescope designed to measure anisotropy in the CMB at angular scales down to 0.1 deg

VLA—Very Large Array, a radio interferometer in New Mexico consisting of 27 antennae spread out over 35 km and operating with 0.1-arcsec resolution

VLBA—Very Long Baseline Array, an array of radio telescopes operating as an interferometer with a transcontinental baseline and resolution less than a thousandth of an arcsecond

VLBI—Very Long Baseline Interferometry, a technique whereby a network of radio telescopes can operate as an interferometer.

VSA—Very Small Array, a project to make images of the CMB radiation
on angular scales of around 1 deg
VSOP—VLBI Space Observatory Program, a mission led by the Institute
of Space and Astronautical Science in collaboration with the National
Astronomical Observatory of Japan
WFPC—Wide-Field Planetary Camera, an instrument on the Hubble
Space Telescope

5

Report of the Panel on Solar Astronomy

SUMMARY

The study of the Sun has revealed fundamental physical puzzles that have resisted understanding for generations of astronomers. The Sun is a typical star, with other stars being at least as complex. Solving mysteries on the Sun—among others, the dynamo process, the intermittency in the surface magnetoconvection, and the heating of the active corona—is important for all of astronomy and astrophysics. The Sun offers unique opportunities for physical insight that go far beyond just resolving astrophysical processes on their intrinsic scales. These opportunities include (1) using the Sun as a plasma physics laboratory, (2) understanding and predicting the impacts of the Sun on Earth's climate and on "space weather" in the near-Earth environment, and (3) understanding the role of solar evolution in the evolution of life in planetary systems. The successes achieved in solar research since the 1991 survey report[1] lead us to expect that many of these mysteries can be resolved by the new projects prioritized in this report. However, it should be kept in mind that at this time, key solar mechanisms are poorly understood even as they are applied in other astrophysical contexts. Or, fascinating new phenomena might be discovered that will give rise to new puzzles to challenge new generations of physicists.

STRATEGY FOR THE DECADE 2001 TO 2010

The progress of the past decade was made possible by investments made in the 1980s that led to revolutionary observational capabilities in space and on the ground, including simultaneous multiwavelength observations of dynamics, precision vector magnetic field measurements, and helioseismology. Breakthroughs in numerical simulations of two- and three-dimensional magnetohydrodynamical (MHD) processes allowed for tailoring solarlike scenarios on the computer. All these advances have led to the formulation of a new strategy—a systems approach—for solar physics in the next decade:

---

[1]Astronomy and Astrophysics Survey Committee, National Research Council. 1991. *The Decade of Discovery in Astronomy and Astrophysics* (Washington, D.C.: National Academy Press).

- The domains of the solar interior, photosphere, chromosphere, corona, and heliosphere should be treated as a single system.
- Diverse data sets should be integrated, as demonstrated in the NASA Solar Data Analysis Center (SDAC).
- The connection to the operational branches of space weather research should be exploited much as weather research received from the National Weather Service is exploited.
- International efforts should be integrated as much as possible.

OBSERVATIONAL EFFORTS

CURRENT

- *Observational facilities in operation.* The Dunn solar telescope with the Advanced Stokes Polarimeter and its adaptive optics (AO) program, the McMath-Pierce telescope with its infrared program, and the Fourier-transform spectrograph should be operated until the Advanced Solar Telescope (AST) becomes available. The seismology network GONG and the Mauna Loa Solar Observatory should be continued. The various university observatories should be maintained at a level that will ensure a broad educational base. The space-based observatories—Yohkoh, SOHO, Ulysses, and TRACE—should be maintained and given adequate funding for data analysis.
- *Observational facilities under construction.* SOLIS (on the ground) and HESSI, Solar-B, STEREO, and Solar Probe (in space) are of utmost importance for expanding some findings of the last decade in critical areas.

FUTURE

- *Primary recommendation, ground-based, medium size: the Advanced Solar Telescope (AST).* In view of solar physics' growing relevance to the climate research and space weather communities, AST should be built and become operational within this decade. Key astrophysical processes will be directly observable with the AST and AO. Half of the $64 million investment would come from international partners.[2]
- *Secondary recommendation, ground-based, medium size: the*

---

[2]The estimated costs for ground-based initiatives include costs for instrumentation, grants, and operations, as described in the preface.

Frequency-Agile Solar Radiotelescope (FASR). This state-of-the-art radio observatory would be primarily for solar observations that can be readily scheduled in coordinated campaigns. Cost: $15 million.

• *Primary recommendation, ground-based, small size: the expansion of SOLIS to a three-station network around the globe.* This would give nearly continuous coverage in full-disk solar vector magnetic field monitoring and would form the backbone of an assessment of the solar magnetic flux budget over the solar cycle. Cost: $4.8 million.

• *Primary recommendation, space-based, medium size: Solar Dynamics Observatory (SDO).* SDO would pursue in particular the newly discovered tomography of subsurface structures through time-distance analysis of running waves at the solar surface and impulsive helioseismology from oscillations of loops in the corona. Cost: $300 million.

THEORY AND DATA MINING

The panel recommends a broadened Solar Magnetism Initiative (SMI) as a comprehensive research framework for theory and data mining for all of the above projects. Understanding in solar physics can be advanced through detailed multidimensional numerical modeling. SMI will provide the coordination between observational activities and numerical experiments in forward modeling, to be done by modelers in solar physics as well as plasma physics and turbulence theory. SMI will be a community-wide research program that has been broadened from its original scope to become a multiagency enterprise. The cost of the program is estimated to be $3 million per year for 5 years, with the option of extension for another 5 years. Since SMI is proposed as a multiagency enterprise, it is not ranked with respect to ground-based or space-based projects but stands on its own.

NEW TECHNOLOGIES

An adaptive optics system for a 4-m-class AST needs to be pursued based on recent dramatic progress in the existing project at the National Solar Observatory (NSO) and international cooperation. The development of lightweight mirrors (like Solar-Lite) to achieve high resolution in the medium and far IR will break the cost curve for future space missions. The development of Stokes polarimeters in the UV will allow for measurements of the magnetic field in the chromosphere.

POLICY ISSUES

NSO should be enabled to take its lead role in developing the AST through changes in its managerial structure. Broad community participation needs to be ensured. The allocation[3] between the National Science Foundation's (NSF's) university grants program (about 63 percent) and its funding to centers in the U.S. solar and solar terrestrial community (37 percent) is appropriate and healthy. Increases in overall funding are necessary, however, given the increased need to understand the Sun for space weather forecasting and for the driving of climate. Additional educational outreach activities will also be required.

WHY DO SOLAR PHYSICS RESEARCH?

The complexities of the Sun—its internal structure, rotation, and convection and the resulting cyclic and random generation of its magnetic fields and the magnetoactive, hot, explosive, extended solar atmosphere and solar wind—are fascinating and challenging (see Figure 5.1). Because these solar phenomena occur over physical scales that cannot be simulated in laboratories on Earth, their study tests and expands our understanding of magnetofluid dynamics and plasma physics. Solar physics is key to much of astrophysics and central to the Sun-Earth connection, and it bears on the quest to determine the origin and extent of life in the universe.

KEY TO THE MAGNETODYNAMIC UNIVERSE

Dynamic magnetic fields are widespread throughout the universe; they are an active ingredient of many astronomical objects, from dwarf stars to accretion disks to clusters of galaxies. Our understanding of the origins and effects of these distant astrophysical magnetic fields is rudimentary at best. The Sun has the most intense magnetic field in the solar system. The entire corona and solar wind and diverse explosive events (many producing bursts of high-energy particles, x rays, and gamma rays

---

[3]From the Task Group on Ground-based Solar Research, National Research Council. 1998. *Ground-based Solar Research: An Assessment and Strategy for the Future* (Washington, D.C.: National Academy Press). Also known as the Parker report for committee chair Eugene Parker.

226

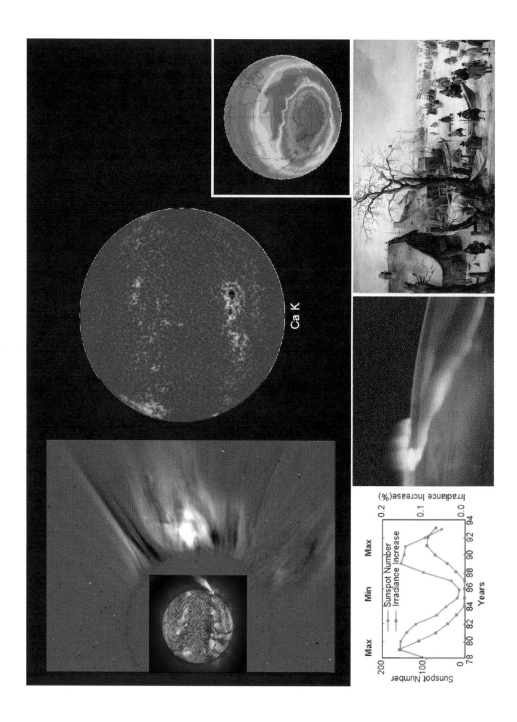

Ca K

and other large events blasting magnetized matter out past the planets) are all magnetodynamic effects. The Sun is a unique laboratory that will lead to an understanding of the dynamic behavior of cosmic magnetic fields.

SOLAR-TERRESTRIAL PHYSICS

Life on Earth depends on the Sun's heat and light. Earth's climate, the state and extent of the upper atmosphere and magnetosphere, and space weather inside and outside the magnetosphere are determined and driven by the Sun's luminosity, by its UV and x-ray spectrum, by the solar wind, and by explosive events on the Sun. Solar irradiance variations appear to be correlated with the level of the Sun's magnetic activity. Extrapolated luminosity changes due to changes in the Sun's production of magnetic field over decades and centuries are large enough (>0.1 percent) to significantly affect Earth's temperature, contributing to global warming and "little ice ages." Changes in the Sun's magnetic activity change the output of UV and x rays by factors as large as 10 or more. These radiations control Earth's thermosphere, ionosphere, and protective ozone layer. Coronal mass ejections (CMEs) on the Sun blast out massive magnetic clouds that plow through the solar wind and impact Earth, causing magnetic storms that can disrupt power systems. In near-Earth space and throughout the solar system, high-energy particles from these events often reach levels that can be lethal to spacecraft and astronauts. To better understand and predict global change and space weather, we need to understand and predict the mechanisms and behavior of their driver, the magnetic Sun.

FIGURE 5.1 Superposed images of aspects of solar variability. **Top left:** composite of an event with a modest flare, the brightest eruptive prominence seen so far by SOHO/EIT, and a 400 km/s CME seen by SOHO/LASCO. **Top right:** the Sun from RISE/PSPT in CaK. **Bottom from left to right:** plot of solar luminosity and the sunspot cycle; auroral curtain during a magnetic storm; H. Averkamp painting of skaters during the Little Ice Age, when solar activity was low during the 17th century (Hendrik Averkamp, **Winter Scene on a Canal,** c. 1615, oil on panel, $18^7/_8 \times 37^5/_8$ in., Toledo Museum of Art, Toledo, Ohio; purchased with funds from the Libbey Endowment, Gift of Edward Drummond Libbey, acc. no. 1951.402). Top left images courtesy of the SOHO/EIT and SOHO/LASCO consortia. SOHO is a project of international cooperation between ESA and NASA. Top right image courtesy of Radiative Inputs of the Sun to Earth/Precision Solar Photometric Telescope Project.

ORIGIN AND EVOLUTION OF LIFE ON PLANETS

Acting over the age of the solar system, the solar wind may well have played a major role in the evolution of planetary atmospheres. It has been suggested that Mars currently has little atmosphere because it has long had, as now, little magnetic field, allowing its early atmosphere to be blown away by the solar wind. To evaluate the magnitude of such effects, astronomers need to understand the Sun's production of magnetic field, the mechanisms underlying the acceleration of the solar wind, and their variation over solar cycles and longer times. Similar considerations apply to the Sun's output of UV and x rays over its history—with this output being controlled by the solar magnetic cycle. Some stars with activity cycles exhibit much greater variability than the present Sun, suggesting that the Sun might have had very active phases in the past. An understanding of the magnetic Sun will form a basis for estimating how stellar magnetism could influence the possibility of life arising on other planets throughout the universe.

THE MOST SIGNIFICANT ADVANCES IN THE LAST DECADE

GOALS ACHIEVED

Many of the goals for solar physics laid out in the 1991 survey report have been met or surpassed, as can be seen below.

THE SOLAR INTERIOR

• Thin flux-tube calculations have been successful in reproducing the synoptic properties of flux emergence over the solar cycle, thereby placing stringent bounds on the magnetic field strength at the base of the convection zone (several 10^4 gauss).

• The radiative core of the Sun rotates as a solid body, while the observed surface differential rotation persists to the base of the convection zone.

• A thin boundary layer, the tachocline (thickness less than 0.05 solar radii), exists at the radiative core convection zone interface and is a propitious site for the solar dynamo. The rotation amplitude between 0.68 solar radii (just below the convection zone) and 0.72 solar radii (just above it) varies locally up to 25 percent over a period of 1.3 years.

- Solar model structures validated by helioseismology rule out an astrophysical solution to the solar neutrino problem and underscore the necessity of elemental diffusion and settling in the Sun's radiative core.
- Local helioseismology (Hankel decomposition, time-distance analysis, and acoustic holography) images to a depth of ~20,000 km subsurface signatures of extant and emerging active regions and has detected steady poleward, near-surface meridional flows of 10 to 30 m/s.

THE SOLAR SURFACE

- Magnetic fields emerge in strong-field concentrations with significant electric currents and helicity. Through convective collapse and the buoyancy of the magnetized plasma, the fields rapidly orient themselves perpendicularly to the solar surface and are enhanced to superequipartition levels of approximately 1500 G.
- The rate of appearance (and disappearance) of the surface magnetic flux, particularly in the form of small-spatial-scale ephemeral regions, is such that the average observed unsigned flux in the quiet Sun would be doubled in approximately 40 h.
- The fact that the average unsigned flux varies by a factor of only 3 to 5 over the entire solar cycle implies that these emerging fluxes must be rapidly "recycled" under the action of a local surface magnetic dynamo. Model calculations indicate these local dynamo processes provide sufficient magnetic energy for the heating of the outer solar atmosphere.
- A significant fraction of a sunspot's magnetic flux is contained in the penumbra, which has a deep fluted structure. Radial spokes of nearly horizontal magnetic field alternate with spokes in which the field is inclined some 40 deg.
- Evidence indicates that the emergence of active regions leads to excess facular emission that exceeds the deficit of the sunspot umbrae and penumbrae. Numerical simulations are beginning to address the question of how deep in the convection zone these irradiance variations first arise. The total change in solar irradiance over a solar cycle, however, remains unexplained.

THE OUTER SOLAR ATMOSPHERE AND HELIOSPHERE

- Both theory and observation now show that the chromosphere and the transition region cannot be regarded as nested physical atmospheric layers with a distinct identity. Rather, their characteristic spectral

signatures arise from radiatively weighted averages of nonlinear magnetohydrodynamic processes that are highly variable in both space and
time and that force the tenuous plasma to be far from radiative equilibrium.

• There is a nascent appreciation that the signatures of wave
propagation, magnetic reconnection, and nanoflares are imprinted on
the line profiles of UV and EUV emission features, allowing the relative
contribution of these processes to the heating of the solar atmosphere to
be determined from data with sufficient resolution in wavelength, space,
and time.

• The reconnection of post-CME loops has been detected through
their continuous glow in soft x-ray emission. The synthesis of radio, x-ray,
and white-light coronal images has begun to reveal the intricate manner
in which the corona ejects magnetized material (carrying magnetic
helicity) in the guise of CMEs while liberating magnetic free energy
through flares (magnetic reconnection events) possessing a continuous
spectrum of sizes.

• The diffuse x-ray irradiance of the corona shows a pronounced
variation with the solar cycle. This variation exceeds the variation
expected from the number and size of active regions.

• The first-ionization-potential (FIP) effect is absent in high-speed
solar wind streams, implying that the structure and dynamics of the upper
chromosphere are fundamentally different in coronal hole regions and
the rest of the Sun.

• Heating in the coronal acceleration region of the high-speed solar
wind leads to large ion-temperature anisotropies and very large perpendicular ion temperatures, implying that ion cyclotron heating is a major
source of the energy required to drive these streams.

THE SOLAR-STELLAR CONNECTION

This report cannot be a comprehensive review of all of stellar physics. Hence, the panel concentrates on two examples where the synergy
between solar and stellar work is particularly beneficial and includes the
sharing of instrumentation. During the last decade, studies of stellar
magnetic activity have made significant progress in some respects, but in
others they have been handicapped by the lack of adequate tools for
some of the observations. Ground-based spectrographic and photometric studies of bright-field stars, including the Mt. Wilson HK (hydrogen
and calcium line) photometry program, the High Altitude Observatory

(HAO)/Lowell Solar-Stellar Spectrophotometer, and the Tennessee State University network of photometric telescopes, have delineated the broad features of stellar magnetic activity. The broad dependencies of stellar activity on stellar temperature and rotation rate are starting to be understood, as are the connections between these properties, a star's age, and the photospheric abundances of lithium and beryllium. The Sun displays an unusually small ratio of photometric to chromospheric variability; this fact is central to the understanding of sunspot and facular contributions to time variations in the solar flux, but it also has the consequence that accurate solar analogues are difficult to find. In order to gain samples of stars that are larger, more homogeneous, and better defined with respect to their mass, age, and composition, access to larger telescopes to observe fainter stars is required.

Astroseismology of Sun-like stars would make critical contributions to solar-stellar problems, better defining the fundamental parameters of the stars under study and helping to reveal their internal processes. So far there are no methods to reliably measure the tiny oscillating signals produced by stars similar to the Sun. Progress has been made in radial velocity observations of a few of the brighter stars. Further efforts in calibration methods, combined with suitable high-resolution echelle spectrographs, can be expected to bring a hundred or so nearby solar-type stars within the reach of this technique. A more far-reaching avenue for the application of seismic methods is photometry from space. Unhindered by atmospheric absorption and scintillation, a modest-size telescope would be able to analyze stellar pulsations in many Sun-like stars in the nearer open clusters. The first steps toward such precise photometry missions are now being taken, involving observations of a few bright-field stars; recent results have come from the star tracker on the WIRE spacecraft, while small photometry missions have been selected for flight by France, Canada, and Denmark.

A SYSTEMS APPROACH TO SOLAR PHYSICS—TOWARD A DECADE OF UNDERSTANDING

The scope of solar research and the methods for performing it are expanding in several respects: (1) the Sun is being treated as one physical system, (2) solar research is being systematized, (3) diverse datasets are being integrated into a framework, (4) connections are being forged to operational forecasting of space weather, and (5) international cooperation is being sought.

The Sun's variable output—radiation, particles, and fields—is controlled by the structure and evolution of its magnetic field. The solar magnetic field is present throughout the solar convection zone and the tachocline immediately below it and the photosphere, chromosphere, and corona above it. Thus, all of these domains are magnetically linked. Historically, the dynamics and magnetohydrodynamics of these domains of the Sun were studied and modeled in distinct, relatively unconnected efforts. Much progress has been made in understanding the physical processes at work in each domain. The time is right to try to understand the whole of solar magnetism, by looking at the Sun as a single system from the convection zone out through the corona. This perspective is supported by a combination of well-established facts about the Sun. Hale's polarity law of sunspots—which says that leader and follower spots have opposite polarities in the north and south hemispheres—is augmented by the observation that the Sun's magnetic field patterns show a predominance of left-handed twist in the north and right-handed twist in the south. This observation points to a global organization of the field despite its structured and filamentary characteristics. Helioseismology has revealed that the tachocline at the base of the solar convection zone is the likely location for dynamo action. The tachocline may be unstable to global MHD disturbances that in turn act as templates for solar activity seen at the photosphere. Mechanisms for the injection of magnetic flux into the bottom of the convection zone have been identified and modeled, and the rise of flux through the convection zone due to magnetic buoyancy has been demonstrated theoretically, explaining several features of sunspot groups. Measurements of vector magnetic fields at the surface have allowed detailed study of the physical interaction between the magnetic field and the thermally radiating plasma (see Figure 5.2). MHD models have captured the process of convective collapse, by which magnetic flux tubes form. Understanding chromospheric heating through radiation-hydrodynamics has also advanced. The evolution of coronal structures throughout all solar cycle phases has been well described, including the sudden changes due to CMEs. Some of these processes have also been modeled successfully. White-light, UV, and x-ray observations of evolving coronal structures are being integrated into a unified picture.

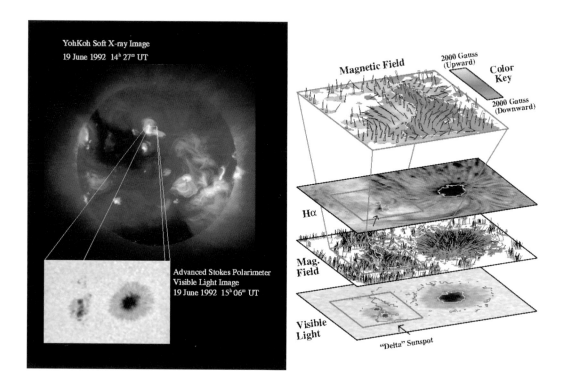

FIGURE 5.2 Vector magnetic-field measurements and visible-light and x-ray observations of a small active region composed of a quiescent unipolar sunspot and a smaller delta spot, which evolved significantly over a few days. This evolution is associated with a coronal brightening observed in soft x ray over the delta spot. In contrast, the larger unipolar spot shows no detectable x-ray signature in the corona above it. Courtesy of B. Lites, High Altitude Observatory, University Corporation for Atmospheric Research. Reprinted by permission from *Reviews of Geophysics* 38 (2000): 1-36; copyright by the American Geophysical Union.

THE CONCEPT BEHIND THE SOLAR MAGNETISM INITIATIVE

An understanding of the Sun's entire magnetic field requires an integrated program of observations and the incorporation of diverse datasets into a common database for community use. This effort has to be closely coupled to theoretical modeling that uses existing models and develops new ones. Since the domains of the Sun are physically linked by the solar magnetic field, an understanding of the Sun and its variability will require that all domains be considered together in a consistent physical model. The solar physics community has defined an initiative to do this: the solar magnetic initiative (SMI), which is described in the section on theory and data mining.

GLOBAL SOLAR DATABASES

Starting with the Japanese Yohkoh mission, launched in 1991, solar physicists worldwide adopted the concepts of distributed software development and integrated ground- and space-based data access and analysis. Concurrent software development at the Solar Data Analysis Center (SDAC) at NASA led to the evolution of the solar software tree. Separate master sites exist for the Yohkoh, SOHO, and TRACE branches. Via a Web interface, users can configure an installation package, download, and install it. The archive consists of data from the 12 SOHO experiments, synoptic data from 14 ground-based instruments (optical and radio), and synoptic data from Yohkoh and TRACE.

OPERATIONAL FORECASTING

Variations in the space environment near Earth that adversely affect mankind and technological systems are driven by variations in solar output. The National Oceanic and Atmospheric Administration's (NOAA) Space Environment Center, the nation's provider of space weather services, uses a variety of means to predict solar activity on timescales as short as a day to as long as the solar cycle. If the global generation and eruption of solar magnetic flux were better understood and modeled, as suggested in the recommendations of this study, earlier watches, accurate warnings, and long-term activity profiles could be issued. Basic solar research will benefit from the feedback from operational forecasting, much as basic weather research did.

INTERNATIONAL COOPERATION

European Space Agency (ESA)-NASA cooperation on the SOHO spacecraft and Japanese-U.S.-U.K. cooperation on the Yohkoh spacecraft have surpassed even the most optimistic expectations in the richness of their scientific returns. Similarly, ground-based networks around the world allowing for near-continuous observations have had great success.

Two telescopes are under construction on the Canary Islands: the Swedish 1-m telescope and the German 1.5-m GREGOR telescope.

Adaptive optics (AO) has been demonstrated successfully on the Dunn solar telescope at Sacramento Peak and on the 50-cm Swedish telescope at La Palma (Figure 5.3). A realistic roadmap for using AO in a 4-m-class AST would involve using the new telescopes as stepping-stones in a collaborative effort.

The panel met with S. Solanki from the Eidgenössische Technische Hochschule in Zürich, now the director of the Max-Planck-Institut in Lindau; with T. Kosugi from the Japanese Space Research Agency (ISAS), head of the Solar-B project; and with O. von der Lühe, director of the Kiepenheuer-Institut für Sonnenphysik in Freiburg, to assess prospects for international cooperation. From the presentations of these

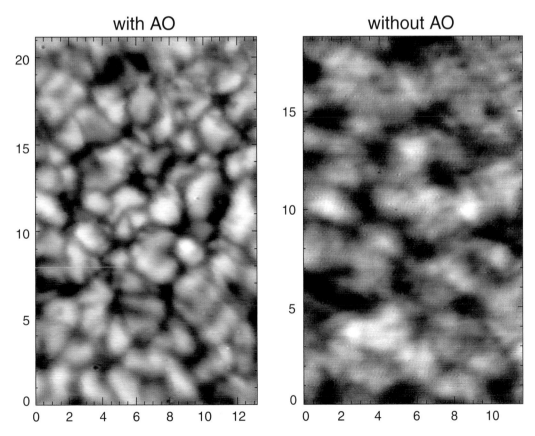

FIGURE 5.3 Snapshots of observations of solar granulation at the Dunn solar telescope without and with adaptive optics. Courtesy of T. Rimmele, National Solar Observatory.

experts and the discussions with them and other international contacts, there is a good basis for international collaboration beyond the projects already mentioned. In particular, there is strong interest in substantial participation in the AST.

EXISTING PROGRAMS

GROUND-BASED OBSERVATIONAL EFFORTS

NATIONAL SOLAR OBSERVATORY (NSO)

- *Evans Facility*—40-cm coronagraph at Sacramento Peak;
- *Dunn solar telescope*—0.76-m vacuum telescope at Sacramento Peak with a prototype AO and the High-Altitude Observatory (HAO)/ NSO Advanced Stokes Polarimeter;
- *Kitt Peak vacuum telescope*—synoptic instrument for full-disk solar magnetograms and He I 1083-nm spectroheliograms;
- *Kitt Peak McMath-Pierce telescope*—1.5-m open telescope for IR and optical observations;
- *Synoptic Optical Long-term Investigations of the Sun (SOLIS)*— currently under construction. The three SOLIS instruments are (1) a full-disk vector spectromagnetograph, (2) a full-disk imager for high-fidelity spectral images of the solar disk, and (3) a solar spectrometer for measurements of line profiles of the Sun as a star. SOLIS is expected to become operational in 2001. Two additional vector spectromagnetographs should be built and installed at much different longitudes to obtain nearly continuous time coverage (see below at "Extension of SOLIS to a Network"); and
- *Global Oscillation Network Group (GONG)*—worldwide network of six seismology instruments. GONG has recently been upgraded to higher spatial resolution. It will continue to operate over a full solar cycle. There will be increased costs ($750,000 per year, beginning in FY2001) associated with operation and data analysis of the enhanced GONG network.

HIGH ALTITUDE OBSERVATORY (HAO)

- *The Advanced Coronal Observing System*—synoptic instrument set to study coronal dynamics;

- *The Low-degree/Experiment for Coordinated Helioseismic Observations (LOWL/ECHO) Oscillations Experiment*—operated on Hawaii and on Tenerife (jointly with the Instituto de Astrofísica de Canarias); and
- *Precision Solar Photometric Telescope (PSPT)*—one of a network of three such telescopes (all built by NSO) to measure solar radiative variability as part of the NSF RISE program.

UNIVERSITY OBSERVATORIES

University-based solar observatories in the United States are critical for the training of the next generation of solar experimental scientists:

- *Big Bear Solar Observatory*—65-cm telescope operated by the New Jersey Institute of Technology (NJIT);
- *Mees Solar Observatory*—operated by the University of Hawaii on Haleakala;
- *San Fernando Observatory*—61-cm vacuum telescope operated by California State University at Northridge to study solar irradiance variability;
- *Mt. Wilson*—the 60-ft tower telescope operated by the University of Southern California as part of a worldwide helioseismology network. The 150-ft tower telescope, operated by the University of California at Los Angeles, investigates long-term changes of solar magnetic activity and large-scale flow systems;
- *Wilcox Solar Observatory*—operated by Stanford University. The observatory began daily observations of the Sun's global magnetic field in May 1975; and
- *Owens Valley Radio Observatory (OVRO)*—array for imaging and spectroscopy operated by NJIT.

Additional facilities are run by NASA in support of space missions and by the Air Force for monitoring space weather events (SOON and ISOON).

INTERNATIONAL GROUND-BASED OBSERVATORIES

Other countries have made significant investments and have seen significant successes with new instruments. These efforts complement U.S. facilities globally, providing good coverage for monitoring dynamical events on the Sun and sometimes coordinating their measurements with space-based observatories. Large facilities around the globe include the following:

- Swedish vacuum solar telescope on La Palma, Spain,
- Dutch open telescope on La Palma, Spain,
- German vacuum tower telescope on Tenerife, Spain,
- German vacuum Gregory telescope on Tenerife, Spain,
- Franco-Italian THEMIS telescope on Tenerife, Spain,
- Solar telescope of the Indian Institute of Astrophysics,
- Huairou solar station in China,
- Hida observatory in Japan, and
- Nobeyama 17- and 34-GHz radio telescope in Japan.

SPACE-BASED OBSERVATIONAL EFFORTS

The last decade saw unprecedented successes in solar observations from space-based observatories. It is imperative that now-active missions be sustained as long as possible to cover the solar cycle. The guest investigator programs under which the United States participated in the Japanese-U.S. Yohkoh mission, the ESA-NASA-SOHO mission (see Figure 5.4), and the TRACE mission (see Figure 5.5) have led to vigorous data analysis efforts at universities and other U.S. research institutions. Sustaining and expanding a support program for data analysis is essential to fully exploit these missions. Four additional missions (HESSI, STEREO, Solar-B, and Solar Probe) are in preparation. The panel considers these missions as approved and/or on their way and therefore does not rank them in this report; however, it draws attention to their significant scientific value. The scientific return of all these missions would be increased significantly if the participation of guest investigators could be increased.

MISSIONS IN FLIGHT

Yohkoh

Yohkoh is a Japanese mission in cooperation with the United States and United Kingdom to observe high-energy radiation in the solar atmosphere. Launched in the autumn of 1991, Yohkoh has provided continuous coverage of solar coronal activity throughout nearly a complete solar cycle. As of the time of this writing, the data have re-

FIGURE 5.4 Solar rotation and polar flows of the Sun as deduced from measurements by the SOHO/MDI instrument. The left side of the image represents the relative rotation speed of various areas on the Sun. Red-yellow-orange is faster and blue is slower than average. The light-orange bands extend down approximately 20,000 km into the Sun. The cutaway reveals rotation speed inside the Sun. The large red band is a massive fast flow beneath the solar equator. A much more subtle stream can be seen in the cutaway at the poles as the light-blue areas embedded in the slower moving dark-blue regions. The blue lines in the cutaway represent the surface flow (10 to 20 m/s) from the equator to the poles, which extends to a depth of at least 26,000 km. The return flow at the bottom of the convection zone is from a simple model and has not been observed yet. Courtesy of the SOHO/MDI consortium.

FIGURE 5.5 Sample picture of coronal loops from TRACE observed in Fe IX/X at 171 Å. Courtesy of NASA and the Stanford-Lockheed Institute for Space Research.

sulted in approximately 500 refereed papers, 38 master's theses, and 34 doctoral theses. Yohkoh is expected to continue operations until 2002-2003.

SOHO

The Solar and Heliospheric Observatory (SOHO), launched in December 1995, is an international cooperation between ESA and NASA. Located at the L_1, SOHO has enough fuel to fully cover a solar cycle. The SOHO mission has three principal goals—to gain an understanding of the mechanisms responsible for the heating of the Sun's outer atmosphere; to determine where the solar wind originates and how it is accelerated; and to measure the properties of the solar interior and flows into it. By flying at L_1, which is ≈ 1 percent of the distance to the Sun on the Sun-Earth line, SOHO is ideally situated to continuously monitor the Sun, the heliosphere, and the solar wind particles streaming toward Earth. The SOHO instruments function together as a coordinated system.

The helioseismology instruments on SOHO measure the surface magnetic fields, the surface flows, and flows and plumes below the surface. The coronal instruments provide both high-spatial-resolution maps of the locations of the energy releases and spectral diagnostics to determine the mechanisms of the release processes. The all-sky Lyman-alpha imager shows the extent of the wind in the heliosphere, and the particle and field detectors measure the energy and constitution of the particles accelerated toward Earth.

Missions have often been planned as coordinated systems. In this case particularly, a unique combination of an excellent instrument selection; pre-mission coordination of data formats, analysis software, and instrument operational methods; an agreement to share a basic data set in near real time; a joint central experimental operations center; topical workshops; a regular schedule of joint planning sessions; and daily coordination of these plans have made SOHO function as a science system rather than a collection of instruments.

TRACE

The Transition Region and Coronal Explorer (TRACE) telescope carries a Cassegrain design with a 30-cm aperture and a field of view of 8.5 arcmin $\times$ 8.5 arcmin. The TRACE instrument employs heritage from

SOHO MDI flight spares, making it much more capable than one would expect from a NASA Small Explorer (SMEX) mission. TRACE is in a polar orbit, allowing for continuous uninterrupted solar observing for approximately 9 months each year. TRACE data and analysis software are freely available to the whole community. The TRACE spacecraft continues to operate nominally; the estimated orbital lifetime is about 5 years.

MISSIONS UNDER CONSTRUCTION

HESSI

The High Energy Solar Spectroscopic Imager (HESSI) is slated for launch in March 2001. The combination of arcsec imaging and spectroscopy will allow HESSI to study impulsive energy release, particle acceleration, and particle and energy transport in solar bursts. Bursts with sufficient count-rate can be imaged in as short as 10 ms, although the best imaging will be obtained in 2 s, set by the spacecraft rotation rate. HESSI will locate the energy release site of flares, trace the propagation of particles from the release site, locate and determine the role of secondary particle acceleration, determine the composition of accelerated ions, and examine the role of long-term storage and acceleration of ions in flares.

HESSI mission plans include direct support of key ground-based observations, needed to place the high-energy processes into the overall context with other processes not observable by HESSI. The HESSI data are to be completely open for unrestricted use, and both data and analysis software will be disseminated through the HESSI European Data Center.

Solar-B

Solar-B is a Japan-U.S.-U.K. mission led by ISAS. The cost to NASA is about $65 million, a quarter to a third of the total mission cost. NASA's contribution includes the focal plane package, including the vector-spectromagnetograph for Solar-B's 50-cm optical telescope as well as parts of the x-ray telescope and the EUV imaging spectrometer. Solar-B is planned to launch in 2004 and to operate for at least 3 years.

Solar-B has a coordinated set of optical, EUV, and x-ray instruments. Together, these instruments allow the interaction between the Sun's magnetic field and its heated outer atmosphere to be investigated as a system, by observing the direct response of the chromosphere and corona to changes in the photospheric magnetic field. Solar-B will, with the perfect seeing of space, provide quantitative measurement of the full vector magnetic field in the photosphere on small enough scales (down to 0.2 arcsec) to isolate elemental flux tubes with a sensitivity to the transverse component of the field vector of about 100 G.

STEREO

The Solar-Terrestrial Relations Observatory (STEREO) is a Solar-Terrestrial Probe mission. STEREO will consist of two SMEX-size spacecraft carrying identical scientific payloads. The spacecraft will be launched into orbits near 1 AU that will allow increasing angular separation between the two spacecraft, with one trailing Earth and one leading.

The STEREO payloads will include both in situ and radio instruments for probing the interplanetary medium and a set of solar imaging instruments to characterize the structure and development of CMEs in the plane of the ecliptic. In addition, when the separation of the two spacecraft is still relatively small, the stereoscopic view from the two sets of imagers will allow new insight into the three-dimensional structure of active region loops. The scientific return from STEREO would be significantly enhanced by the presence of similar imaging and in situ instrumentation along the Sun-Earth line.

Solar Probe

The Solar Probe, evaluated in *A Science Strategy for Space Physics*,[4] will fly from pole to pole through the solar atmosphere, as close as 3 solar radii above the surface at perihelion, and will perform the first close-up exploration of the Sun. Scheduled for launch in February 2007, Solar Probe will travel along a polar trajectory to the Sun, where it will arrive in 2010. The second flyby will be near the solar minimum in 2015. Embedded in the acceleration region of the solar wind, Solar Probe will address the basic questions surrounding the origin of the fast and slow solar wind. By flying over the poles, it will sample the slow and fast solar

---

[4]Committee on Solar and Space Physics, National Research Council. 1995. *A Science Strategy for Space Physics* (Washington, D.C.: National Academy Press).

wind directly where their acceleration takes place, and it will image the solar atmosphere with better spatial resolution than is currently possible. It will also determine for the first time the solar surface properties of the polar regions of the Sun. The payload will consist of miniaturized imaging and in situ instruments. Together they will provide the first three-dimensional view of the corona, high-resolution spatial and temporal observations of the magnetic fields, helioseismic measurements of the solar polar regions, and local sampling of plasmas and fields at all latitudes.

NEW INITIATIVES

Figure 5.6 illustrates the spatial resolution and wavelength coverage of major instruments on the ground and in space. The new ones will be ranked separately below.

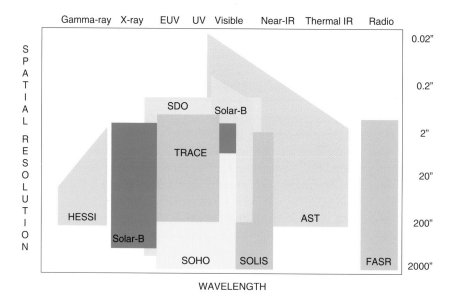

FIGURE 5.6 Schematic overview of the coverage in wavelength and spatial resolution provided by some of the observational facilities discussed in this report. Note that some of the overlap is necessary because of the different time resolution of the various instruments.

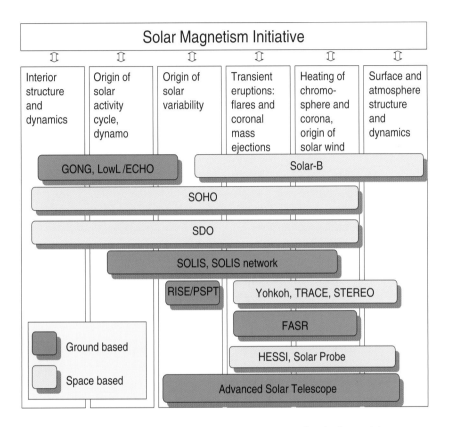

FIGURE 5.7 Schematic overview of a global approach to understanding the Sun and the coverage of the various physical regimes by existing and future observational facilities.

FROM THE GROUND

The priority of the projects below, as ranked by the panel, is in agreement with the conclusions formulated in the Parker report. The third project, a small one, was part of the original proposal for SMI (see Figure 5.7). It has since been separated out as a stand-alone astronomical project. Several groups have brought other plans for new instrumentation to the attention of the committee (without entering them formally into the ranking competition). These projects are summarized under "Other Projects—Not Ranked."

THE ADVANCED SOLAR TELESCOPE

The Advanced Solar Telescope (AST) will enable the observation of plasma processes with unprecedented resolution in space and time and will provide a unique opportunity to probe cosmic magnetic fields and test theories of their generation, structure, and dynamics. Progress in these issues requires a solar telescope with (1) an angular resolution of better than 0.1 arcsec to resolve the pressure scale height and the photon mean free path, (2) a high photon flux at high spatial resolution for precise magnetic and velocity field measurements, and (3) access to a broad set of diagnostics for a wavelength range from 0.3 to 35 μm. No current or planned ground-based or space-based solar telescope meets these requirements. Recent major advances in technology, such as the successful development of solar adaptive optics; the construction of the Dutch open-air solar telescope on La Palma, which produces superb images without a vacuum sustained in the light path of the telescope; and the development of large-format infrared detectors are making it possible to realize the AST within the decade. The AST will replace NSO facilities at Sacramento Peak and Kitt Peak.

Solar physics has advanced to a point where existing solar telescopes are no longer sufficient to conduct critical observational tests of models for the underlying physical mechanisms. For this reason, two European telescopes on the Canary Islands are being replaced by telescopes with larger apertures. The AST is proposed as a joint project of the United States and its international partners, to be centered at NSO. It will complement other solar facilities in space and on the ground. A major design driver for the AST is the capability to perform precise and accurate measurements of solar magnetic fields over a large wavelength range. The AST will be sited at a location that offers superb seeing and clear weather for sustained periods of time.

The Origin of the Sun's Magnetic Cycle

The 11-year sunspot cycle and the 22-year magnetic cycle are still a mystery. Dynamo models assuming a large-scale dynamo at the base of the convection zone were promoted for many years, but it was finally realized that they are not self-consistent. Recent observations and modeling indicate that there may be two dynamos: a high-field-strength dynamo at the base of the convection zone and a weak-field, turbulent dynamo near the surface. Indeed, hints of a weak magnetic field component that covers the entire Sun have been discovered in several recent observations. This global phenomenon may be of crucial importance for the magnetic cycle and its variability. The AST is the ideal tool for quantitative measurements of these weak, turbulent fields.

The Solar-Stellar Connection

The observation of the solar cycle at high precision with modern instrumentation is only a few decades old. Thus, our knowledge of the full range of solar variability that may occur is extremely limited. The observation of solar-type stars can be productively exploited to overcome the temporal confines of the solar database, thereby revealing the potential range and nature of solar variability over timescales that are now inaccessible to the solar database of only a few decades. In particular, a program of high-precision spectroscopic and photometric observations of the numerous solar-type stars in the solar-age and solar-metallicity cluster M67 can reveal all the potential modes of variability in both magnetic activity and radiative output in Sun-like stars and, by implication, in the Sun itself. The study of stars similar to our Sun is a practical approach to understanding and forecasting the long-term behavior of the Sun. The AST will be used at night as a dedicated 4-m-class facility to observe the faint solar-type stars in distant clusters such as M67 at high precision. Studying solar-type stars with ages, chemical compositions, or rotation rates different from those of the Sun will tell us how these stellar parameters relate to the physical principles responsible for them.

Stellar Chromospheres

Measurements of CO around 4.7 μm show extremely cool clouds that appear to fill much of the volume in the low chromosphere. Only a small fraction of the volume is filled with hot gas, as expected from static models that exhibit a homogeneous temperature rise in the chromosphere. The observed spectra appear to be explained only by dynamic models of the solar atmosphere. Numerical simulations indicate that the temperature structures occur on spatial scales that cannot be resolved with current solar telescopes. A test of these models therefore requires a large telescope that provides access to the thermal infrared. Stellar magnetic fields can be measured most accurately in the infrared. Thus such observations should be a major component of the nighttime program for the AST.

The Sources of Global p-Mode Oscillations

A process that occurs on very small scales (<100 km) causes global solar oscillations: narrow, supersonic downdrafts in intergranular lanes continuously produce "acoustic noise" that powers the oscillations. Numerical simulations provide detailed predictions on how convective energy is converted into acoustic energy that powers the p-modes. Acoustic events may also contribute to the heating of the lower chromosphere through the formation of acoustic shock waves. While the resolution of current facilities is sufficient to verify the existence of acoustic events, it is insufficient to determine in any detail the underlying physical mechanisms.

Structure of Sunspots

Strong photospheric magnetic fields are concentrated in flux rope units in which local fields are strong enough to control the local environment but whose collective behavior is controlled by the photospheric convection patterns. In sunspots the total magnetic field is large enough to dominate the hydrodynamic behavior of the solar atmosphere. To test numerical simulations of sun- and star-spots in general, 0.05- to 0.1-arcsec-resolution vector polarimetry with low-scattering optics is required.

Confronting Models with Observations

Figure 5.8 shows magnetic fields in dynamic pressure equipartition (~400 G) with convective motions from numerical simulations. Magnetic field elements are small scale (<70 km), of mixed polarity, intermittent, and mostly concentrated in the narrow intergranular lanes where strong downflows are present. To understand the importance of these fields in the dynamo process astronomers need to observe how and at what rate these weak fields form and to determine the spectrum of field strength. Polarimetric measurements show that the "quiet" Sun appears to be covered with weak magnetic fields. IR observations of sufficient spatial and temporal resolution would allow observing the formation of kilogauss flux tubes, the building blocks of solar and stellar activity, by

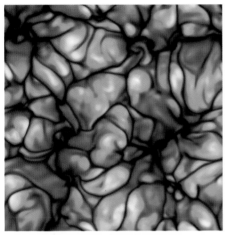

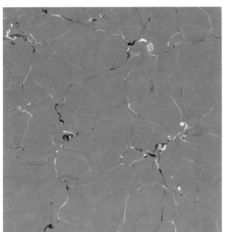

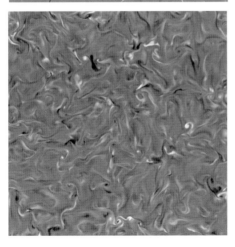

FIGURE 5.8 Snapshot of the appearance of temperature and magnetic field from a three-dimensional MHD simulation. Color plot of temperature fluctuations (top; bright indicates hot material) and gray-scale plots of the vertical component of the magnetic field (middle panel). The middle panel corresponds to a layer near the upper boundary; the bottom panel corresponds to a deeper layer. Courtesy of F. Cattaneo, University of Chicago.

convective collapse. No direct observational evidence for this process exists to date.

AST Baseline Parameters and Their Science Drivers

Basic design guidelines for the AST and their link to science requirements are given below. Conceptual design efforts are preliminary at this time.

- *Spatial resolution.* A 4-m diffraction limit using AO is needed to (1) resolve the photon mean free path and the pressure scale height in the photosphere and (2) probe the IR signature of cool clouds in the chromosphere and test models of their radiative cooling.
- *Sensitivity.* A 4-m photon collection aperture is needed to (1) study the ubiquitous weak magnetic field and test models of a turbulent dynamo in the upper convection zone and (2) measure waves in magnetic flux tubes and test models of chromospheric and coronal heating.
- *Field of view.* A field of view of 5 arcmin is needed to (1) test models of the eruptions of flux that form active regions from the strong-field dynamo in the lower boundary layer of the convection zone, (2) test models of large-scale coherent processes that lead to flares and CMEs, and (3) observe large-scale oscillations in prominence and compare with models.
- *Wavelength range.* A wavelength range of 0.3 to 35 μm is needed to (1) provide access to a broad range of diagnostics from the photosphere to the corona and (2) observe the widest variety of diagnostic spectral lines to constrain atmospheric properties. The Mg lines at 10 mm allow for magnetic field measurements with large Zeeman splitting, and they form in local thermodynamic equilibrium as opposed to the UV and visible lines formed in the same region, which do not. The range from 20 to 30 μm is unknown territory. The telescope technology to go beyond 20 μm does not change.
- *Polarization accuracy.* A polarization accuracy of 10^{-4} intensity is needed to (1) precisely measure the vector magnetic field and test models of wave generation in magnetic flux tubes by the surrounding granulation and (2) use the Hanle effect to test models of extremely weak magnetic fields in the photosphere, chromosphere, and prominences.
- *Scattered light.* Large sunspots with field strengths in excess of 3 kG often have residual intensities of less than 10 percent of the mean

TABLE 5.1 AST Investment Strategy (thousand dollars)

Cost Increment	Year									Total
	1	2	3	4	5	6	7	8	9	
Adaptive optics	1,500	1,500	1,500	500						5,000
Site tests	550	150	150	150						1,000
Conceptual design	500	1,100	1,500							3,100
IR technology	100	300	300	300						1,000
Focal-plane instrument packages			200	500	1,000	3,000	3,000	2,000	2,000	11,700
AST construction					14,000	13,000	12,000	3,000		42,000
Total	2,650	3,050	3,650	1,450	15,000	16,000	15,000	5,000	2,000	63,800

granular intensities, even in the red. To isolate effects intrinsic to the umbra, the umbral signal should be greater by a factor of 10 than the signal from the stray and scattered light. The scattered light from the instrumentation must then be on the order of 1 percent or less.

• *Location.* The best possible site in terms of seeing and sunshine hours is needed to maximize telescope performance and minimize the cost of AO.

• *Costs.* Table 5.1 gives the incremental costs of designing and building the AST. About one quarter of the cost for the construction will go into the focal plane instrumentation packages (FPIP). Costs include a 15 percent contingency. Both Germany and Japan have indicated interest in the project, and plans are being made to share the cost of developing of the AO system, which is a critical technology for the AST. It is planned to seek sufficient international and domestic partnerships to bring the cost to the United States to approximately one half of the total cost, or about $32 million.

Development Plan

The AST will represent the first major U.S. investment in ground-based solar physics in over 40 years. As such, it must involve a substantial fraction of the solar community. Much of the instrumentation development, site testing, and design will be done by universities and international partners. Once completed, the international operating organization must also run the AST facility and ensure continued instrument development. Figure 5.9 gives a time line for the various ingredients and phases of the AST development. Since the AST should involve a wide spectrum of the solar community and since it will require international partners, an AST board consisting of representatives of the national

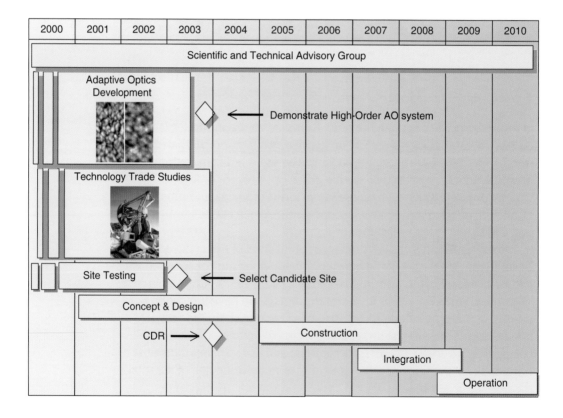

FIGURE 5.9 Roadmap for the development of the Advanced Solar Telescope. Courtesy of S. Keil, National Solar Observatory.

centers, the university community, other interested government agencies, and non-U.S. partners should be formed. This board will provide overall project guidance and oversee project execution through advisory subcommittees. It will also oversee the financing of the project. Funds will be provided by the NSF, by other interested agencies, and by non-U.S. partners. The board will organize the distribution of these funds to the various AST functions, AST design, site testing, construction, instrument development, and operation. These functions will be distributed among the partners, with the NSO taking overall responsibility for developing the telescope.

THE FREQUENCY-AGILE SOLAR RADIOTELESCOPE

Solar radio instrumentation worldwide lags far behind instrumentation for other wavelength regimes. The Frequency-Agile Solar Radiotelescope (FASR) would represent a major advance over existing facilities, with capabilities specifically for observing the Sun. It is designed to produce high-quality images of solar processes over a core frequency range of 0.3 to 30 GHz, with sufficient angular, spectral, and temporal resolution to fully exploit radio emission as a diagnostic of the wide variety of physical processes that occur on the Sun. It improves on the 84-antenna Nobeyama Radioheliograph by working at a multitude of frequencies rather than just two, and its spatial resolution is better by a factor of 10.

The science goals of the FASR include the following: (1) transient energetic phenomena: energy release, plasma heating and electron acceleration, electron transport, and formation and destabilization of large-scale structures, (2) the nature and evolution of coronal magnetic fields: precise measurement of coronal magnetic fields, temporal evolution of coronal magnetic fields, the role of coronal currents, and the storage and release of magnetic energy, and (3) the solar atmosphere: coronal heating, structure of the quiet solar atmosphere, origin of the solar wind, and the formation and structure of filaments.

The approach to attaining these goals is through the measurement of broadband, spatially resolved microwave, and decimeter-wave spectra. Such spectra are the key to unraveling the multiparameter dependence inherent in radio emission, which would make possible the precise measurement of physical quantities such as magnetic field strength and direction, temperature, electron density, and the shape of the electron energy distribution. The FASR specifications, listed in Table 5.2, give the

TABLE 5.2 Specifications for the Frequency-Agile Solar Radio Telescope
(FASR)

Parameter	Specification
Frequency range	0.3-30 GHz
Frequency resolution	
	~3%, 3-30 GHz
	~1%, 0.3-3 GHz
Time resolution	
	<1 s, 3-30 GHz
	<0.1 s, 0.3-3 GHz
Antenna size	2-5 m
Number of antennas	~100
Number of baselines	~5000
Polarization	Dual
Number of IF pairs	4-8
Angular resolution	20″ (1 GHz) to 0.5″ (30 GHz)
Field of view	Full Sun

requisite spatial and spectral resolution, image quality, and temporal
resolution to address the following topics:

• Spatial, temporal, and spectral characteristics of the site of energy
release;
• Measurement of magnetic field strength at coronal heights in flares
and active regions;
• Determination of the electron energy distribution;
• Detection of CMEs, both off the limb and on the solar disk;
• Elucidation of causes of coronal heating (nanoflares or destabiliza-
tion of coronal currents); and
• Synoptic measurement of both thermal and nonthermal activity.

The FASR design can be entirely based on existing technology,
although innovative concepts should be explored to cut costs and extend
capabilities. Many aspects of the system (antenna size, broadband
frequency coverage, signal transmission, and perhaps even digital signal
processing and correlation) overlap with the One Hectare Telescope
(1HT), now under way at the University of California at Berkeley and the
Search for Extraterrestrial Intelligence (SETI) Institute.

Extension of SOLIS to a Network

A single SOLIS facility is severely limited by the day-night cycle because many important phenomena connected with magnetic fields occur on timescales of a day. To study and understand the magnetic origin of solar variability, two additional vector spectromagnetographs should be built and installed at widely separated longitudes to obtain nearly continuous time coverage. Precision spectropolarimetry with a SOLIS network affords the possibility of monitoring the evolution of solar surface vector magnetic fields as toroidal flux rope systems, created by dynamo action at the base of the solar convection zone, rise through the solar surface. The full history of the birth, evolution, and decay of solar magnetic flux during the disk passage of active regions (10 to 14 days) is now recognized to be essential to an understanding of the physical processes underpinning flux emergence and activity. Fortunately, most active regions pass through their critical early emergence phase within a few days to a week, so that one can be reasonably optimistic that the passage of emerging regions across the visible solar hemisphere will illuminate much of the crucial physics. A SOLIS network will provide continuous coverage of such an event. Operation of such a network over the solar cycle will provide critical insight into the magnetic origins of solar activity and variability. Expansion of SOLIS from one station to a three-station network will cost about $4.8 million, with an additional $200,000 needed annually for operation of the two additional stations and the associated increase in data volume.

Other Projects—Not Ranked

Large-Aperture, Ground-Based Coronagraph

There is evidence from a recent eclipse expedition that a Si I line is present prominently in solar coronal structures. This spectral line could be the best tool for direct measurements of coronal magnetic fields. To gather enough photons to quantitatively exploit this line and other infrared lines in the solar corona, much bigger coronagraphs than the existing ones are needed. A group at the University of Hawaii is pursuing this with a proof-of-concept telescope of about 50-cm aperture. A coronagraph of several meters aperture is targeted by this group in the context of replacing the NASA infrared facility on Mauna Kea. Such a telescope with an aperture of 6.5 m would have tremendous value for nighttime observations in planetary astrophysics as well as for observations of faint objects around bright sources.

LOFAR

A project for a long-wavelength array for general astronomical use is under way. It is a merger of two previous concepts—the Naval Research Laboratory's (NRL's) Long-Wavelength Array and the Dutch Low-Frequency Array (LOFAR). This array would have significant impact on space weather research. It will operate in the frequency range below 150 MHz and will obtain high-resolution (1- to 10-arcsec) images with excellent sensitivity in the region of the corona approximately 0.5 to 2 solar radii above the surface. In this region, the imaging of large-scale thermal and nonthermal coronal structure will be possible, including likely signatures of CMEs, shock waves (type II bursts), noise storms (type IV), and electron beams (type III), all of which directly affect the space weather environment. The solar science to be addressed by the array needs to be more fully investigated, and full consideration for solar science is urged in the design of the instrument, to ensure fast imaging over the entire spectrum.

SRBL

The Solar Radio Burst Locator (SRBL) is a project in the prototype stage and slated to become an operational four- or five-station system operated by the U.S. Air Force. In addition to the operational use of the system for locating bursts, it will provide broadband spectral coverage (in the 2- to 18-GHz range) of solar bursts. The panel urges that provision be made for proper calibration and wide dissemination of the spectral data to the scientific community.

Ballooning

Recently, a group led by Johns Hopkins University flew a balloon experiment in Antarctica with the goal of long-duration observations of solar processes. A second flight has just been completed. NASA developed the concept of a long-duration balloon facility offering the option of solar observations largely undisturbed from seeing in the Earth's atmosphere and with fairly long observation cycles.

IN SPACE

The Solar Dynamics Observatory is the recommended mission for the next decade. All the other missions are either part of the Solar Terrestrial Probe line (category "small" for NASA) or will be evaluated by the Space Studies Board of the NRC.

SOLAR DYNAMICS OBSERVATORY

The Solar Dynamics Observatory (SDO) is a mission to explore the Sun from the subsurface layers of the convection zone to the outer atmosphere. Breakthroughs in techniques for analysis of solar oscillations using MDI (see Figure 5.10) data have opened a new frontier. Acoustic imaging reveals that active regions generate a complex pattern of wave absorption. However, existing instruments cannot provide the combination of temporal and spatial resolution and coverage necessary to exploit these techniques to study the eruption and evolution of active-region magnetic structures. SDO will explore the complete life cycle of solar active regions using 1-arcsec-resolution, full-disk Doppler velocity and vector-magnetic-field observations.

Because of its location at L_1, SOHO was limited to a data rate of 40 kbps for most observations (see Table 5.3). For 8 hours a day, the MDI data rate was increased to 640 kbps. Also, for a 2-month period every year, the MDI rate was 640 kbps. The SOHO imagers used a total of five 1024^2 charge-coupled devices (CCD) detectors. SDO operates 24 hours a day in a geostationary orbit at a data rate of about 160 Mbps, which is greater by a factor of 250 than the 8-h data rate of SOHO and greater by a factor of 4000 than the normal operational mode (see Table 5.4). The SDO uses twelve 4096^2 CCD detectors, which allow simultaneous observation in all the EUV channels to separate temporal changes from evolutionary ones (SOHO requires sequential operations). In its normal mode, SDO can make 4096^2 images 250 times as fast as SOHO makes 1024^2 images.

SOHO/EIT and TRACE observations have demonstrated a requirement for high temporal and spatial resolution over the full disk to understand the interconnected dynamic structures of the transition region and the low corona and to conduct coronal helioseismology. EUV spectroscopic imaging with 1-arcsec resolution will determine the atmospheric magnetic connectivity of active regions. SDO should be launched early in the rising phase of solar cycle 24 to observe relatively isolated active

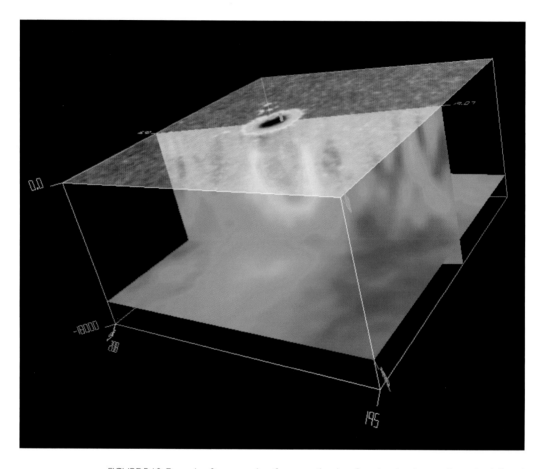

FIGURE 5.10 Example of tomography of a sunspot's subsurface structure in sound speed as inferred from a time-distance analysis of running waves. Courtesy of the SOHO/MDI consortium.

regions and follow the rapid increase in complexity as the solar activity maximum is approached.

SDO carries the Atmospheric Imaging Assembly, an array of telescopes that image the Sun in the temperature range 4000 to 9,000,000 K; an advanced Doppler and vector magnetogram instrument package (Helioseismic and Magnetogram Imager, or HMI) that images subsurface structures, detects spots on the opposite side of the Sun, and produces vector magnetograms; and a pair of coronagraphs (the Coronal Imaging Assembly, or CIA) that make precise measurements of the Sun's white-light corona from 1.05 to 18 solar radii. In addition, SDO carries an array of precision radiometers that measure the solar irradiance from the far UV to the IR.

Measurement Strategy

All imaging instruments will observe the full Sun continuously, and there will be one basic observing mode to simplify operations and data analysis. Continuous full-disk observations will provide the ability to follow the location of every active region as it transits the solar disk, whenever it emerges. Measurements that simultaneously achieve high spatial and temporal resolution and wide spatial and temporal range are the critical elements of SDO mission strategy.

Science Objectives

SDO should help to answer the following questions:

- Why are there sunspots and solar active regions?
- How do magnetic regions emerge, evolve, and decay?
- How do the active-region fields interact with the small-scale fields?
- Do local dynamo processes occur?
- How does the large amount of magnetic energy created at small scales dissipate?
- How do small- and large-scale coronal magnetic field reorganizations occur?
- What are the surface and subsurface magnetic configurations that lead to CMEs and flares?
- How important is the inverse cascade of small-scale flux emergence to large-scale flux expulsion?
- To what extent are CMEs and flares predictable?
- Does activity affect solar convection and irradiance?
- How are the dynamics of the interior and the quiet and active solar corona linked?

SOHO/MDI has shown that we can make detailed maps of magnetic fields before they reach the surface by the technique of acoustic imaging. Then the surface magnetic fields and flow systems can be tracked using polarized spectral images. TRACE has shown that it is necessary to take data with a cadence of less than 10 s to follow the small- and large-scale

TABLE 5.3 SOHO Imagers

Instrument	Measurement Type	Field of View (arcmin)	Spatial Resolution (arcsec)
MDI	Dopplergram	34 × 34	30
(Full Sun)	Dopplergram	34 × 34	4
(HiRes)	Dopplergram	10.5 × 10.5	1.2
(Full Sun)	Line-of-sight magnetogram	34 × 34	4
(HiRes)	Line-of-sight magnetogram	10.5 × 10.5	1.2
EIT	Intensity	40 × 40	5
	Intensity	40 × 40	5
	Intensity	40 × 40	5
	Intensity	40 × 40	5
LASCO			
C1	Intensity	1.1-3 solar radii	11
C2	Intensity	1.5-6 solar radii	25
C3	Intensity	3.7-30 solar radii	110

NOTE: MDI is Michelson Doppler Imager, EIT is Extreme Imaging Telescope, LASCO is Large-Angle and Spectrometric Coronagraph Experiment, and C1, C2, and C3 are detectors on LASCO.

reconnections of magnetic fields in the corona. SOHO/EIT has shown that waves can propagate at least over a visible hemisphere.

Analysis of the high-cadence TRACE data has revealed waves in the corona excited by flares, CMEs, filament activation, and magnetic reconnection. The new science of coronal helioseismology has already revealed preliminary estimates of the magnetic field in the corona and has started a fundamental reevaluation of the basic magnetohydrodynamic wave damping processes. Like earthquakes, coronal explosions are episodic, and the measuring system must be in place before the event

Spectral Resolution	Temporal Cadence	Data Rate (kbps)	Time Coverage (h/d)	Detector
94 mÅ at 6768 Å	60 s	5	24	One 1024^2
94 mÅ at 6768 Å	60 s	160	8 (24)	
	60 s			
	90 min			
	90 min			
171 Å	23.5 min	1	23.5	One 1024^2
195 Å	4.7 min	5.6	0.5	
284 Å				
304 Å				
				Three 1024^2
Broadband	25 min			
Broadband				
Broadband				

in order to study it. Solar researchers expect that coronal helioseismology will teach them as much about the corona as traditional helioseismology has taught about the solar interior.

Technology Requirements

SDO will use large-format, low-power, fast-readout CCD detectors, and it will rely on fast spacecraft data compression hardware with large, low-power, fast, smart memories. The spacecraft will use artificial

TABLE 5.4 Solar Dynamics Observatory Imagers

Instrument	Measurement Type	Field of View (arcmin)	Spatial Resolution (arcsec)
HMI	Dopplergram	34 × 34	1
	Vector magnetogram	34 × 34	1
AIA-EUV	Intensity	36 × 36	1.1
	Intensity	36 × 36	1.1
	Intensity	36 × 36	1.1
	Intensity	36 × 36	1.1
	Intensity	36 × 36	1.1
	Intensity	36 × 36	1.1
	Intensity	36 × 36	1.1
AIA-UV	Intensity	36 × 36	1.1
	Intensity	36 × 36	1.1
	Intensity	36 × 36	1.1
	Intensity	36 × 36	1.1
	Intensity	36 × 36	1.1
CIA-Inner	Intensity, polarization	33.6-72	2.2
CIA-Outer	Intensity, polarization	2-18 solar radii	17

NOTE: HMI is the Helioseismic and Magnetogram Imager, AIA is the Atmospheric Imaging Assembly, and CIA is the Coronal Imaging Assembly.

intelligence for self-operation and monitoring along these lines. Technology developments can increase the scientific return of the mission and lower its cost.

Education and Public Outreach

The SDO data set will be available in real time at the science data center for distribution over the Internet. The data will be of great use to scientists worldwide. In addition, national and international agencies concerned with space weather forecasting will have an invaluable asset—continuous, real-time solar data from the interior to the outer atmosphere. By reformatting the data and using the spacecraft as a repeater, as was done with GOES satellites, a video signal showing the

Spectral Band	Temporal Cadence	Data Rate (Mbps)	Time Coverage (h/d)	Detector
94 mÅ (6768)	45 s	30	24	One 4096^2
	5 min	30	24	One 4096^2
171 Å	10 s	10	24	One 4096^2
195 Å	10 s	10	24	One 4096^2
284 Å	10 s	10	24	One 4096^2
304 Å	10 s	10	24	One 4096^2
211 Å	10 s	10	24	One 4096^2
133 Å	10 s	10	24	One 4096^2
304 Å	10 s	10	24	One 4096^2
1900 Å	10 s	10	24	One 4096^2
1700 Å				
1600 Å				
1550 Å				
1216 Å				
Broadband	30 s	3		One 4096^2
Broadband	30 s	3		One 4096^2

Sun in action can be made available to the entire world for use in schools, by forecasters, by scientists, by the general public, or by the media. It would not be surprising if every major science museum in the world had a high-definition TV showing "Live from the Sun."

Costs

The costs for SDO are well understood. The instruments are straightforward enhancements of those on SOHO, TRACE, and Yohkoh, and accurate cost estimates exist. A geosynchronous orbit requires a Delta-class launch vehicle. By using a single ground station at the science data center, operating costs are minimized. The preliminary estimate through launch and the first 2 years of operation is $300 million.

SOLAR TERRESTRIAL PROBE MISSIONS

Particle Acceleration Solar Orbiter (PASO)

PASO will address the fundamental question of how the Sun accelerates particles to high energies in solar flares and CMEs. It is designed to use solar sailing to achieve a near-synchronous orbit at 0.16 to 0.2 AU, that is, an orbital period about equal to the solar rotation period. This would allow continuous observation of particle acceleration from active regions and CME-related solar features from their birth through their rise to a maximum and decay. PASO would provide hard x-ray/gamma-ray imaging of flares with 25 to 36 times the sensitivity and 5 or 6 times the linear spatial resolution of observations from 1 AU. Neutrons of energies below tens of MeV and down to ~1 MeV, which can provide direct evidence for acceleration of low-energy ions, would only be detectable by getting this close, since they decay in flight (e-folding decay time of ~1000 s). Getting this close is also the only way to obtain measurements of the energetic particles freshly accelerated by CME shocks before they have been significantly modified by scattering and energy changes. Finally, PASO will provide the first systematic exploration of the inner (<0.16 to 0.3 AU) heliosphere.

Reconnection and Microscale Probe (RAM)

The RAM is designed to investigate the structure and dynamics of the magnetized coronal plasma with continuous broadband solar observations from the L_1 orbit. It aims to understand the microscale instabilities that lead to reconnection and, ultimately, to flares and CMEs. The probe will be equipped with an ultrahigh-resolution telescope imaging the Sun at 195 Å with a resolution of 0.02 arcsec. It will perform high-resolution spectroscopy (0.2 arcsec from 0.3 to 10 keV) and high-resolution EUV spectroscopy at 170 to 220 Å. The probe is projected to be a Solar-Terrestrial Probe (STP) mission. It is a follow-up mission on TRACE, SOHO-EIT, and Yohkoh, going for the specific problem of reconnection. It will have to be equipped with large-format cryogenic imaging detectors, which are under development.

THEORY AND DATA MINING: THE SOLAR MAGNETISM INITIATIVE

The SMI concept of a comprehensive research program involving the various facilities and the solar community was introduced earlier. SMI would include a series of focus programs on particular aspects of the problem; these would be carried out by groups of scientists collaborating in extended workshops at a single location supported by appropriate computing, observing, and data analysis resources. Candidate topics include the solar dynamo and interior dynamics; magnetic flux transport through the convection zone; an observational description of emerging magnetic flux; the history of magnetic flux; the solar magnetic cycle at the solar surface; and the physics of coronal mass ejections, their causes, and heliospheric effects. It is expected that each focus program would take about 1 year, with intensive workshops every 6 months at which progress would be compared and coordinated. At the end of each focus program, results would be presented to the broader community in a workshop or at a scientific meeting.

The SMI would be overseen by a steering committee that could be modeled after the CEDAR steering committee. In addition to defining the focus programs described above, it would advise on scientific priorities throughout the life of the program, coordinate observing campaigns, and keep the community informed and involved though newsletters and presentations at scientific meetings and workshops.

Two- and three-dimensional simulations of MHD processes mimicking some of the processes on the Sun have been undertaken with great success by many groups in the United States and elsewhere. There are major numerical efforts at the following universities: University of Colorado at Boulder, Harvard University, Michigan State University, Stanford University (Lockheed), University of California at Berkeley, University of Alabama at Huntsville, University of Chicago, University of Rochester, and Yale University, as well as at the Bartol Institute, NASA Goddard Space Flight Center, National Center for Atmospheric Research/High Altitude Observatory, Naval Research Laboratory, and the San Diego Supercomputer Center/Science Applications International Corporation.

To take advantage of the rapidly increasing quantities of solar observational data (from, for example, the SOLIS instruments) and to expand the effort in numerical modeling, SMI will need about $2 million per year in funding for an expanded university grants program. Grants would be awarded following standard NSF procedures for directed programs, such as CEDAR and GEM.

To entrain new scientists with recent Ph.D.s into SMI, a program of at least two SMI postdoctoral fellowships should be established, to be

TABLE 5.5 Estimated Annual Investments for the Solar Magnetism Initiative

Initiative	Annual Investment ($)
University grants program and SMI postdocs	2,200,000
Focus programs, workshops, and coordination	90,000
Centralized scientific support and community service	500,000

hosted by any institution involved in SMI. The appointees would be chosen in a community-wide competition according to a process to be determined by NSF.

Some centralized support would also be needed to develop the community Stokes inversion program, which is essential for utilizing the new and greatly expanded vector magnetograph data expected from SOLIS, and to support observing campaigns, focused programs, and the SMI database, which provides both observational and modeling data (Table 5.5). There would also be some one-time costs for hardware for the SMI database, proposed to reside at the National Center for Atmospheric Research (NCAR)/HAO. It is expected that supercomputing requirements for SMI, particularly for numerical modeling and Stokes inversions of data, while not small, can be accommodated within current plans to upgrade the NCAR Scientific Computing Division's supercomputing capability and by some of the NASA centers.

TECHNOLOGIES FOR THE FUTURE

ADAPTIVE OPTICS

The AST project will develop a visible adaptive optics (AO) system. A multiconjugate adaptive optics (MCAO) system based on this technology will achieve diffraction-limited resolution over fields of view (FOVs) significantly larger than the isoplanatic patch size of, typically, a few arcseconds. The Sun is an ideal object for the development and application of MCAO since the multiple wavefront measurements as a function of the FOV required for MCAO can be performed using solar structure as the wavefront sensing target. The complexity involved in having to use multiple laser guide stars can be avoided. MCAO should therefore be developed for the Sun first. The AO development is proposed as a collaborative effort of NSO and NJIT/Big Bear Solar Observatory, the Center for Adaptive Optics, and international partners.

SOLAR-LITE

Solar-B will have 150-km resolution, 10-G sensitivity to the line-of-sight field component and 100-G sensitivity to the transverse component. Solar Probe will provide the first glimpse of the line-of-sight magnetic field with 10-G sensitivity at 50- to 25-km resolution. Solar-B will have the resolution to isolate separate elementary flux tubes (150 km in diameter); Solar Probe will have the resolution to look within a tube but will not measure the transverse component. Measurement of the transverse component is essential for determining the three-dimensional configuration of the field. Now is the time to begin defining and developing the science and technology for the next-generation, high-resolution solar mission. The development of a lightweight mirror larger than the 50-cm mirror of Solar-B, as begun in the Solar-Lite technology studies, will lead to less expensive instruments than those envisioned for OSL/SOT, which was planned for in the 1991 survey report but never built, for financial reasons.

HIGH-RESOLUTION VECTOR MAGNETOMETRY OF UV LINES

Measurements are needed of the three-dimensional vector of the magnetic field in lines formed above the photosphere, in the field-dominated, force-free domain of the solar atmosphere. This requirement is motivating the development of new filters and polarimeters for vector magnetography of UV lines formed in the chromosphere and low transition region.

CONNECTION TO LABORATORY ASTROPHYSICS

ATOMIC/MOLECULAR/NUCLEAR PHYSICS

- *Identification of EUV line spectra.* Of more than 800 observed lines, SOHO has revealed that 300 or so between 500 and 1600 Å are unidentified.
- *Accurate laboratory wavelengths.* Laboratory measurements for Mg X and Ne VII lines found in the corona/transition region are of insufficient accuracy or missing altogether. The NIST spectrograph should follow up with measurements accurate to 1 part in 200,000.
- *Photoionization resonances.* The OPACITY project data for He I show resonances close to the Fe IX/X and Fe XII lines emitting in the TRACE bandpasses, with energies uncertain to 1 eV. High-resolution measurements of the photoionization resonance structure of neutral or singly (doubly) charged ions of abundant elements are needed.
- *Collision cross-sections for particle impact.* A new area of research recently opened up using the Hanle effect in Stokes spectra obtained inside the solar limb. Many details of the atomic physics involved in the collisional depolarization need to be measured to much greater accuracy than can presently be achieved in the laboratory.
- *Landé g factors.* The Landé g factors of absorption lines of complex atoms (e.g., Fe I and Fe II) depend on the configuration mixing. As the infrared becomes accessible to high-resolution Stokes measurements, precise g factors for Fe I and Si I are needed.
- *Neutrinos.* The LOWL instrument provided proof that the neutrino deficit as it is measured for the solar neutrino flux on Earth is not due to a deviation of the solar structure from the standard model. The key lies in the regime of particle physics. A continuation of measurements of neutrinos is, however, essential to determine finite mass and possible magnetic moment of the neutrinos.

PLASMA PHYSICS

Basic plasma physics and magnetohydrodynamic processes, which are thought to be central to solar physics, can be studied in the laboratory. The Magnetic Reconnection Experiment (MRX) device at Princeton has been used for a series of magnetic reconnection experiments. One of the main issues studied by MRX is the relationship between the reconnection region, which is extremely small, and the global magneto-hydrodynamic equilibrium. The results have been successfully interpreted as magnetic reconnection at the Sweet-Parker rate. The basic device is a modified plasma fusion reactor design of a type known as Spheromak. There have been other contributions from laboratory plasma physics in the areas of anomalous thermal conductivity and anomalous resistivity, relaxation of plasma to a force-free state, and dynamo activity. Such experiments have an important role in solar physics, providing a basis for theory and interpretation of observation.

POLICY AND EDUCATIONAL ASPECTS

The panel examined issues surrounding the standing of solar physics in the U.S. university community and the overall balance between the NSF grants program and the two NSF solar physics centers—NSO and HAO. It also considered how a major development project like the AST should be optimally organized within the United States as well as with international partners.

THE UNIVERSITY-BASED SOLAR PHYSICS COMMUNITY IN THE UNITED STATES

Driven by the successes of solar space missions and by helioseismology, there has been a rejuvenation of solar physics at U.S. universities, with two new departments having been established. There is a vigorous university research community in solar physics built mostly on research faculty positions rather than on regular faculty. However, several traditional chairs for solar physics at some major universities were not refilled as they became vacant. This development continues to be of concern in view of the growing importance to society of understanding the Sun in the context of space weather and climate change. It also contrasts starkly with the scientific opportunities and the apparent strong interest of the many excellent young researchers who work in the field supported by soft money. The reasons for this development lie partly in the shift in emphasis from solar (and stellar) physics in astronomy to solar physics in the geophysical context. Neither the funding agencies nor the universities have yet been able to address the challenge posed by the changing astrogeophysical framework for solar physics. The panel urges them to do so.

FUNDING ASPECTS

Appendix G of the Parker committee report documents the balance of funding between NSF grants to universities and funds spent on NSF-supported centers (NSO and HAO). In solar physics research, about 63 percent of NSF funding goes to grants and 37 percent to centers. This is very similar to the balance between NCAR (the largest NSF center) and grants in atmospheric science. The panel regards the balance of funds in solar support by NSF as healthy. The overall demographics in the solar community should be adequate to support and fully exploit the missions and programs planned for the next decade, except that more faculty positions are needed. If NASA provides funding for strong guest investigator programs and NSF provides funds sufficient to run and exploit new ground-based observing capabilities, there will be new incentives for universities to hire faculty.

THE NATIONAL SOLAR OBSERVATORY

The panel endorses the recommendation of the Astronomy and Astrophysics Survey Committee's Panel on Education and Public Policy that the NSO should be separated from the nighttime parts of the National Optical Astronomy Observatories as soon as is reasonable. This would allow the best possible posture for NSO and the solar community to advocate and develop the AST, which should be (and is becoming) the primary future focus for NSO. NSO should then establish structures that ensure broad community participation in preparing and building the AST. The panel also recommends that postfocus instrumentation for the AST be developed in collaboration with the community, with instrument packages outsourced but developed under overall guidance from a central authority for AST. In addition, the panel sees many advantages to having international partners in the AST project, provided adequate control remains with the United States.

EDUCATION

For the broader educational outreach aspects the panel refers the reader to Chapters 4 and 5 of the survey committee report. Solar physics can contribute considerably to the educational effort in astronomy. In particular, the highly dynamical nature of solar processes—like CMEs, which can be observed with high time and spatial resolution—make solar observations an attractive means of portraying astrophysical concepts to the public. There is a need to educate the growing community of space weather forecasters in solar physics as well as decision makers in government and in industry.

ACRONYMS AND ABBREVIATIONS

1HT—One Hectare Telescope
ACOS—Advanced Coronal Observing System
AIA—Atmospheric Imaging Assembly
AO—adaptive optics
ASP—Advanced Stokes Polarimeter
AST—Advanced Solar Telescope
AU—astronomical unit
CCD—charge-coupled device
CDR—concept and design review
CEDAR—Couplings, Energetics and Dynamics of Atmospheric Region, a part of the NSF solar influences program
CIA—Coronal Imaging Assembly (on SDO)
CME—coronal mass ejection
EIT—Extreme Imaging Telescope (part of SOHO)
ESA—European Space Agency
EUV—extreme ultraviolet
FASR—Frequency-Agile Solar Radiotelescope
FIP—first-ionization potential
FOV—field of view
FPIP—focal-plane instrumentation packages
GEM—Geospace Environment Modeling, a part of the NSF solar influences program
GOES—Geostationary Operational Environmental Satellite, a series of meteorology satellites
GONG—Global Oscillations Network Group
HAO—High Altitude Observatory
HESSI—High Energy Solar Spectroscopic Imager
HMI—Helioseismic and Magnetogram Imager (on SDO)
IF—intermediate frequency produced when a radio receiver mixes the input signal with a local oscillator
IR—infrared
ISAS—Institute of Space and Astronautical Sciences (Japan)

ISOON—Improved Solar Observing Optical Network
KPVT—Kitt Peak Vacuum Telescope
LASCO—Large Angle and Spectrometric Coronagraph Experiment
LOFAR—Low-Frequency Array
LOWL/ECHO—Low-degree/Experiment for Coordinated Helioseismic
    Observations
MCAO—multiconjugate adaptive optics
MDI—Michelson Doppler Imager, an instrument on SOHO
MHD—magnetohydrodynamics
MPT—McMath-Pierce Telescope (on Kitt Peak)
MRX—Magnetic Reconnection Experiment (at Princeton University)
NASA—National Aeronautics and Space Administration
NCAR—National Center for Atmospheric Research
NJIT—New Jersey Institute of Technology
NOAA—National Oceanic and Atmospheric Administration
NRL—Naval Research Laboratory
NSF—National Science Foundation
NSO—National Solar Observatory
OPACITY—A solar physics project conducted by the Institute of Physics,
    Bristol
OSL—Orbiting Solar Laboratory (never-built NASA project)
OVRO—Owens Valley Radio Observatory
PASO—Particle Acceleration Solar Orbiter
PSPT—Precision Solar Photometric Telescope
RAM—Reconnection and Microscale Probe
RISE (SunRISE)—Radiative Inputs of the Sun to Earth, a part of NSF
    solar influences program
SAIC—Science Applications International Corporation
SDAC—Solar Data Analysis Center
SDO—Solar Dynamics Observer
SETI—search for extraterrestrial intelligence
SMEX—Small Explorer (NASA)
SMI—Solar Magnetism Initiative
SOHO—Solar and Heliospheric Observatory
Solar-B—NASA mission to measure magnetic field and luminosity of the
    Sun
SOLIS—Synoptic Optical Long-term Investigation of the Sun
SOON—Solar Observing Optical Network
SOT—Solar Orbiting Telescope (never-built NASA project)
SRBL—Solar Radio Burst Locator

STEREO—Solar-Terrestrial Relations Observatory
STP—Solar-Terrestrial Probe
THEMIS—Heliographic Telescope for the Study of Magnetism and
    Instabilities on the Sun (French/Italian project)
TRACE—Transition Region and Coronal Explorer
UV—ultraviolet
WIRE—Wide Field Infrared Explorer

6

Report of the Panel on Theory, Computation, and Data Exploration

SUMMARY

The Panel on Theory, Computation, and Data Exploration was charged with surveying two separate branches of astronomy and astrophysics: theoretical astrophysics and data exploration. "Theoretical astrophysics," in the convention of this report, includes both analytic theory and numerical simulation. The term "data exploration" is introduced to describe the newly emerged discipline of mining insight from large and complex astronomical databases using sophisticated modeling tools. The panel reviews the status of these two branches of astronomy and astrophysics and provides separate sets of recommendations for prioritized initiatives and policy directives.

THE SCOPE OF THEORETICAL ASTROPHYSICS

Unlike many astronomy communities, which identify themselves by wavelength or mission, the community of theoretical and computational astrophysicists defines itself by synthetic tasks that cross disciplinary boundaries:

- *Defining the frontier.* The community invents concepts that create frontiers and a framework for observational discovery—new ideas about the universe—from the extremes of space-time to physics in exotic environments to the new universe of captured digital knowledge.
- *Model building.* It creates intelligible descriptions of physical systems with precise quantitative connections to reality, including both sophisticated simulations that incorporate a comprehensive range of physical processes and compact mathematical constructions that identify and represent the key physical effects.
- *Synthesizing a world view.* It knits physical science into a coherent narrative of our place in the universe, one that is accessible, interesting, and edifying to society at large. This scientific view of the universe competes in the free market of ideas; theorists tell and sell the astronomers' story of the universe.

In these tasks, theoretical and computational astrophysicists combine leadership for and service to the larger astronomical community. Their specific activities are defined by both their core intellectual values and their interactions and collaborations outside their community. Three important themes recur in this report:

- The rapid pace of discovery in this golden age of observational astronomy has created for the first time a comprehensive view of what is happening in the universe over much of observable space-time and is quickly expanding the complexity and range of accessible phenomena.

- The continuing advance in digital technology is redefining and expanding the character of knowledge. In astronomy, the digital revolution is creating explosive growth in the quality and quantity of data and our ability to model complex phenomena, resulting in a new "digital universe."

- Paradoxically, the cultural gap between the frontier science community and much of the rest of American society continues to widen at precisely the time that new technologies allow us to create tools for broader and more rapid dissemination of knowledge outside the science community. We must work to resolve this paradox.

The panel believes that theory defines the context within which most of astronomy operates. Observers may answer the questions what and where, but theory addresses the how and why and seeks explanation and synthesis. No modern observation would make sense, or could be properly interpreted, without the pioneering theoretical work that gave us white dwarfs, neutron stars, black holes, atomic physics, relativity, radiative transfer, hydrodynamics, mechanics, statistical physics, high-energy radiation processes, and so on, and without the theorists who are the developers and users of these concepts and tools. One need only think of Penzias and Wilson without Gamow, Dicke, and Peebles or Bell and Hewish without Gold and Wheeler to understand the centrality of fundamental theory.

Given this history, the panel asserts that the theoretical core of astronomy must be nurtured and strengthened in the next decade in order to optimize the scientific return from the coming explosion of astronomical discoveries.

THEORY INITIATIVES PROPOSED BY THIS PANEL

The theoretical foundation of any discipline is most secure when the broadest possible range of research is supported. The panel proposes three new initiatives for theoretical astrophysics, believing that they will bring significant, tangible benefits to the entire astronomy community. One of these initiatives advances a new model for supporting theory efforts aligned with observational missions. The other two initiatives are designed to enhance the current broadly based theory programs that support the inspirational theoretical research that complements mission-oriented research. The panel summarizes its initiatives and then supplies details of each in a later section.

THEORY CHALLENGES TIED TO PROJECTS AND MISSIONS

The most important output of the Astronomy and Astrophysics Survey Committee is its prioritized list of facilities and missions for the next decade. The panel proposes that most of these prioritized initiatives should be accompanied by, and continuously interact with, one or more coordinated theory challenges. The challenges should describe theoretical problems that are ripe for progress and either relevant to the planning and design of the mission or key to the interpretation and understanding of its results in the broadest context.

The theory challenges should be planned and budgeted as an integral part of the project or mission. The funds should be allocated through open competition in the national community rather than as add-ons to observational or instrumentation grants or contracts, and under no circumstances should they divert funds from existing grants programs for broadly based theory. Both individuals and consortia should be supported. Panels drawn from the theoretical community, broadly construed, should select the award recipients. Support should cover the broadest possible range of theory, from the basic theoretical foundations for the mission to detailed modeling.

Appropriate challenges will evolve during the life cycle of the project or mission. In the early stages, the challenges contribute to mission definition, identify opportunities, and add enthusiasm and ideas. As the data flow in, theory contributes to the interpretation of the results and sets the context for subsequent initiatives. At the end, theory helps produce a synthesis of the results.

The cost of theory challenges might typically be 2 to 3 percent of the project or mission cost, although the scope of the challenges should be determined individually for each project, and much larger fractions—or no theory challenges at all—could be appropriate for some projects. The panel believes that this small cost will be repaid many times over by the contributions of theorists to mission design, analysis of newfound phenomena, and the vision that inspires future missions.

A National Postdoctoral Program in Theoretical Astrophysics

All of astronomy and astrophysics thrives under constant reinvigoration by young researchers. Postdoctoral fellows fill a unique role as innovators, owing to their combination of scientific independence, freedom from administrative and teaching duties, and ambition to establish a personal scientific identity. The presence of talented young people enhances research productivity and cross-field collaborations. Theoretical astrophysics, in particular, has seen its directions and technical methods driven largely by the efforts of postdoctoral fellows, and theory postdoctoral fellows do much of the highly innovative nonprogrammatic research that inspires new missions and research directions far into the future. Nor should we forget the important role of postdoctoral fellowships in training the astrophysicists of the future.

The present support mechanisms for theoretical postdocs are inadequate and unstable. Grants to individual theory researchers rarely are sufficient to fund a full-time postdoc. The few who are supported in this way are tied to a specific project, with limited freedom to pursue untested or potentially revolutionary ideas. A handful of U.S. institutions award fellowships that occasionally support theorists. However, several foreign institutions have large and strong theory postdoctoral programs that exceed in scope the programs available at almost any single U.S. institution. Some support for theorists has also been available from the Hubble, Compton, and Chandra fellowship programs; however, this support is programmatically selective and fragile.

To meet the need for a healthy postdoctoral research corps in the United States, the panel proposes a national program of freestanding postdoctoral fellowships. The panel envisions the program being administered in much the same way as the successful Hubble postdoctoral program, with postdocs distributed at institutions throughout the country and selected through competitive review. Such a program will provide an indispensable base for fundamental, creative theoretical research; it will identify the most outstanding young theorists and foster their development in a cost-effective way; it can encourage ethnic and gender diversity; and it will enhance the vitality of research at universities across the country and the talent pool available for this research.

The panel recommends a program that would award 10 or so 3-year theory fellowships a year, at an annual cost of about $2 million.

RIGHT-SIZING THEORY SUPPORT

Federal support for theoretical research is central to the continuing health of astronomy. However, despite the best efforts at the National Science Foundation (NSF) and the National Aeronautics and Space Administration (NASA), support for theory has not kept pace with the considerable growth in astronomical data in the last 10 years, creating an imbalance in the funding profile at a time of exceptionally rapid discovery.

To help remedy this imbalance, the panel recommends that major observational facilities, projects, and missions share the responsibility of funding both mission-related and broadly based theoretical research. Moreover, because the direct benefits of theoretical research are difficult to quantify, the funding agencies should develop guidelines for its support. The panel believes that a suitable preliminary guideline is that at least 30 percent of the costs for research personnel in grant programs, academic departments, and research institutes should normally be directed toward theoretical research activity.

The most cost-effective mechanism to address the challenge of right-sizing support for theory in this era of discovery is the targeted expansion of existing grants programs at the NSF and NASA that support broadly based theoretical research. In particular, the panel recommends a substantial augmentation of NASA's successful Astrophysics Theory Program (ATP). Enhanced support through such programs would complement the theory challenge program and the national theory postdoc program to establish a thriving and balanced research effort in theoretical astrophysics.

DATA EXPLORATION INITIATIVE PROPOSED BY THIS PANEL: THE NATIONAL VIRTUAL OBSERVATORY

Astronomy will experience a major paradigm shift in the next few years, driven by large, systematic sky surveys at multiple wavelengths. The panel believes that these digital archives will soon be the astronomical community's main avenue for accessing data. Systematic exploration and discovery in these databases will play a central role in the day-to-day research activities of most astronomers. This data avalanche—the flood of terabytes of data—is happening, whether or not we plan effectively for it. The first megasurveys are already in progress, including 2MASS, SDSS, and MACHO.

This transition is driven by advances in technology. The last decade witnessed a thousandfold increase in computer speed, a significant increase in detector size and performance, a dramatic decrease in the cost of computing and data storage capabilities, and widespread access to high-speed networks. Despite these advances, the environment to exploit these huge data sets does not exist today. In order to handle terabytes of data efficiently, one needs database engines with fast input/output speeds and advanced query engines that can access databases spread throughout the country. Existing analysis tools do not scale to terabyte data sets. In combination, these factors make a new initiative, the National Virtual Observatory (NVO), both feasible and compelling.

The NVO will help the astronomer preparing for the next observing run, the theorist analyzing large-scale structure, or the phenomenologist searching for extremely rare objects. The NVO will link the major astronomical data assets into an integrated—but virtual—system that enables a qualitatively new type of astronomical research: automated multiwavelength and multiple-epoch exploration and discovery among all known catalogued astronomical objects. The NVO will initially provide access to tens of terabytes of catalog and image data, growing to multiple petabytes by the end of the decade. It will influence all disciplines of astronomy and astrophysics, from x rays to optical and infrared through the radio wavelengths, and it will be essential for confronting sophisticated theories with observations.

The NVO is "national" because it serves the needs on a national scale; it is "virtual" because it supports observations on digital representations of the sky, and it is an "observatory" because it is a general-purpose facility, just like a traditional observatory. The four major elements of the NVO are (1) integration of major data archives, (2) advanced services for the astronomical community, (3) standards and tool development, and (4) education. The NVO will involve a coordinated—but distributed—effort of universities and national centers to develop an integrated data architecture for astronomical research. Such standards and coordination will play a key role in linking the multiple archives and service providers; without them, astronomy will be unable to exploit its data fully and efficiently. The NVO will be a powerful resource for public education and outreach at many levels. A digital representation of the sky, easily accessible via the Web, has the potential to excite the imagination of future scientists and allow them to participate in the astronomical discovery process. Educators will draw on the NVO to develop educational materials, and public institutions such as planetariums will use the NVO to develop presentations and displays.

The software and standards developed as the core of the NVO will be relevant to many other fields of research and should attract the attention of researchers in computer science, statistics, bioinformatics, earth sciences, and other fields. The NVO supports many of the goals of the recent federal Information Technology for the 21st Century Initiative (IT^2) and is particularly appropriate for multiagency funding.

The panel envisages that the management of the NVO will be similar to that of the NSF Science and Technology Centers or NASA's Astrobiology Institute. The NVO should be a multistaged effort, consisting of definition, demonstration, development, and deployment phases. Cost estimates for the NVO project are preliminary. Firm estimates will require a broad consensus on how the project is to be organized as well as on the scope and schedule of the project. Current projections for the definition and demonstration phase in years 1 to 3 are around $5 million total, with development scoped at $10 million total and deployment in years 3 to 5 at around $30 million total. Public outreach, grants to observers, and research in related areas of computer science and astronomy, including theory, could amount to an additional $15 million over 5 years.

SUMMARY OF PANEL FINDINGS AND RECOMMENDATIONS

The specific findings and recommendations of the Panel on Theory, Computation, and Data Exploration are summarized in this section. Details and supporting arguments are found, in each case, elsewhere in this report.

PROPOSED INITIATIVES

The panel's principal recommendations take the form of three initiatives in theoretical astrophysics and one in data exploration:

• The panel recommends that most prioritized projects or missions recommended by its parent committee, the Astronomy and Astrophysics Survey Committee, be accompanied by one or more coordinated theory challenges. The theory challenges should be budgeted and programmed as an integral part of the project or mission. However, the funds should be allocated through periodic peer-reviewed open competitions in the national community rather than as add-ons to project grants or contracts. Both individuals and consortia should be supported. Support should cover the broadest possible range of theory, from speculative scenario building to detailed modeling.

• The panel finds that present support mechanisms for theoretical postdoctoral fellows are inadequate and unstable, and it proposes a national program of "portable" theoretical postdoctoral fellowships. The program would be administered much like the successful Hubble postdoctoral program. Postdocs will be distributed at institutions throughout the country and selected through competitive review. The program will identify the most outstanding young theorists and foster their development in a cost-effective way. It will enhance the strength and vitality of university-based research and encourage ethnic and gender diversity.

• The panel believes that there is an imbalance between the considerable growth in astronomical data in the last 10 years and the support of theory that is essential to a proper understanding and context for that data. The panel recommends that steps be taken to correct this imbalance. Specifically, the panel recommends that at least 30 percent of the support for research personnel in grant programs, academic departments, and research institutes normally should be directed toward theoretical research activities; that major observational facilities, projects, and missions should fund both "harvest" and "seed-corn" theoretical research; and that funding agencies should develop overall guidelines for right-sizing their levels of theory support. A high priority should be given to the expansion of NASA's Astrophysics Theory Program.

• The panel recommends the creation of a National Virtual Observatory to accomplish the integration of the nation's priceless astronomical data, to facilitate its archiving, and to lead in the development and dissemination of tools for new scientific discovery from archival data. The NVO will undertake activities in standards, archive services, basic analysis tools, and advanced analysis tools. In toto, these efforts will require resources comparable to those of a small satellite mission. Very roughly, a 3-year definition and prototyping stage will be budgeted at $5 million. Development and testing would pick up in the second year and cost $10 million over 4 years. Deployment and operations would ramp up from the second through fourth years, with a 5-year cost of $30 million. Public outreach, grants to observers, and research in related areas of computer science and astronomy, including theory, would be budgeted at $15 million over 5 years. Unless specifically rechartered, the project would terminate or descope after 5 years of operation.

OTHER RECOMMENDATIONS

• *DOE support of theoretical astrophysics.* While the survey committee is not specifically tasked to examine Department of Energy (DOE) support of astrophysics, it would be impossible to give a complete picture of U.S. theoretical astrophysics without doing so to some extent. The panel offers the following suggestions for optimizing the effectiveness of DOE's contribution to astrophysics:

— DOE's Office of Science should leverage its efforts in low-energy nuclear physics with appropriate research in nuclear astrophysics; a particular example is radioactive ion beam research.

— Research in cosmology and matter under extreme conditions is similarly relevant to major new DOE facilities such as the Relativistic Heavy Ion Collider (RHIC) and the Continuous Electron Beam Accelerator Facility (CEBAF), the Stanford Linear Accelerator Center's (SLAC's) B-factory, the European Laboratory for Particle Physics' Large Hadron Collider (LHC), and the Fermi National Accelerator Laboratory's (FNAL's) collider and fixed target programs.

— DOE's Defense Programs should recognize more explicitly the close synergy between their national security missions and research in astrophysics and (noting past high returns on investments made, particularly in terms of quality personnel brought into the defense program) should support with programmatic funds certain areas of relevant theoretical research in astrophysics. The ASCI program, in particular, is both the beneficiary and the benefactor of research in theoretical astrophysics, and this connection could be strengthened.

• *Institutes with opportunities for visiting theorists.* While conferences allow sharing of research results, they seldom provide opportunities for actual collaborative work. The panel endorses the support by funding agencies of the several institutes that provide opportunities for extended working visits by scientists away from their home departments. The panel recommends that such institutes continue to receive healthy support.

• *High-performance computing.* The panel notes that several national initiatives in computing will allow for the participation of, and allocation of resources to, computational astrophysicists—but only if every effort is made within NASA and NSF to be responsive to these initiatives at an agency level. The panel recommends that the funding agencies position themselves to benefit from such initiatives by supporting algorithm development, Grand Challenge applications, and the development and dissemination of community codes and by conveying at agency levels the message that they support the continued health of the national supercomputer centers.

DESCRIPTION OF THEORETICAL ASTROPHYSICS

The initiatives that the panel recommends represent only the part of theoretical astrophysics that readily matches the AASC process. The panel represents a much broader community than can be adequately served by "initiatives." More than any other area of astrophysics, theory depends on continuous healthy support of broadly based science rather than large discrete investments, although history has shown that theory has a remarkably strong influence on large scientific investments. To help the reader understand the field of theoretical astrophysics, the panel describes here its vision for the field, and in doing so looks at who theoretical astrophysicists are, where they have been, and where they are going.

THE NEW THEORIST

Traditionally, a theoretical astrophysicist studied astronomical phenomena primarily by analytical and conceptual (pencil-and-paper) reasoning. Today's theorist performs on a larger stage. In particular, the 1980s saw the emergence of numerical modeling and simulation as a distinct and powerful subdiscipline of theoretical astrophysics. In a wide variety of astrophysical problems—cosmological structure formation, globular-cluster evolution, star formation, supernova explosions, accretion disks and jets, galactic dynamics, formation and evolution of planetary systems—the synergy of numerical modeling and analytic theory led to far greater progress than either tool could achieve on its own. Numerical modeling has also blurred the distinction between theory and observation: to properly understand and model selection biases, experiments and observations are now often numerically simulated, and theoretical simulations are observed using the same acceptances as the actual observations.

Two developments in particular have driven this evolution. The first has been a tremendous expansion in both the quality and quantity of observational data as the result of the deployment of new facilities. For example, the angular resolution afforded by the Hubble Space Telescope (HST) and the exquisite sensitivity of instruments such as the High-Resolution Echelle Spectrometer on the Keck I telescope demand far more detail from theoretical modeling than was necessary in the past. New data sets are of sufficient size and complexity that new discoveries will require creative new strategies for organizing the data. Second, and equally important, has been the spread of computing resources of unprecedented power. Calculations that were recently impossible anywhere can now often be done on consumer-level computers. This expansion in computing speed has made it possible to construct physical models with sufficient detail to confront the new observational data. Astrophysicists are now able to test fundamental theories for a wide range of complex problems, from the growth of structure in the universe to the physical mechanisms of supernovae.

All areas of science rest on the three legs of theory, observation, and experiment (some colleagues regard numerical simulation as a fourth leg). The relative strength of these legs varies with time in any discipline: in the 19th century, experiment and theory in biology were clearly subordinate to observation, but such is no longer the case. The panel believes that the importance of the theory leg in astronomy and astrophysics is growing, not only because of the expansion of the scope of theoretical activities to include numerical modeling and data mining, but also because of the increasing sophistication of our conceptual understanding, our rapidly improving numerical algorithms, and the explosion in computer speed and memory.

Theory plays several distinct roles in astronomy research:

• Theory uses laboratory-tested physical laws to interpret and explain astronomical observations, providing a coherent and satisfying framework for the interpretation of diverse observational data and showing how they fit together to yield an integrated picture of the universe as a whole. The most exciting observations are not those that confirm past predictions but rather those that are surprising or apparently paradoxical—and astronomy has no shortage of these.

• Theory guides the choice of new observations and the development of new instruments and observational strategies. For example, the wide variety of ongoing and planned ground-based, balloon, and satellite experiments to measure fluctuations in the cosmic background radiation are a direct response to theoretical predictions that these fluctuations provide an exquisitely sensitive probe of cosmological parameters; another example is the immensely successful gravitational-microlensing experiments of the past decade, which were stimulated by the theoretical prediction of this effect.

- To a growing extent, physicists use astronomical observations and the associated theoretical calculations to probe fundamental physics in regimes that cannot be reached by traditional Earth-based experiments—for example, properties of neutrinos, including mass and lifetime; constraints to the existence of hypothetical particles including the axion, neutralino, and gravitino; and the behavior of matter at nuclear density and above.

The panel believes that disciplines without powerful theoretical underpinnings tend to become stagnant pools of undigested information.

SUCCESSES OF THE PREVIOUS DECADE

It is difficult to enumerate concisely the successes of theoretical astrophysics, given its breadth and diverse roles. Confirmation of explicitly predicted phenomena provides the most dramatic successes of theoretical astrophysics. Often, the observational verification comes decades after the theoretical prediction. Thus, we count an old prediction as a "recent" success of theoretical astrophysics if it was confirmed only in the past decade.

THEORETICAL PREDICTIONS VERIFIED BY OBSERVATION

Observation has confirmed the following theoretical predictions:

- *Blackbody spectrum of the cosmic microwave background.* The COBE-FIRAS measurement of the spectrum of the cosmic microwave background (CMB) confirmed that it was indeed the cosmological blackbody predicted nearly 50 years ago in the hot Big Bang model. The frequency spectrum would have been considerably different if the CMB were the integrated flux from a number of discrete unresolved objects.
- *Big Bang and stellar nucleosynthesis.* Increasingly precise measure-

ments of light-element abundances (particularly the deuterium abundance at high redshifts) are in excellent agreement with the predictions of Big Bang nucleosynthesis. The recently observed patterns of heavy-element abundances in halo stars agree with those expected from the r-process in the late stages of stellar nucleosynthesis.

- *Galactic black holes and quasars.* The prodigious luminosity of quasars was interpreted almost 30 years ago by theorists to be the accretion of material onto a black hole of mass 10^6 to 10^9 times that of the Sun. The inevitable consequence of this dramatic prediction—that many inactive nearby galaxies should host massive black holes, even if they are quiet from lack of matter to accrete ("dead quasars")—has been verified directly during the past decade with a wealth of kinematic evidence for the existence of compact dark objects in the centers of many galaxies, including our own.

- *Helioseismology.* Models of the solar interior predict the temperature, density, and pressure in the Sun on the basis of a handful of parameters and physics, under conditions far from those that can be achieved on Earth and far from those at the solar surface. Nevertheless, exquisitely sensitive probes of the solar interior from helioseismology have shown that the standard solar model predicts the sound speed throughout the solar interior to within 0.1 percent and the density to within 1 percent.

- *Neutrino astrophysics.* Neutrino astrophysics has now matured to the point where it provides unique and important information on neutrino properties. Measurement of the width of the Z^0 boson at the European Laboratory for Particle Physics confirmed the upper limit on the number of light-neutrino species from Big Bang nucleosynthesis. The success of solar models now implies that the solar-neutrino problem almost certainly points to new physics, probably in the form of nonzero neutrino mass.[1]

- *Gamma-ray burst afterglows.* Theorists realized that if gamma-ray bursts are at cosmological distances, then injection of such a huge amount of energy would drive a relativistic shock into the interstellar medium. The temporal and spectral behavior of the x-ray, optical, and radio afterglows observed provide excellent qualitative agreement with the predictions of such "fireball" models.

- *Star formation.* The theoretical paradigm for star formation has

---

[1]Recent detection of flavor-changed neutrinos from the Sun by the Sudbury Neutrino Observatory accounts for the missing neutrino flux and indeed implies a nonzero mass for neutrinos.

long involved infall to an intermediate state with a protostellar disk. Recent observations have verified the presence of such a disk, as well as the predicted inside-out dynamical infall pattern in the surrounding envelope.

OTHER SUCCESSES OF THEORY, COMPUTATION, AND DATA EXPLORATION

Many of the most important contributions of theoretical disciplines to astronomy and astrophysics cannot be characterized as simple theoretical prediction/observational confirmation stories. For example, modeling can make sense of preexisting observations that were at first incomprehensible. Although some discoveries are serendipitous, an increasingly large number of major breakthroughs in astronomy come from observations motivated by theory. Finally, during the past decade theorists have begun to exploit massive data sets to address fundamental issues in astronomy. Some examples of these successes follow:

• *Magnetohydrodynamic instabilities and accretion disks.* During the past decade, it was discovered that saturated magnetohydrodynamic (MHD) instabilities driven by shear in rotating flows provide the long-sought physical mechanism for driving outward angular-momentum transport and, hence, inward matter accretion in a wide variety of astrophysical disks. The identification of this linear instability was immediately followed by numerical simulations that verified the linear analysis and showed that a saturated state develops in which turbulence is steadily driven by the interaction of magnetic tension with orbital motion and in which the effective viscous stress scales with the magnetic pressure (see Figure 6.1).

• *Cosmological simulations and the origin of the Lyman-alpha forest.* In the last few years, the incorporation of hydrodynamic effects into cosmological simulations has invigorated the N-body simulations that preceded them and allowed for the first time detailed comparison between the simulations and the visible large-scale structure in the universe. The most dramatic achievement of these simulations is the striking similarity between simulations and observations of the Lyman-alpha forest of absorption lines produced by intergalactic gas clouds.

• *Mergers and galactic structure.* In the 1970s, theorists proposed that many features of peculiar galaxies were due to gravitational tidal forces acting between colliding galaxies. The roughly concurrent discov-

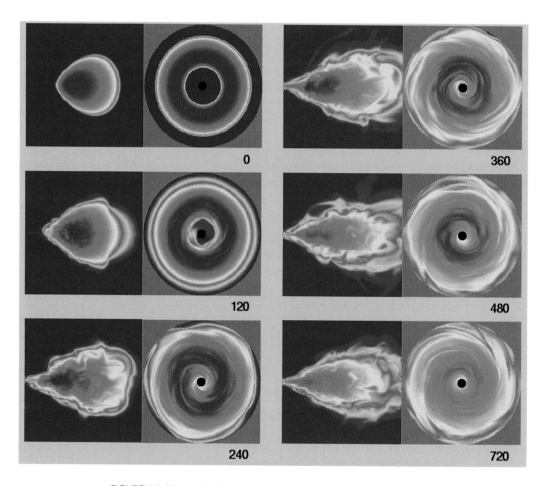

FIGURE 6.1 Magnetohydrodynamic simulation of an accretion torus surrounding a black hole in a galactic nucleus. Each snapshot shows the density distribution as a (one-sided) cross section and as a top view. The simulation shows that laminar flow in the torus is unstable and rapidly develops turbulence. Courtesy of J. Hawley, University of Virginia.

ery of flat rotation curves (that implied massive dark halos) and the observation that galaxies tend to congregate in clusters led theorists to surmise that many, if not all, peculiar galaxies could be attributed to mergers. Detailed numerical simulations of interacting galaxies, made possible by innovations in computer algorithms and the explosive growth of computing power during the past decade, now produce many features in remarkably good accord with those observed in interacting galaxies.

- *Cosmic microwave background anisotropy.* Following the spectacular success of COBE, theorists began to study what could be learned by mapping the CMB with better angular resolution and improved sensitivity. Theorists showed how a suite of intriguing structure-formation theories (e.g., inflation and topological-defect models) could be tested with the CMB. By additionally illustrating that some cosmological parameters could be determined precisely, they articulated clearly the science case for NASA's MAP mission and the European Space Agency's Planck mission, scheduled for launch in summer 2001 and 2007, respectively, and helped to define these missions (polarization optimization on Planck, number and position of frequency channels, etc.).

- *Gravitational microlensing.* The observation of gravitational microlensing provided dramatic confirmation of this half-century-old prediction of general relativity. The observed rate of microlensing toward the Galactic bulge is in rough agreement with what one would expect from models of the Galaxy and the stellar populations therein. Moreover, the success of microlensing surveys demonstrates the feasibility of astrophysics experiments that require the collection, search, and analysis of huge data sets.

- *Three-dimensional simulations.* The decade of the 1990s was a watershed for computational astrophysics because, for the first time, computer speed and memory allowed three-dimensional simulations to be performed with respectable resolution and physics. This capability enabled a number of impressive computational successes, including two mentioned above: the elucidation of the nonlinear behavior of MHD instabilities in accretion disks and the understanding of the origin of the Lyman-alpha forest (see Figure 6.2).

- *Algorithm development.* In the 1990s, powerful and efficient algorithms were developed for following the evolution of large N-body stellar systems, planetary systems, magnetohydrodynamics, reactive flow, radiation hydrodynamics, special and general relativistic hydrodynamics, and Einstein's field equations in four dimensions. The development of multiscale algorithms for astrophysical problems, such as tree codes and adaptive mesh refinement, enabled theorists for the first time to resolve a vast range of length scales at reasonable computational cost.

- *Practical applications of theoretical astrophysics.* In a surprising twist of technology, two of the most abstract realms of astrophysics theory, orbital dynamics and general relativity, have achieved prominent and ubiquitous practical application in the Global Positioning System, with diverse uses ranging from earthquake prediction to navigation to recreation.

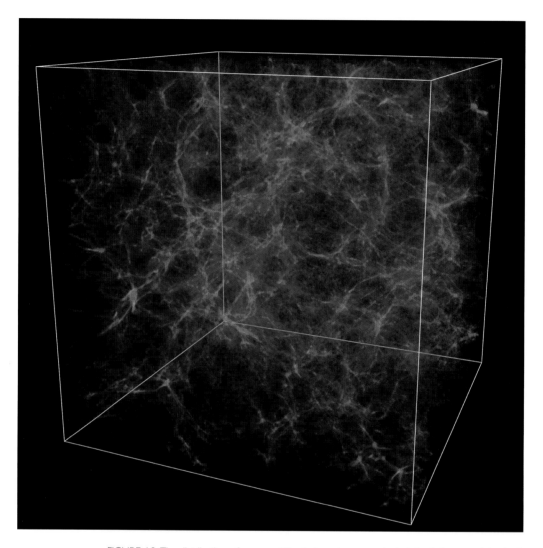

FIGURE 6.2 The distribution of x-ray-emitting hot gas in a cosmological simulation containing both dark matter and ordinary matter. Courtesy of G. Bryan and M. Norman, National Center for Supercomputing Applications, University of Illinois at Urbana-Champaign.

THEORY CHALLENGES TIED TO PRIORITY MISSIONS AND PROJECTS

INTRODUCTION

We are living in a golden age of astronomical discovery, unprecedented in its breadth across the electromagnetic and particle spectrum. Major advances—some serendipitous, others anticipated by theoretical work—seem to be reported almost weekly. These remarkable ongoing achievements in observational astronomy cannot be fully understood, nor can future directions in observational astronomy be intelligently planned, without commensurate advances in theoretical astrophysics. The panel believes that in times of rapid observational progress, theorists should be intimately involved in all stages of a mission or project, from the earliest planning to the ultimate interpretation of the scientific results.

To accomplish this, theorists should be involved in the agency, institutional, and collaborative apparatuses (committees, collaborations, teams, working groups, etc.) that direct each mission and project. Moreover, theoretical astrophysics should be encouraged to expand, especially in areas aligned with prospective observational projects. This encouragement should begin at the earliest possible stage of the mission planning and continue throughout the mission life cycle.

Diverse processes shape the intellectual progress of theoretical astrophysics. Sometimes progress is triggered by a single theoretical idea, sometimes it is a result of national programs focused on a carefully selected area that is ripe for theoretical advances, and sometimes it comes from an unexpected observational discovery. For example, our new understanding of the gravitational dynamics and state of primordial gas at significant redshifts (the problem of Lyman-alpha clouds) can be traced to the success of one of the projects in the NSF computational Grand Challenge program. This example (and there are others) shows that theoretical efforts of the highest quality can be elicited and focused by the conscious and thoughtful intervention of a funding agency.

The panel proposes that most new initiatives prioritized by the survey committee should be accompanied by, and continuously interact with, one or more coordinated theory challenges. The challenges should describe theoretical problems that are ripe for progress, relevant to the planning and design of the mission, and/or key to the interpretation and understanding of its results in the broadest context and should provide support for research on these problems over the broadest possible range, from speculative scenario-building to detailed modeling. Broadly based missions may have many suitable challenges, not all of which will be supported. The theory challenges should be budgeted and programmed as an integral part of the project rather than through existing grant programs. However, funds should be allocated through periodic peer-reviewed open competitions in the national theory community rather than as add-ons to observational or instrumentation grants or contracts. Under no circumstances should they be allowed to divert or diminish the funds available for broadly based theory. Both individuals and consortia should be supported as appropriate. theory challenges are important at all stages of a mission but are perhaps most critical in the early planning stages, when theorists contribute to mission definition and identify novel opportunities, and near the end, when theory synthesizes the results and helps to define the big questions that inspire future missions.

There is no simple and persuasive algorithm to determine the cost of appropriate theory challenges for any given project. The scope of the challenges should be determined individually for each project. A typical cost might be 2 to 3 percent of the project or mission cost, although much larger fractions—or no theory challenges at all—could be appropriate for some projects. A guideline for the minimum funding required for a worthwhile theory challenge is $2 million.

EXAMPLES OF THEORY CHALLENGES

The specific theory challenges tied to each mission and project should be determined by the informed astronomical community—probably through ad hoc panels drawn from the theory community and convened for this purpose. However, to illustrate the concept, it is worthwhile to suggest some possible challenges for two existing major projects (SIM and ALMA) and several of the recommended initiatives of the survey committee.

SIM

The Space Interferometry Mission (SIM) will determine the positions and distances of stars several hundred times more accurately than they are now known. SIM will determine the distances and velocities of stars throughout the Galaxy and in nearby galaxies and will probe nearby stars for evidence of planets. Following are examples of theory challenges that could be posed by SIM:

• *To understand what determines the phase-space distribution of stars, star clusters, satellite galaxies, and dark matter in giant spiral galaxies like our own.* The theoretical challenge raised by the SIM observations is to improve the resolution and fidelity of models of structure and galaxy formation to enable us to compare these models to SIM's survey of the Galaxy.

• *To establish what determines the frequency, size, and mass of planetary systems around solar-type stars.* The challenge posed by SIM, and later by the Terrestrial Planet Finder (TPF), is to develop our understanding of planet formation to explain how the orbital radii and masses of planets are determined.

ALMA

ALMA is a large millimeter array that will look with high sensitivity and angular resolution for molecular line and continuum emission from galaxies forming at large redshift and probe star-forming regions. A possible theory challenge for ALMA is *to formulate a comprehensive theory of global star formation, aiming to answer such questions as how the star formation rate and initial mass function depend on the properties of a galaxy.* This will require developing theories and algorithms to address such questions as the following: How do turbulence and magnetic fields affect the masses and clustering properties of the stars? How is the binary star or planet formation rate related to the properties of the parent cloud prior to its collapse and fragmentation?

NGST

The Next Generation Space Telescope (NGST), intended as an infrared successor to the HST, has a much larger aperture and is cooled to much lower temperatures. In the 1- to 5-μm infrared waveband, NGST would be over 1000 times more sensitive than any existing or planned facility, while retaining the image quality achieved by HST at visible wavelengths. Major theory challenges posed by NGST could include the following:

• *To develop an integrated theory of the formation and evolution of large-scale structure, Lyman-alpha clouds, galaxy clusters, and galaxies.*

• *To understand the formation of planetary systems in the context of star formation and protostellar disks.*

These tasks will involve radiation/hydrodynamic simulations on multiple scales, with detailed atomic physics.

CON-X

The Constellation-X Observatory (Con-X) will provide high-resolution x-ray spectroscopy over a wide bandwidth. It will analyze broadened iron emission lines from the region near the event horizon of supermassive black holes in active galactic nuclei, thereby probing the black hole masses and spins and testing strong-field general relativity. It will measure the properties of clusters of galaxies at high redshifts and thereby constrain models of early galaxy formation.

A possible theory challenge posed by Con-X is *to develop accurate, documented, general-purpose codes to model multidimensional, time-dependent radiation hydrodynamics in full general relativity.* High-spectral-resolution observations of the iron K-alpha emission line by Con-X should inform us about the properties of space-time near a spinning black hole, testing general relativity and constraining the mass and spin of the hole. The time is right to envisage constructing sophisticated and validated codes that could handle the relativistic MHD—including radiative transfer in a Kerr metric—needed to simulate the inner portions of these accretion disks (see Figure 6.3).

TPF

TPF is a near-infrared, space-based interferometer with a baseline of up to 1 km that will allow detailed imaging at 1 milliarcsec resolution with the sensitivity of NGST. TPF will conduct a complete census of planets as small as Earth around nearby stars. TPF will also allow us to image the centers of our own and nearby galaxies.

A possible theory challenge of TPF is *to understand the unique objects and processes that occur at the centers of galaxies, such as stellar collisions, tidal disruption of stars, supermassive black holes, accretion discs, and relativistic jets,* and to understand how the interplay of these objects leads to the complex phenomena of active galactic nuclei.

GSMT

The Giant Segmented Mirror Telescope (GSMT) is a giant (30 m)

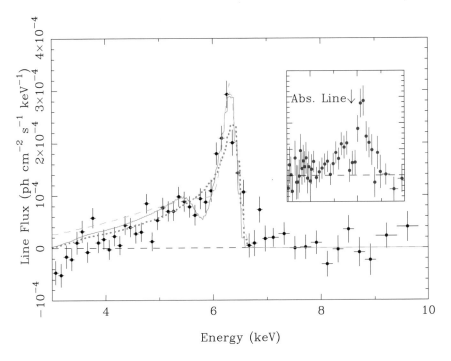

FIGURE 6.3 The broadened iron K-alpha line in the Seyfert 1 galaxy NGC3516. The solid green line shows the predicted spectrum from an accretion disk around a rotating black hole. Courtesy of P. Nandra, Goddard Space Flight Center.

optical/infrared (0.3 to 25 μm) telescope equipped with adaptive optics in order to achieve diffraction-limited resolution down to ~1 μm.

A possible theory challenge for GSMT is *to develop models of star and planet formation, concentrating on the long-term, dynamical coevolution of disks, infalling interstellar material, and outflowing winds and jets.*

The modeling of disks, which incorporates both (magnetized) gas and solid materials (dust and ices), is required to understand the sequence and details of disk dynamical processes, which high-resolution GSMT images and spectral studies may reveal for the first time.

Unresolved theoretical questions include the following: What small-scale processes drive protostellar disk accretion, and which conditions lead to episodic versus steady accretion? Is coagulation starting from micron-sized particles required to initiate planet formation, or does gravitational instability play a role? At what stage of evolution does fragmentation produce binary stars, how does accretion proceed afterwards, and how do final mass ratios depend on initial conditions? How do the orbits of planets and binaries evolve in the presence of disks of varying masses, with varying ratios of gas/solids and varying solid-body size distributions?

EVLA

The Expanded Very Large Array (EVLA) will achieve microjansky sensitivity and 10-milliarcsec angular resolution. It will produce images of high-redshift galaxies with sufficient detail to determine whether active galactic nucleus (AGN) activity associated with a supermassive black hole precedes, is contemporaneous with, or follows starbursts of the first generation of stars.

A possible theory challenge for the EVLA is *to understand from a theoretical perspective the respective roles of star formation and supermassive black holes in powering luminous galactic nuclei.*

LSST

The Large-Aperture Synoptic Survey Telescope (LSST) is a large ground-based telescope with a wide field of view, designed to map the entire accessible sky in a few nights. LSST will provide a unique new window on rare and transitory astronomical phenomena.

A possible theory challenge for LSST is *to understand the origin and fate of small bodies in the solar system, and to interpret them as fossils of solar system formation.* LSST will discover and follow vast numbers of Earth-crossing and main-belt asteroids, Centaurs, Kuiper Belt objects, and comets, enabling the development of large catalogs with accurate orbital elements and well-understood selection effects. These catalogs can be used in conjunction with long-term orbit integrations to address such issues as the extent and mass of the Oort comet cloud, the evolution of Jupiter-family comets, the impact hazard from long-period comets, and the population of inactive comets in the inner solar system.

LSST will open an enormous discovery space for rare, faint, or short-lived astronomical events. Possibly the most important theory challenge for this project could be *to discover, interpret, and explain unexpected rare phenomena buried in the LSST database.* Responding to this challenge will require both sophisticated algorithms for data mining and speculative model-building.

GLAST

The Gamma-ray Large Area Space Telescope (GLAST) is a high-energy gamma-ray observatory. It will probe the 20 MeV to 300 GeV region of the electromagnetic spectrum with far better sensitivity and angular and energy resolution than earlier missions. It will observe gamma rays from supermassive black holes, Galactic gamma-ray sources, and gamma-ray bursts, and it will explore the star-formation history of the universe and the nature of dark matter.

A possible theory challenge posed by GLAST is *to model the relativistic jets that emanate from the central black holes in active galaxies, to elucidate the radiation and acceleration mechanisms in jets.* This challenge will require modeling the inner parts of the accretion disk and the "launching" of the jets and explaining particle acceleration within relativistic shocks. This work will also benefit studies of galactic superluminal sources and gamma-ray bursts, where much of the same physics appears. Most jet phenomena vary in time, so that time-dependent hydrodynamic studies are also required.

LISA

The Laser Interferometer Space Antenna (LISA) is a spaceborne gravitational-wave detector aimed primarily at studying strong-field gravity, the coalescence of supermassive black holes, and Galactic binary-star systems, including those with black holes and neutron stars. A major theory challenge posed by LISA could be *to compute the expected gravitational waveforms from black-hole mergers.* This will require developing robust three-dimensional general relativistic codes with adaptive mesh refinement. Many of the required algorithms were developed by the NSF-funded Binary Black Hole Grand Challenge project in the second half of the 1990s.

SDO

The Solar Dynamics Observatory (SDO) will image the Sun in many wavelengths from the visible to the extreme ultraviolet. Combining spectroscopy and helioseismic inversions, SDO will measure flows and temperatures above and below the surface and observe photospheric vector magnetic fields with high resolution. The goal is to understand the origin and evolution of active regions and how they affect the structure, dynamics, and heating of the corona. Possible theory challenges include *modeling the interaction of turbulent convection and magnetic flux tubes and the interaction and reconnection of magnetic structures.*

AST

The Advanced Solar Telescope (AST) is a large-aperture, adaptive-optics, ground-based solar telescope designed to achieve high resolution in the infrared (0.1 arcsec resolves features on the Sun's surface of ~75 km). Its aim is to obtain observations at high spatial, temporal, and spectral resolution, in order to resolve the microstructure of solar magnetic fields, determine the evolution of magnetic structures, and clarify the basic physics of the solar magnetic activity cycle. It will advance our understanding of the dynamo origin of solar magnetic fields and their emergence through the solar surface, the recycling of magnetic flux, the relation of global to small-scale magnetic field organization, and the effect of magnetic fields on convective energy transport and its variation over the solar cycle. A possible theory challenge for AST is *to model the solar activity cycle.* This will require three-dimensional magnetohydrodynamic simulations spanning multiple length and time scales.

VERITAS

The Very Energetic Radiation Imaging Telescope Array System (VERITAS) will observe TeV gamma rays from the jets of active galactic nuclei (AGN) and possibly also gamma ray bursts (GRBs). These highly variable, energetic signals reflect violent processes occurring in the active inner regions of their distant sources.

A possible major theory challenge posed by VERITAS is *to understand the origin and characteristics of the energetic signals from AGN and GRB.* Although there are a number of models for high-energy emission from AGN and GRB, comparison of the predictions of the different models with present observations has not been able to identify either the type of primary particle that is accelerated (hadrons or electrons) or the acceleration mechanism (shocks, MHD processes, electric fields). A closely related challenge is to compute the expected very-high-energy spectra of AGN and GRB for a variety of acceleration mechanisms and particle types to try to identify unique observational signatures. Modeling the high-energy spectral shape and cut-off location will also be critical in differentiating intrinsic cutoffs and absorption effects from the expected external absorption due to pair production with cosmic infrared background radiation.

COMPUTATIONAL THREADS IN THEORY CHALLENGES

Several computational threads run through the theory challenges. The growth in computing power over the next decade will enable these challenges to be attacked with a far greater level of physical detail and realism than has hitherto been possible. By 2010, it is expected that the national supercomputing centers will provide access to individual systems capable of executing 100 Tflop (10^{14} floating point operations per second), and distributed aggregations of supercomputers approaching 1000 Tflop will be accessible from users' desktops in a seamless fashion. The desktops themselves will be in the 10 to 100 Gflop range—comparable to the supercomputers currently housed in NSF and NASA centers. Research groups will have at their disposal dedicated clusters with intermediate capabilities. Assuming a factor of 10^2 from the exponential growth of processor speed (Moore's law), a factor of 10 from parallelism (i.e., from $\sim 10^3$ to 10^4 processors), and a factor of 10 from improved algorithms, we can look forward to a factor of $\sim 10^4$ growth in computing power over the next decade, enabling us to attack the large range of spatial and temporal scales present in astrophysical systems with unprecedented success. Some of the areas that will benefit the most from these increases in computational power are discussed below:

• *Stellar magnetohydrodynamics*. Magnetic fields are ubiquitous in astrophysical systems. The Sun has shown us that unexpected phenomena occur. Current models of solar convection and magnetic turbulence solve MHD coupled to crude radiative transfer over a narrow range of length scales. In 2010, solar physicists will be able to simulate the entire convective layer in the Sun, for scales ranging from granulation to giant cells and for timescales approaching the solar cycle. Speedup by a factor of about 3 is needed for improved radiative transfer, by a factor of 3 for the application of adaptive grid techniques to handle the large range of spatial scales, and by a factor of 1000 for long temporal integrations. With such improvements, much-desired goals such as developing direct stellar dynamo models may finally be achieved. Codes of this type will also find applications to star formation, magnetic stars, and the Galactic dynamo.

- *Cosmological structure formation.* The twin requirements driving this field are the needs for a large spatial dynamic range and more realistic models of star formation. A spatial dynamic range of 10^5 is presently achievable with adaptive mesh refinement and tree codes. A factor of 10^4 hardware speedup, with balanced increases in memory and storage sizes, would allow increasing this number to $10^{6.5}$, sufficient to resolve the internal structure of forming galaxies within a cosmological volume. Codes of this kind will be able to study the entire history of cosmic structure formation from the first bound objects to large-scale structure.

- *Relativistic astrophysics.* Three-dimensional, fully general relativistic hydrodynamics is currently being done on $\sim 200^3$ grids to study neutron star mergers. A factor of 10^4 would permit the construction of a more general code for studies in relativistic astrophysics, including MHD, radiative transfer, and high-energy plasma processes. Such a code would have applications to black hole accretion, relativistic jets, and mergers of neutron stars and black holes. Long time integrations would be possible, permitting the study of topics such as AGN variability and the decay of binary orbits from gravitational radiation.

- *N-Body dynamics.* Current N-body modeling of the structure and evolution of star clusters employs particle numbers up to 10^4 or 10^5. While impressive, these calculations still fall far short of the actual number of stars in a globular cluster (up to 10^6) or galactic nucleus (up to 10^8). A factor of 10^4 increase in computing power would enable simulations of stellar systems in which stars are represented by computational particles on a one-to-one basis. Simulations of this magnitude are essential for understanding the dynamical processes driven by two-body relaxation that determine the evolution of these stellar systems.

- *The interstellar medium.* Because of its enormous range of physical conditions and spatial and temporal scales, the interstellar medium is one of the most complex astrophysical systems. Realistic models require incorporation of dynamically strong magnetic fields, nonequilibrium heating/cooling and ionization/recombination, optically thick radiative transfer, and dynamical representation of the cosmic-ray distribution. Using adaptive meshes, it is now possible to follow the collapse and fragmentation of a protostellar cloud in its early optically thin stages; computational power enhanced by a factor of 10^4 would enable the simulation of the subsequent evolution under the optically thick conditions that accompany disk and protostellar core formation.

THE NATIONAL VIRTUAL OBSERVATORY

MOTIVATION FOR THE NVO

THE NEW ASTRONOMY

A data revolution is occurring in astronomy. In the past, most astronomical data consisted of individual observations of a small sample of objects, usually over a narrow wavelength range. The Palomar Sky Survey plates served as a reference catalog of the sky, with visual inspection as the access method.

The situation is changing dramatically. New dedicated surveys, such as SDSS and 2MASS, are producing uniform, high-quality photometric and spectroscopic observations of millions of objects, and these observations will be available in digital form on the desktop of every astronomer. The science returns from these surveys will be enormous: from the discovery of large numbers of high-redshift quasars, to the characterization of the moderate redshift universe, to the structure of our own Galaxy, to definitive studies of low-mass stars and brown dwarfs.

The new surveys typically have 0.5 arcsec × 0.5 arcsec pixel size over the 40,000 deg^2 of the whole sky, resulting in 2 trillion pixels, or about 4 TB in each waveband. Ongoing surveys are currently generating tens of terabytes of data, and by observing many different epochs, the LSST will be generating petabytes by the end of the next decade. The volume of data from these surveys will be increasing every year, and there will be enormous incentive to integrate the separate archives into a seamless entity that allows true multiwavelength astronomy to be performed on entire classes of objects.

In addition, astronomers will need integrated access to the archives of individual observations from ground- and space-based observatories and seamless connection to information/metadata archives such as NASA's Astrophysics Data System (ADS), the NASA/Infrared Processing and Analysis Center Extragalactic Database (NED), and the Set of Identifications, Measurements, and Bibliography for Astronomical Data (SIMBAD), as well as legacy data archives such as the High Energy Astrophysics Science Archive Research Center. The panel recommends the creation of a National Virtual Observatory to accomplish the integration and facilitate the archiving of the nation's priceless astronomical data.

The NVO is only practical because of recent developments in database technology, high-speed networking, and storage systems. An NVO would be a true national observatory facility capable of supporting a large number and range of astronomical investigations initiated by individual researchers. It would give astronomers unprecedented multiwavelength access to large areas of the sky. However, unlike any current observatory, the NVO would be virtual, supporting astronomical observations and investigations via digital representations of the sky and associated electronic archives. It would be distributed in nature, using the next generation of high-speed networks for infrastructure.

The NVO will build upon and help integrate existing data and information projects such as NED and SIMBAD, as well as the various astronomical archives. One major goal of NVO is to discover new astronomical objects by comparing millions of objects at various wavelengths. If the new objects are galaxies, NVO would provide an interface to NED or its successor to record and catalogue these newly identified objects (i.e., produce metadata on the objects). Likewise, NED or SIMBAD could be used to generate a list of objects of a particular type. The astronomer could then use the data mining and cluster analysis tools of the NVO to identify new candidate objects of the same class.

SCIENCE RETURN

The NVO would enable a wide range of unique and important astronomical research: multiwavelength identification of large candidate samples of objects such as brown dwarfs, high-redshift quasars, gravitational lenses, and ultraluminous infrared galaxies; multiwavelength cross-identification of sources discovered in new surveys and observations; and searches for rare and exotic new objects among the billion or so catalogued sources. The NVO would be an important new tool for several main-line fields of astronomical research: large-scale structure, galaxy evolution, active galaxies, galaxy clusters, galactic structure, and stellar populations.

Most important, the NVO would enable science of a qualitatively different nature. The panel imagined exploiting the revolution in computing and networking to carry out a new and different type of astronomical research: multiwavelength exploration and discovery over the entire sky using all known catalogued astronomical objects and background radiation maps. The exploration process will include discovery and identification of unique astronomical objects through unusual colors;

discovery of patterns revealed from the analysis of statistically rich and unbiased image and catalogue data; and gaining of new insights into complex astrophysical systems through confrontation between sophisticated numerical simulations and the data. The discovery process will be accelerated through the application of advanced visualization, data mining, and statistical tools.

The NVO or some similar facility is indispensable for the full scientific utilization of wide-angle variability experiments such as those proposed for LSST. The NVO will be the keystone of the user interface for the LSST database, and in turn the LSST project will drive the scope and requirements for the NVO in the next decade.

Expected science returns include the following:

- *Discovery and identification of unusual astronomical objects.* Early examples of this type of science include detection of L and T dwarfs and high-redshift quasars. Rare objects occurring at a rate of 1 in 10 million will be detectable with the NVO in significant numbers. With automated discovery tools, astronomers will be able to search for exotic objects on an unprecedented scale. These will stand out by virtue of their unusual colors, variability, peculiar morphology, and/or unusual spatial coincidence.
- *Definitive population studies of galactic and extragalactic objects.* Unbiased surveys are essential for determining luminosity functions, mass functions, and evolutionary characteristics. Studies of large-scale structure, galactic structure, and galaxy evolution will all benefit significantly from the NVO.
- *Enabling new science.* The NVO will become the premier tool for designing observing programs for large ground- and space-based telescopes. The broad, multiwavelength coverage will allow efficient construction of target lists to be followed up by detailed spectroscopic and imaging observations.
- *Rapid turnaround in the discovery process.* Digital access to the whole sky at multiple wavelengths can dramatically compress the idea-observation-analysis-result-publication sequence. Researchers at smaller institutions will benefit in particular, since the NVO will provide equal access to all.

WHY NOW?

The time is right. All of the components—storage, computer, and networking technology—are here, and the large-format detectors are pouring out data, but these parts are not yet integrated into a coherent system. Similar integration efforts in other areas of science have not always worked well, but there are strong reasons to believe that astronomers can do better, partly because they (and space scientists) form a relatively small community. They have been able to create and adhere to standards in the past—a prominent example is the well-established FITS format, which cuts across many subdisciplines of astronomy. Perhaps most important, astronomers all have access to the same objects in the sky.

Major Aspects of the Virtual Observatory

The following sections contain a brief initial design for the organization, tasks, administrative structure, and budget for the NVO. All of these are expected to evolve rapidly over the next few years, due to technological and conceptual advances in computer science and database management, the increasing maturity of the NVO concept, and synergies with the LSST and other survey projects. The description in this report should be regarded as a snapshot of NVO development as of mid-1999 rather than as a rigid template for the future. In fact, the panel recommends that one of the first steps toward the NVO should be the establishment of a working group to define the tasks and organizational structure of the NVO more precisely.

The NVO would consist of the main repositories of astronomical data in the United States, together with significant computational resources, all connected by high-speed networks. It would include information services such as ADS and NED and compute servers such as those at the NSF and NASA supercomputing sites. It would be responsible for serving data sets to the whole community in a transparent and efficient manner. NVO participants would collaboratively manage the available resources and ensure that data sets were up to date and online. The NVO would also provide links to astronomy name servers and legacy data. Searches for astronomy sources would be executed across the data sets in a fashion transparent to the users.

The NVO would be based on four layers:

• *Standards*. This layer defines the means of communication between the layers, by establishing a set of interfaces and data exchange formats.

• *Archive services*. This layer contains the individual data and information archives, their connections to the virtual system, and their respective basic data delivery services.

• *Basic analysis tools*. This layer consists of the tools that integrate the archives and includes the basic cross-archive query tools, browsing tools, and basic statistical tools.

• *Advanced analysis tools*. This layer consists of the advanced visualization, classification, and data mining tools that will be used to exploit NVO data.

Standards

A critical and immediate activity of the NVO will be the development of standards for information exchange. Specific areas for standardization include those for metadata, metaservices, streaming formats, object relationships, and object attributes (e.g., position, flux, and band). These standards will allow individual data archives to develop services that integrate seamlessly with other components of the NVO.

Archive Services

Data Content

The essence of the NVO is the interconnection of distributed astronomical data sets, compute servers, and information providers. The NVO will link data from the major astronomical surveys, such as 2MASS, SDSS, DPOSS, Radio-FIRST, NVSS, MACHO, ROSAT, LSST, EXIST, and GALEX, into an integrated data system together with data from major ground- and space-based telescopes such as HST, Chandra, SIRTF, Gemini, VLA, and ALMA. The NVO will also provide access to information/metadata services such as NED, ADS, and SIMBAD. As an example, an astronomer interested in discovering new galaxy clusters will be able to ask for all compact associations of objects in SDSS that also show x-ray and radio emission from the cluster of objects. A subset of these that has ROSAT or Chandra imaging could be selected and a list of those with existing literature/NED/ADS references generated automatically.

The panel envisages that data would reside with the respective groups, which know their own data best. The groups would receive NSF or NASA support to maintain their own data. There is also a need to support continued access to important archival data sets after the active mission phase, as well as to support information/metadata services such as ADS, NED, and the Los Alamos e-print archive. The funding of the surveys and data/information archives is envisioned as separate from NVO as it is currently conceived. NVO primarily provides the infrastructure and services needed to access, cross-compare, and analyze data from multiple information sources. In this sense, NED, SIMBAD, 2MASS, SDSS, and so on are all seen as sources of data that the astronomer needs to access and manipulate in an automated fashion in order to deal efficiently with large numbers (millions) of astronomical objects.

An important feature of the NVO would be a program of grants to support data analysis and associated theory. Just as with other novel facilities, grants encourage the community to invest its energy in a new observational tool, facilitate the development of shared expertise, and ensure that the up-front investment in a new facility yields the maximum scientific return.

Interconnections

The interconnection of the different archive/service sites could be accomplished via one of the testbed programs connected to the next generation of high-speed networks. Extensive use of data mirroring and data caching would also be employed to allow efficient, high-speed access to the data sets. The NVO could thus become a prime example of creative network usage.

Query and Computation Support

Much of the general astronomy community will use the NVO only on a casual look-up basis and will utilize a www interface, generating a large number of rather simple queries. These can be easily supported via a central Web site with a modern, simple-to-use query engine. Intermediate users will want to access the archives in a more elaborate fashion, using a more advanced query engine. Their usage should be free, but limited in scope, in order to manage the available resources.

The greatest difficulty arises from the requirement to serve "power users," who will undertake multiple searches through terabytes of data, extracting hundreds of gigabytes for further processing. This task resembles accessing supercomputer resources and could be handled in a similar manner; that is, database server time would be allocated in a fashion analogous to supercomputer CPU time. Centers and individual researchers would get funding to support these database activities. As computing technology and networks evolve, more of the query engine activity could be accomplished on so-called commodity supercomputers such as Linux clusters.

TOOL DEVELOPMENT

Effective use of the NVO will require the development of basic analysis tools that provide the ability to look at data and perform queries across the different catalogs and to navigate through and manipulate terabyte data sets. Advanced tools will be needed to look at multidimensional objects and discover unexpected patterns in an automated fashion (data mining), as well as for visualization and classification, and numerical simulations will be needed to efficiently confront the large data sets.

OUTREACH

Education

The NVO will be a wonderful resource for education at all levels, including K-12. Planetariums and public science museums could utilize the resources of the NVO in exhibits and presentations. Through Web-based resources, the excitement of astronomy can be conveyed to every interested student on the planet. The panel also anticipates benefits for upper-level education in astronomy and other disciplines. The intellectual work of the NVO will involve astronomers, computer scientists, statisticians, and even mathematicians; it should enhance linkages between astronomy and these other disciplines and thereby broaden the range of training and career options for both undergraduate and graduate students.

Information Technology

The NVO must solve challenging problems using state-of-the-art techniques from computer science. The solutions will be applicable to other scientific disciplines and to business. Among the most critical problems are creating appropriate data structures, storing the data on physical media in a fashion that anticipates and adjusts dynamically to subsequent queries, and carrying out computations on multiterabyte databases. Implicitly, these require using huge input/output bandwidth and massive computational power; developing improved query engines that enable users to formulate sophisticated search and recognition queries efficiently; and developing advanced statistical and visualization tools for massive databases.

The technology to access and mine the data can best be developed by a wide collaboration that involves not only astronomers but also computer scientists, statisticians, and participants from industry. The NVO would be a very credible interdisciplinary project that could be paid for by funds for long-range information technology (IT) research.

PROJECT SCOPE, STRUCTURE, AND TIME LINE

PROJECT TASKS

The following components of NVO will require support:

- *System development.* Database technologies, query estimation and optimization, standards development, and network technology;
- *Maintenance of NVO-specific databases.* For example, cross-identification information and custom data subsets;
- *User tools and interfaces.* Applications environment and tool kit, query languages, statistics tool kit, and visualization tool kit;
- *User services.* Documentation, user feedback, and bug-fixing;
- *Advanced research.* Automatic search algorithms, data mining, statistics, and visualization; and
- *Operations.* Data and compute servers for NVO-related data and services.

PROJECT STRUCTURE

The panel suggests that the NVO program fund a range of research activities, including smaller grants to individuals and research groups as well as larger efforts at universities and national centers. In the panel's view, the variety of skills, their location in multiple institutions with different cultures, and the need for agile deployment of resources to match skills to new opportunities mean that NVO will need to have a *distributed structure* with the following components:

- A small core group having the skill mix to work with the community to plan and manage NVO activities. This includes a central management office;

• Project teams located at a number of institutions throughout the nation and the world, each charged with working on one or more of the key NVO project tasks; and

• Project teams at universities, the national data centers, and the national laboratories, each project being charged with working with the NVO to develop protocols, standards, and quality assurance for carrying out surveys and populating archives.

Useful models for distributed management structures spanning multiple institutions can be found in the NSF Science and Technology Centers or the NASA Astrobiology Institute.

PROJECT COSTS AND PHASING

To be successful, the NVO will require resources comparable to those of a small satellite mission. In this section, the panel makes a first attempt to scope an NVO effort. This estimate is limited to the scope and cost of the new elements of the NVO and does not attempt to summarize the total cost of maintaining the national astronomy data resources, which includes the individual surveys and data archives.

The development cycles for the NVO's four layers (standards, archive services, basic analysis tools, and advanced analysis tools) will go through four phases: (1) definition, (2) prototyping, (3) development and testing, and (4) deployment/operations. Initially, the system will include only survey data, with individual observations and information services being integrated at a later stage.

The panel estimates that the definition and prototyping phases will cost approximately $5 million total and be staged over the first 3 years of the program. Development and testing would begin in year 2 and continue through the program's end, with an estimated total cost of approximately $10 million. These two phases will require an estimated 80 person-years of effort. Deployment and operations would start during the latter half of year 2 and ramp up to full level during year 4. The cost is estimated at approximately $30 million total, assuming seven major operations sites are funded. Public outreach, grants to observers, and research in related areas of computer science and astronomy, including theory, could cost an additional $15 million over 5 years. It is important to recognize that the cost estimate primarily consists of funding for software development for the integration of existing data sets and does not include funding for the archives themselves. It is also important to recognize that LSST will be an essential component of the NVO and will influence its scope and cost. The data efforts within the LSST and the NVO should be closely coordinated so that astronomers will have optimal access and use LSST data when it comes online.

The panel anticipates that if NVO is successful as a managed development project, it will terminate after the 5-year period, leaving a legacy of standards and software that can be further developed by the community. The level and nature of continued operations for the resulting integrated astronomical data system should be assessed at the end of the 5-year period.

NATIONAL POSTDOCTORAL FELLOWSHIPS IN THEORETICAL ASTROPHYSICS

A strong postdoctoral program is essential for the long-term health of all areas of astronomy and astrophysics. The postdoctoral period is a critical and ubiquitous step in the development of an independent and creative researcher. Even more important, postdocs are mature and independent researchers who remain free from teaching and administrative duties, allowing them to play a central role in establishing new theoretical concepts or research thrusts as well as new observational programs and experiments. This role of postdocs is becoming more important as missions and facilities grow and, along with them, the administrative duties of senior scientists.

There are numerous cases where theory postdocs have made fundamental contributions to the development, execution, and analysis of experiments, observations, or missions. Theory postdocs are particularly effective because of their flexibility: they can change fields and focus their energy on the most exciting new results or areas more rapidly and effectively than either senior researchers or postdocs whose support is closely tied to data analysis. An equally important role of theory postdocs is to provide the visionary, speculative research that sets the direction of the field and enhances the discovery potential of future missions over timescales of decades.

Current support for postdoctoral fellows in astrophysics is inadequate. Postdoctoral researchers are the most expensive single line-items in most theory research proposals and hence are the first to be cut if funds are tight. More important, the few postdocs who *are* supported from research grants are usually tied to a specific project and have no (or only limited) freedom to pursue independent and creative research. A handful of institutions can support postdoctoral fellows in astronomy from internal endowed or operating funds, but the number of such postdocs is far too small to sustain the creative spark for the entire national astrophysics program. Many foreign institutions (e.g., Cambridge University, the Canadian Institute for Theoretical Astrophysics, and the Max-Planck-Institut) have stronger and more flexible postdoc programs than their U.S. counterparts of comparable size and reputation. This situation becomes more acute as the fraction of research support that is tied to specific observing or instrumentation projects grows.

The panel applauds the success of the Chandra, Compton, and Hubble postdoctoral fellowship programs. These fellowships are awarded to young researchers working on mission-related problems and can be held at institutions throughout the country. The competition is stiff, with the competitors generally regarded as among the most prestigious and desirable postdocs in the world. Indeed, there is a much higher success rate in placement for faculty jobs from the holders of these fellowships than from the postdoctoral pool at large. Though a significant fraction of these fellowships have been awarded to theorists, most have (quite properly) gone to observers.

The support for theorists from these important programs is fragile, and it is not enough. The Compton fellowships were funded out of Guest Investigator support and were phased out as the mission-operations and data analysis (MO&DA) dollars shrank for that mission. The Chandra fellows are funded in much the same way and will likely decrease as that mission ages. Selection committees for these fellowships have generally interpreted "mission-related" research broadly, which allows them to support theorists, but it cannot be assumed that this interpretation will persist over the next decade. Moreover, as emphasized above, the work of theorists whose research is not specifically related to current missions is likely to be crucial to the planning of future missions and is certainly crucial for the long-term vitality of astronomical research.

These considerations lead the panel to propose a national program of postdoctoral fellowships in theoretical astrophysics tied not to specific missions but rather to elucidating the long-term vision of astrophysical research. The program can be administered in much the same way as the Hubble postdoctoral program; that is, the postdocs will be selected through a competitive peer review and distributed at institutions throughout the country. Such a program has demographic and institutional benefits as well as scientific ones:

• Centralized selection and peer review would identify the most promising young theorists more accurately and with less administrative cost than a host of independent searches tied to individual grants; moreover, the process would allow the best young researchers to thrive in the environment of their choosing.
• The centralized selection of competitive fellowships could encourage ethnic and gender diversity, and the portability of the fellowships would benefit two-career couples.
• The presence of talented young people could dramatically enhance the research productivity and collaborations of others at the host institution; moreover, a national postdoc program could distribute this talent to a large number of institutions.
• In an era of increasing specialization, such fellowships would enable young researchers to shift their research area in response to new opportunities or to synthesize previously disjointed topics.

The panel recommends a program that would give out about ten 3-year theory fellowships a year, for a steady-state number of about 30 at any one time, managed with rules similar to those for the Hubble fellowship. The annual cost would be around $2 million. This program should not be seen as discouraging the Hubble, Compton, and Chandra programs from awarding fellowships to theorists in the respective subfields.

RIGHT-SIZING THEORY SUPPORT

One of the central issues faced by this panel and by the survey committee is how to determine the appropriate level of support for research in theoretical astrophysics. Prioritizing and budgeting support for theory research faces several challenges: innovative theoretical programs usually cannot promise to achieve a well-defined set of scientific goals within a fixed schedule and budget; when budgets are squeezed or capital projects have cost overruns, theory support is usually the most expendable budget item, and the full impact of theoretical research is often not visible until decades after the research is complete. A further complication is the need to distinguish two broad classes of theoretical research: (1) "harvest" research, which reaps the observational product of a mission or facility and is closely tied to data analysis, and (2) "seed-corn" research, which enables the community to develop new concepts and innovative strategies and targets for the next generation of major projects.

A variety of statistics can be used to indicate the impact of theoretical research within astronomy. In the past 25 years, theorists have comprised roughly 30 percent of the winners of the Henry Norris Russell Lectureship, the most prestigious award of the American Astronomical Society. Theorists have won 20 percent of the Hubble postdoctoral fellowships awarded since the program began (this figure would probably be higher except that fellowship holders must be doing mission-related research). Of the 50 astronomers ranked highest in the Institute for Scientific Information's list of most-cited physicists, 45 percent are theorists. At the top five U.S. universities in astronomy and astrophysics, as ranked by the 1995 Goldberger report, 40 percent of the faculty are theorists. The report *Federal Funding of Astronomical Research*[1] estimates that 24 percent of the members of the American Astronomical Society are theorists active in research.

All of these statistics suggest that roughly 35± 10 percent of the most influential and visible researchers in astronomy and astrophysics are theorists—at all levels from postdocs to senior faculty. This proportion is remarkably consistent with the proportion in the DOE's program in high-energy physics, which devotes roughly 30 to 40 percent of its university support to theoretical groups (as measured by number of Ph.D. researchers or number of students).

These arguments lead the panel to the following recommendations:

• At least 30 percent of the costs for research personnel in grant programs, academic departments, and research institutes should normally be directed at theoretical research activity.

• Major observational facilities, projects, and missions must share the responsibility for funding both harvest and seed-corn theoretical research.

• Because the direct benefits of theoretical research—particularly seed-corn research—are difficult to quantify, the funding agencies should develop guidelines for its support.

---

[1]Committee on Astronomy and Astrophysics, National Research Council. 2000. *Federal Funding of Astronomical Research* (Washington, D.C.: National Academy Press).

The NSF and NASA were important sources of support for astrophysics theory during the 1980s and 1990s, with the programs of the Division of Astronomical Sciences at the NSF and the Astrophysics Theory, Long-Term Space Astrophysics, and Supporting Research and Technology programs at NASA providing the bulk of the individual investigator grants. The introduction of NASA's Astrophysics Theory Program (ATP) in the 1980s was a major shot in the arm for theoretical astrophysics, so that the NSF and NASA programs now provide comparable support for theory. However, despite the best efforts of program officers and administrators at both NSF and NASA, support for theoretical research has not kept pace with the impressive growth in astronomical data over the last 10 years.

A few numbers taken from *Federal Funding of Astronomical Research* serve to illustrate the current problems. For the ATP, the oversubscription by number increased from ~3.0 in 1987 to ~4.8 in 1997; the oversubscription by funds was higher. The per-grant award (in 1997 dollars) decreased from a high of $190,000 in FY1987 to $85,000 in FY1997, reflecting the trend away from group grants and the attempt to maximize the number of principal investigators supported within a limited budget. The annual awards declined from a peak of $5.8 million (1997 dollars) in 1994 to $3.1 million in 1997. At the NSF, the oversubscription rate for individual proposals increased steadily over the past decade and has recently been ~5.0, while the average award has remained flat in constant dollars.

This unhappy situation is exacerbated by other circumstances. The best students are increasingly turning away from theory to observation, because that is where the money and the exciting new facilities lie. In addition, many senior theorists are joining guest observer programs to obtain funding. Although the panel applauds closer connections between theorists and observers, these should be motivated by science rather than funding imperatives. Whether or not theorists have obtained access to MO&DA money in this way, during budget squeezes support for theory is always the first to be cut.

The panel's concerns about theory support are echoed in *Federal Funding of Astronomical Research*, which concludes that theory and instrumentation are two specific areas in which "support has not been adequate to support the dramatic scientific discoveries of the last decade. . . . [T]his has severely limited the field's ability to understand and interpret the wealth of new data."

By far the most direct and cost-effective single contribution to right-sizing support for theory would be the long-overdue expansion of the individual investigator grants program at the NSF and NASA. In particular, the panel recommends a substantial augmentation of NASA's ATP. Such expansion would benefit broadly based theoretical research and thus would complement the more directed initiatives of the theory challenge program and the career-development initiative of the national postdoctoral fellowships to establish a balanced and thriving effort in theoretical astrophysics research. Not to address the growing crisis in theory research will jeopardize the future health of *both* observational astronomy and theoretical astrophysics. The astronomy community must decide whether theory will be given the modest resources it requires to flourish or whether through benign neglect and inflexibility it will be driven into decline.

As a path forward, the panel suggests that NSF's Division of Astronomical Sciences and NASA's Office of Space Science study the issue of right-sizing theory support, perhaps through the mechanism of the NRC's Committee on Astronomy and Astrophysics, and certainly in consultation with NSF's Division of Physics, which exemplifies the separate theory program.

INSTITUTIONAL ISSUES FOR THEORETICAL ASTROPHYSICS

UNIQUE ROLE FOR THE DEPARTMENT OF ENERGY

The DOE supports astrophysics and cosmology at universities and at its national laboratories in areas where there is intellectual overlap between astrophysics and the missions of the DOE. Much of this work involves theoretical and computational astrophysics; in addition, there is a vigorous and growing interest in astronomical data exploration at the national laboratories. The panel briefly highlights some of this work and then makes recommendations. DOE-supported university research programs in nuclear and particle physics often contain elements of nuclear or particle astrophysics. This work is frequently only part of the research effort of one scientist in a group; support for purely astrophysical research is uncommon. One outstanding historical example of DOE-sponsored research is the discovery of cosmological inflation.

Substantial astrophysics research programs are found at the national laboratories. There are programs at both the Office of Science laboratories (which are discussed first) and the Defense Programs laboratories (see below). At Fermi National Accelerator Laboratory (FNAL), there is an active theoretical astrophysics group. At Oak Ridge National Laboratory, there is extensive research in nuclear astrophysics, and much of the work on the development of new radioactive-ion-beam facilities is motivated theoretically by astrophysical considerations. Similarly, much of the motivation for the RHIC at Brookhaven National Laboratory and CEBAF at the Jefferson Laboratory is to study the quark-hadron phase transition in the early universe. Cosmology is also an important consideration in the experimental programs at SLAC's B-factory, the LHC, and FNAL's collider and fixed target programs (e.g., search for supersymmetry, neutrino mass, and charge-parity violation).

Basic research at the Defense Programs laboratories, especially at Los Alamos National Laboratory (LANL) and Lawrence Livermore National Laboratory (LLNL), is concentrated in areas that support the core national security mission of the labs. This has been interpreted to include a wide variety of astrophysical problems, including modeling the conditions interior to stars and supernovae, modeling explosive processes such as supernovae, and computing the synthetic spectra of high-temperature gases.

The national laboratories have been in the vanguard of large-scale data exploration. The MACHO project originated at LLNL and has now gathered and processed over 6 TB of data. The data processing for SDSS is performed at FNAL. Two projects designed to detect optical counterparts at gamma-ray bursts, ROTSE at LANL and LOTIS at LLNL, routinely image the entire night sky and have gathered several terabytes of image data. These various efforts are intellectually successful, but their visibility and impact could be greatly enhanced if the DOE were to recognize their value more explicitly and encourage research in astrophysics as a policy at DOE headquarters level. Specifically, the panel offers the following suggestions:

- The Office of Science should support research in nuclear astrophysics to complement its efforts in low-energy nuclear physics and, in particular, to support its new programs in radioactive-ion-beam research.
- The Office of Science should support research into cosmology, especially where relevant to its major new facilities (e.g., RHIC, CEBAF) or to theoretical work in particle physics.
- Defense Programs should recognize the close synergy between its national security missions and research in astrophysics and should

support with programmatic funds certain areas of basic theoretical research in astrophysics at Defense Programs laboratories. Examples include the physics and computational methods of stellar structure and evolution, the theory and computation of magnetic accretion disks and jets, the modeling of relativistic blast waves, and data-mining techniques for large astrophysical databases. The ASCI program, in particular, is both the beneficiary and the benefactor of research in theoretical astrophysics, and this connection could be strengthened.

Finally, the panel stresses what is perhaps the most important contribution of DOE to the day-to-day life of astronomers: the celebrated Los Alamos preprint server, which provides a fully automated e-print archive for astrophysics and many other subjects. The Los Alamos server now archives a significant fraction of all new astronomical literature, distributes new e-prints around the world less than 24 hours after submission, and has virtually eliminated the practice (and significant expense to funding agencies) of distributing paper preprints.

INSTITUTES FOR VISITING THEORISTS

There is no substitute for face-to-face interaction and collaboration. Even though physicists pioneered new technologies for electronic collaboration (such as preprint archives and browsing software), personal contact adds an important dimension to creative work. This is true in all of science but especially in conceptually challenging theory.

Astrophysics is a big subject covered by relatively few researchers. While faculty groups are the norm in many large disciplines of physics, astrophysical theorists are thinly spread among institutions. Conferences allow sharing research results but seldom allow for collaborative work or collective exploration of ideas. The panel therefore applauds the success of several institutes that provide opportunities for extended working visits by scientists away from their home departments, allowing them an opportunity to work closely with their collaborators and providing the infrastructure required for research.

One example is the Institute for Theoretical Physics (ITP) in Santa Barbara. It organizes topical programs lasting for several months and encourages and financially supports long visits. The science program is anchored by a strong core of long-term participants. Another example is the Aspen Center for Physics. The summer program is organized around a series of thematic workshops lasting several weeks, with 3-week-long visits encouraged. The science program is anchored by the members of the center, chosen from leaders in many areas of physics, and by workshop organizers, chosen for each summer's program. Housing subsidies are funded on a graded scale that strongly favors young scientists. Both institutes also support shorter conferences organized along the themes of their long-term programs. There are plenty of examples where subfields of astrophysics were given a significant boost by programs such as these; for example, the basic elements of the Cold Dark Matter paradigm for structure formation were substantially worked out during a program at ITP, which is generally acknowledged to have accelerated the development of the subject by at least 2 years.

The panel recommends that institutes such as these continue to receive healthy support.

HIGH-PERFORMANCE COMPUTING

What is in store for high-performance computing and communications in the first decade of the 21st century?

• *Continuation of exponential growth.* The National Technology Roadmap for Semiconductors of the Semiconductor Institute of America foresees that exponential growth in computing power will be sustained by current complementary metal oxide semiconductor (CMOS) technologies and manufacturing processes through 2006. Beyond that, new short-wavelength lithography techniques will be required, as well as overcoming formidable device design and packaging challenges. The industry has historically met such challenges, and the panel believes it will do so again.

• *March to the petaflop.* Spurred on by the DOE's Accelerated Strategic Computing Initiative (ASCI), the NSF supercomputing centers have deployed 1 Tflop computers and are pursuing aggressive upgrade paths to deploy 10 and 100 Tflop computers by 2003 and 2007, respectively. It is expected that NASA will follow suit. These levels of performance are achieved through massive parallelism ($\sim 10^3$ CPUs) and advances in commodity microprocessor architectures that should yield 1 Gflop chips early in this decade.

• *Affordable, ubiquitous computing.* For a few thousand dollars, every researcher will be able to afford desktop computers only about 1000 times less capable than the high-end supercomputers described above. This puts the supercomputer performance of the mid-1990s on everyone's desktop within this decade.

• *Proliferation of commodity PC clusters.* Research groups and departments are currently building high-performance clusters based on commodity PCs that rival the capabilities of high-end supercomputers for far less money. While these systems have distributed memories with slower interprocessor communications than tightly integrated supercomputers, their number, performance, and scientific impact will grow in this decade. A key policy issue will be striking the appropriate balance between federal investments in national supercomputing centers and in group and departmental resources.

• *Information grids, computational grids.* Both the NSF and NASA have programs to link nationally distributed high-value resources such as supercomputers, data archives, and scientific instruments via high-speed networks into what is referred to as a grid. The NSF Partnerships for Advanced Computational Infrastructure Program will probably be in place until 2007. NASA's Information Power Grid project has a more uncertain tenure. Computational astrophysics and remote and interactive observing are key application drivers.

• *Information technology.* The President's Information Technology Advisory Committee, recognizing that IT will be a key factor driving American progress and economic competitiveness in the 21st century, has recommended a broad-based, long-term program of basic research and development in IT across federal agencies, including NSF (lead agency), NASA, DOE, the Department of Defense, the National Institutes of Health, and the National Oceanic and Atmospheric Administration. One of the novel recommendations is the creation of interdisciplinary R&D virtual centers of computer and application scientists making bold assumptions about the future and then asking what information technologies will be required to get there.

How can astronomy and astrophysics take advantage of these technological and societal trends? Because the Information Technology for the 21st Century Initiative (IT^2) initiative is broad based, there is no guarantee that the programs that fund astronomers and astrophysicists will benefit. Therefore, every effort must be made to ensure that existing programs within NASA and NSF are responsive to this and similar follow-on initiatives. Areas that are particularly ripe for increased funding include but are not limited to the following:

- *Algorithmic development and Grand Challenge applications.* To take advantage of these computational advances, the panel recommends that funding be provided for both algorithmic development and scientific applications. This funding should support both small consortia and Grand Challenge efforts.
- *Development and dissemination of theoretical simulation software (community codes).* Such software should increasingly come to be viewed as a standard tool. Just as it has proven centrally useful to the astronomy community to have available standardized software such as FITS, IRAF, and AIPS++, so too would it be productive for theorists to have access to repositories of well-tested, flexible, expandable, and documented production codes. However, the best numerical codes that are the natural products of many modern astrophysical investigations generally are not made available to the broader community. The panel recommends support for code documentation and standardization to facilitate public dissemination of commonly used software.
- *Support for the national supercomputer centers.* The national centers provide a unique resource beyond the capabilities of any single institution for state-of-the-art calculations. The panel recommends that support for these centers be continued. The ASCI program, in particular, is both the beneficiary and benefactor of research in theoretical astrophysics, and this connection can and should be strengthened.

Although not strictly within its scope as a panel on theory, the panel nevertheless notes the importance of experimental efforts that verify and validate the underlying physics. The goal of simulation is reality, not virtual reality.

ACRONYMS AND ABBREVIATIONS

2MASS—Two Micron All Sky Survey
ADS—Astrophysics Data System (NASA)
AGN—active galactic nuclei
AIPS++—Astronomical Image Processing System, software used for image processing and data analysis
ALMA—Atacama Large Millimeter Array
ASCI—Accelerated Strategic Computing Initiative (DOE)
AST—Advanced Solar Telescope
ATP—Astrophysics Theory Program (NASA)

CEBAF—Continuous Electron Beam Accelerator Facility (DOE)

CMB—cosmic microwave background

CMOS—complementary metal oxide semiconductor

COBE—Cosmic Background Explorer, a NASA mission launched in 1989 to study the cosmic background radiation from the Big Bang

Con-X—Constellation-X Observatory

DOE—Department of Energy

DPOSS—Digitized Palomar Observatory Sky Survey

EVLA—Expanded Very Large Array

EXIST—Energetic X-ray Imaging Survey Telescope, to be attached to the ISS

FIRAS—Far Infrared Absolute Spectrophotometer, an instrument on COBE

FIRST—European Far Infrared Space Telescope

FITS—Flexible Image Transport System; format adopted by the astronomical community for data interchange and archival storage

FNAL—Fermi National Accelerator Laboratory

GALEX—Galaxy Evolution Explorer, a space ultraviolet imaging and spectroscopic mission

Gemini—an international project operating two 8.1-meter telescopes, one located on Mauna Kea and the other in Cerro Pachon, Chile

GLAST—Gamma-ray Large Area Space Telescope, a joint NASA-DOE mission

GRB—gamma-ray burst

GSMT—Giant Segmented Mirror Telescope, a 30-m-class, ground-based telescope

HST—Hubble Space Telescope, a 2.4-m-diameter space telescope designed to study visible, ultraviolet, and infrared radiation; the first of NASA's Great Observatories

IPAC—Infrared Processing and Analysis Center (NASA)

IR—infrared

IRAF—Image Reduction and Analysis Facility, a set of computer programs for working with astronomical images

ISS—International Space Station

IT—information technology

IT^2—Information Technology for the 21st Century Initiative (federal program)

ITP—Institute for Theoretical Physics (in Santa Barbara)

LANL—Los Alamos National Laboratory (DOE)

LHC—Large Hadron Collider (European Laboratory for Particle Physics)

LISA—Laser Interferometer Space Antenna

LLNL—Lawrence Livermore National Laboratory (DOE)

LOTIS—Livermore Optical Transient Imaging System, the primary purpose of which is to search for simultaneous optical counterparts of gamma-ray bursts

LSST—Large-aperture Synoptic Survey Telescope

MACHO—massive compact halo object

MAP—Microwave Anisotropy Probe mission

MHD—magnetohydrodynamic

MO&DA—mission operations and data analysis

NASA—National Aeronautics and Space Administration

NED—NASA/IPAC Extragalactic Database

NGST—Next Generation Space Telescope, an 8-m infrared space telescope

NSF—National Science Foundation

NVO—National Virtual Observatory, a virtual sky based on enormous data sets

NVSS—NRAO/VLA Sky Survey (uses the VLA in producing radio images)

R&D—research and development

RHIC—Relativistic Heavy Ion Collider (at DOE's Brookhaven National Laboratory)

ROSAT—Röntgen Satellite, an orbiting x-ray telescope launched in 1990 and named after the German scientist W. Röntgen, the discoverer of x rays; a German-U.S.-U.K. collaboration

ROTSE—Robotic Optical Transient Search Experiment, designed and operated by a collaboration of astrophysicists from the Los Alamos National Laboratory, Lawrence Livermore National Laboratory, and the University of Michigan

SDO—Solar Dynamics Observatory, a successor to the pathbreaking SOHO mission

SDSS—Sloan Digital Sky Survey

SIM—Space Interferometry Mission

SIMBAD—Set of Identifications, Measurements, and Bibliography for Astronomical Data, created and maintained by the Strasbourg Astronomical Data Center

SIRTF—Space Infrared Telescope Facility, NASA's fourth Great Observatory, which will study infrared radiation

SLAC—Stanford Linear Accelerator Center (DOE)

TPF—Terrestrial Planet Finder, a free-flying infrared interferometer designed to study terrestrial planets around nearby stars

VERITAS—Very Energetic Radiation Imaging Telescope Array System

VLA—Very Large Array, a radio interferometer in New Mexico consisting of 27 antennas spread over 35 km and operating with 0.1 arcsec resolution

7

Report of the Panel on Ultraviolet, Optical, and Infrared Astronomy from Space

SUMMARY

The following missions are the priority recommendations of the Astronomy and Astrophysics Survey Committee's Panel on Ultraviolet, Optical, and Infrared Astronomy from Space. All recommendations are a consensus of the panel.

MAJOR MISSIONS

When it prioritized major missions, the panel assumed that the Space Interferometry Mission (SIM), one of the initiatives recommended in the 1991 survey committee report,[1] will be flown and that the Hubble Space Telescope (HST) will operate until 2010.

NEXT GENERATION SPACE TELESCOPE

NGST, ranked by the panel as the top-priority major mission for the decade, will reveal the onset of star and galaxy formation in the early universe. Its combination of scientific breadth and depth make it a compelling successor to the Hubble Space Telescope. It is the first of two logical paths to improved image resolution and sensitivity in space: increase overall aperture size. It should be technologically ready to be launched before 2010.

The panel considered extensions of the core mission, currently 1 to 5 µm, and favors an extension to longer wavelengths, beyond 20 µm, for example, as scientifically more useful than extension to shorter wavelengths.

TERRESTRIAL PLANET FINDER

TPF was ranked as the second-priority major mission for the decade. Designed to observe directly Earth-sized planets near other stars, it is potentially the most scientifically exciting of all the major missions, depending on the breadth of its mission goals. It is the second logical path to improved image resolution and sensitivity in space: distributed aperture interferometry. Because TPF will depend on the successful

---

[1]Astronomy and Astrophysics Survey Committee, National Research Council. 1991. *The Decade of Discovery in Astronomy and Astrophysics* (Washington, D.C.: National Academy Press).

technology developed for both SIM and NGST, the panel saw it as being less technologically ready for the coming decade and therefore gave it lower priority.

MODERATE MISSIONS

SINGLE-APERTURE FAR INFRARED OBSERVATORY

SAFIR, the panel's top-priority moderate-size mission, is a large, filled-aperture telescope sensitive in the far infrared using NGST technology that will enable a distributed array in the decade 2010 to 2020. Such a telescope will be able to determine the total energy output from the first galaxies and will unambiguously determine the separate contributions of stars and accretion onto black holes to the total radiant energy. It will also be able to see through the opaque cores of molecular clouds that are creating new stars, thus providing a window on the first stages of star formation. An 8-m-class telescope will improve observational speed by factors of 10^4 to 10^5 over SIRTF, using the definition of "astronomical capability" in the 1991 survey committee report. The far infrared presently has enormous discovery potential. The single most important requirement is improved angular resolution. The logical build path is to develop a large, single-element (8-m-class) telescope leveraging NGST technology on time scales set by NGST's pace of development. A later generation of interferometric arrays of far-infrared telescopes could then be leveraged on SIM or TPF technologies on correspondingly longer time scales.

The panel considered SAFIR to be a moderate project and rated it as such. The survey committee placed it in the major category because its fairly uncertain costs fall close to the boundary separating moderate from major missions. The panel recognized that this mission could appropriately be included in the class of major missions in the survey committee report, and the committee decided to do so.

SPACE ULTRAVIOLET OBSERVATORY

SUVO, the panel's second-priority moderate-class mission, is a development effort leading to an 8-m-class ultraviolet/optical telescope in the decade starting with 2010. The main science goal is to map the distribution of matter between the galaxies by observing its absorption of light against distant quasars and to search for dark matter by obtaining gravitationally lensed images of galaxy clusters. It is thought that most of the baryons in the universe may reside in this intergalactic matter. Other core science includes studies of protostellar disks in Orion-like environments and starburst galaxies at $z < 1$. An 8-m-class UV/optical telescope will achieve gains of 100 to 1000 over the current capabilities of the Hubble Space Telescope, especially when combined with the next generation of energy-sensitive detectors for the UV/optical bands. Because the costs of a large-aperture UV/optical telescope are presently unknown, the recommendation is for technology development leading up to a new start at the end of this decade or later.

SMALL MISSIONS

ULTRALONG-DURATION BALLOON FLIGHTS

The panel's top priority for small missions is to include ULDB flights in the Explorer line so that they can compete for funding at least at the SMEX level—$75 million. Recent developments in ballooning have resulted in flights lasting on the order of 100 days. Atmospheric models show that the atmosphere is exceptionally transparent and stable at altitudes of 30 km and above. A whole class of missions, ranging from hard x rays to the submillimeter regime, could benefit from missions at cost levels of mid-Explorer-class projects and lower. The panel recommends that NASA allow long-duration ballooning projects to compete within the Explorer program with space missions for funding. With funding similar to that for a SMEX, it should be possible to improve the technology of balloons to make them attractive alternatives to spacecraft for some applications. Such suborbital flights also have great potential for training students who will one day become the technical force in space science.

LABORATORY ASTROPHYSICS

The panel recommends an expanded National Aeronautics and Space Administration (NASA) investment in laboratory research in several areas of astrophysics in support of space missions. The areas requiring fundamental studies and measurements include (1) spectroscopic properties of irradiation-processed ices, (2) the properties of refractory components of interstellar grains, (3) astrophysical and space plasmas, (4) the organic chemistry of the interstellar medium, (5) radiative and dielectronic recombination, and (6) collisional and photoionization studies, including measurements of iron cross sections for both processes. New space missions will explore wavelength regions and temperature regimes not previously observed and study interstellar dust and organic molecules with unprecedented sensitivity, requiring a broader suite of supporting laboratory investigations than have ever before been conducted.

TECHNOLOGY DEVELOPMENT

ENERGY-RESOLVING DETECTORS

The highest priority for technology development is detectors sensitive to ultraviolet/optical wavelengths. Energy-resolving detectors, both superconducting tunnel junctions (STJ) and transition-edge sensors (TES), will completely revolutionize observations at ultraviolet and optical wavelengths when fully developed. The potential is so enormous that it is important to invest heavily in making arrays of flight-qualified detectors with either of the two new technologies. Improving detectors will expand observing capability in the UV-optical range by large factors and is a much less expensive way to achieve gains than building larger telescopes with today's detectors.

FAR-INFRARED DETECTORS

State-of-the-art far-infrared detectors, as represented by the arrays to be flown on SIRTF, have background-limited sensitivity but are of modest format (32 × 32). If SAFIR and subsequent far-infrared missions are to achieve their full potential, NASA should support continued development of these photoconductor and bolometer array technologies to formats of at least 128 × 128. Because there are no commercial or defense-related efforts in this area, as there are at shorter wavelengths, it is essential that NASA provide the support. An investment of $10 million in the next decade would lead to array sizes better matched to the capabilities of future telescopes.

REFRIGERATORS

At the same time, it is important to develop a new class of refrigerators for these new detectors, allowing routine operation at millikelvin temperatures. The techniques for cooling are well known but require investment to produce reliable, long-lived devices for spacecraft.

SPACECRAFT COMMUNICATIONS

It is essential for NASA to improve the bandwidth of space telemetry systems, so that expensive missions will not be underutilized as a result of low data rates. The limit to science capability for many of the future missions will be the rate at which data can be transmitted to the ground. Modern detectors can produce data at rates far exceeding those at which data are transmitted through the Telemetry Data Relay Satellite System (TDRSS) or the Deep Space Network.

ULTRALIGHTWEIGHT OPTICS

Three of the major recommendations from this panel involve large-aperture space telescopes, for which very lightweight optics are highly desirable. The panel supports NASA's initiative to develop gossamer optics to enable the next generation of large telescopes for ultraviolet, optical, and infrared applications.

SCIENCE OPPORTUNITIES

Ultraviolet, optical, and infrared (UVOIR) astronomy, the region between about 0.1 and 1000 μm, is the largest source of information about the universe. Stars and planets emit most of their radiation in this energy range. The electronic transitions of molecules, atoms, and ions as well as the vibration-rotation transitions of molecules fall within the UVOIR; the range includes all chemistry important to life. Approximately 50 percent of the photon energy density in the Galaxy is starlight in the UVOIR; the rest is primarily in the cosmic background radiation. For these reasons, observations at these wavelengths dominate astronomical study and are likely to continue to do so.

UVOIR astronomy flourished in the decade 1990 to 1999. The Hubble Space Telescope was launched, repaired, and enhanced to produce images of unparalleled resolution and depth and spectra that are only now being rivaled by the new class of giant telescopes on the ground. Its public impact has been profound. Hubble alone has put astronomy much more in the minds of nonscientists than any other facility as the number of Hubble discoveries continues to capture the attention of the public. The Hubble Deep Field opened the way to study galaxies when the universe was a small fraction of its current age and focused the world's attention on astronomy. The Hubble ultraviolet spectrographs provided a new way to observe stars, planets, and galaxies. The first of the giant ground-based telescopes, the 10-m Keck and the VLT, went into operation and began to revolutionize ground-based astronomy. The Keck telescopes produced a number of exciting new discoveries, especially about the early universe, for which smaller telescopes cannot collect enough light for spectral analysis of the most distant objects.

Interferometers were developed and used for astrometry and imaging. Instruments were improved. Advances in the precision with which spectroscopic lines could be measured resulted in the first discoveries of planets outside the solar system. Hipparcos, the European astrometric survey satellite flown early in the decade, provided the most accurate map of stellar positions in history. The European Infrared Space Observatory (ISO), flown in mid-decade, returned a wealth of new data on the mid- and far-infrared bands. The Far Ultraviolet Spectroscopic Explorer (FUSE) was launched in 1999. The Space Infrared Telescope Facility (SIRTF) got a new start and will be launched early in this decade.

Research with these new facilities produced two scientific themes. One was to observe ever more distant objects in an attempt to discover the first stars and galaxies as they came into being after the Big Bang. This theme continues to dominate extragalactic astronomy. The primary motivation for constructing 8-m-class telescopes was to collect enough light to make it practical to take spectra of the faint galaxies and quasars at redshifts greater than 1. The second motivation was the growing recognition that planetary systems may be common in the Galaxy, culminating in the first, albeit indirect, discoveries of planets outside the solar system. The discoveries rekindled the age-old interest in the uniqueness of the solar system, an understanding of our origins, and the possibility that the detection of extraterrestrial life is within our reach.

These two themes dominate any discussion of scientific priorities for the next decade. They embody some of the oldest philosophical questions: How did the universe come into being? How did life arise? Are we alone? We are, indeed, fortunate to live at a time when our technology makes possible the answers to these questions, where we see a clear path to addressing these old questions directly with the scientific method. The priorities of the panel reflect the deep urge to understand our roots through measurement and observation, testing of theory, and discovery of those properties of the universe now beyond our imagination.

The most distant objects in the universe are typically less than an arcsecond in size. We need subarcsecond resolution simply to discern their shapes and structures. Even with adaptive optics and artificial guide stars, only a small fraction of the youngest galaxies can be observed with adequate resolution from ground-based telescopes now, and the challenge to provide good sky coverage with theoretical systems of the future is daunting. At redshifts exceeding 2, most of the observable energy of these objects emerges in the near-infrared part of the spectrum, at wavelengths beyond 1 μm. They are so faint that the nighttime sky is usually more than 10,000 times brighter than the objects, even at the best sites on Earth. And the most interesting parts of the spectra are often blocked by the Earth's atmosphere, as the important spectral lines shift from the nearly transparent visual window to wavelengths between 1.5 and 4 μm.

Extrasolar planetary systems pose an even greater challenge to observation. The light from planets either is emitted in the far-infrared parts of the spectrum, where detection from ambient-temperature telescopes on Earth is essentially impossible, or is reflected starlight, in which case the stars are more than a billion times brighter than the planets. Distortion by Earth's atmosphere smears the starlight, causing it to overwhelm the light from a planet in any ground-based observation. Even from space, the optics of a telescope create insurmountable limits to the observation of a faint planet next to a bright star unless the telescope is enormous—100 m in diameter, say—or the starlight is reduced by interference effects. The study of planet building through observations of the circumstellar disks from which planets are born is easier but nevertheless demands much higher angular resolution at infrared wavelengths than anyone has attempted.

Two scientific themes—discovering the formation of stars and galaxies after the Big Bang and searching for planets around other stars—challenge our technical prowess to develop the next generation of astronomical observatories. Many of the wavelengths covered by this panel are inaccessible from ground-based sites and can only be observed from space. Even at optical and near-infrared wavelengths, where the Earth's atmosphere is largely transparent, the enormous gains in resolution and sensitivity made possible by freedom from the distortion and background created by Earth's atmosphere mean that spacecraft are poised to dominate many observations. The large impact of the Hubble

Space Telescope with its aperture of modest size is a good example of how the advantages of space make up for disadvantages in size.

These missions are just beginning to tap the potential of space-based observations of the cosmos in the UVOIR bands. Scientifically, there are important areas not touched by the current generation of space satellites. Technology has now progressed to where gains of several orders of magnitude are possible in each of the areas covered by the current generation of missions. The Hubble Space Telescope sees deeply into the universe but not yet deeply enough to see the "edge of light," where the first stars and galaxies came into existence after the Big Bang. Both HST and FUSE are still inadequate to detect faint quasars needed to map out the intergalactic medium using observations of their absorption lines. For these tasks, more collecting area—a larger aperture—will be needed. Hipparcos measured positions of stars to about 1 milliarcsec, an impressive precision but still more than an order of magnitude shy of the level needed to detect planetary companions to nearby stars by measuring the stellar wobble. Improving on the precision of Hipparcos requires interferometers with baselines on the order of 10 m, many times larger than that of Hipparcos.

ISO and SIRTF use telescopes less than 1 m in diameter, meaning that source confusion sets the limit to the sensitivity throughout the far-infrared band. The high density of sources in the far infrared means that the greatest advances will come from increases in angular resolution or image sharpness. Gains in resolution imply larger structures, since all far-infrared telescopes have been limited entirely by the diffraction of the primary optics. Therefore, larger telescopes and eventually interferometers with telescopes spread across several hundred meters or more are needed to exploit the potential of this rich wavelength regime.

The recommendations of this panel are made so that known scientific problems of great significance can be addressed by the new facilities. But the history of astronomy suggests that the most important and interesting results will be new discoveries unanticipated by those who advance our observational capability. New discoveries often result when capability is increased by a factor of 10 in some dimension. They are common for improvements of 100-fold. Increases by a factor of 1000 or more virtually guarantee discoveries. The panel was especially enthusiastic about initiatives such as NGST that promise 1000-fold (or more) gains, since it believes that such gains are likely to revolutionize our knowledge of the universe.

There are, as a consequence, many opportunities for new UVOIR spacecraft. Some of these fall within NASA's Explorer (<$140 million) and Discovery (<$300 million) programs. These programs are peer-reviewed in regular calls for proposals, so the panel did not attempt to rank any projects that could be proposed as Explorer or Discovery missions, although there are some broad directions that it highlights for special attention. NASA selects its Grand Challenge missions, whose total costs typically exceed $1 billion, by means of a strategic planning process involving much of the astronomical community. The panel independently reviewed the missions in the current strategic plan and produced its own ranking. It also reviewed the complementary character, both scientifically and technologically, of the Grand Challenge missions—SIM, NGST, and TPF—and found that they constitute a compelling package. It discussed several possibilities for moderate missions that are highly desirable and are ranked by priority. Several small missions and technology investments received strong endorsement from the panel. In fact, modest investments in technology are likely to increase astronomical capability in several areas where alternative approaches are very expensive, so these relatively inexpensive recommendations are among the most important.

ASSUMED FACILITIES

THE HUBBLE SPACE TELESCOPE

The panel fully supports the recommendations relating to the Hubble Space Telescope contained in the 1996 report of the Dressler committee, *HST and Beyond*,[2] which emphasized the importance of an extended life for HST.[3] The report's rationale remains as apt today as when it was written.

The first 10 years of HST have been a remarkable period of advance in the science and practice of astronomy. The qualitative superiority of

---

[2]HST and Beyond Committee. 1996. *HST and Beyond* (Washington, D.C.: Association of Universities for Research in Astronomy, Inc.); also known as the Dressler report for the committee's chair, Alan Dressler.

[3]Information about HST can be found online at the Space Telescope Science Institute's Web site at <http://www.stsci.edu>.

Hubble's imaging and spectroscopy, combined with its sophisticated data management, has enabled astronomers worldwide to make observations that further reveal the remarkable complexity of the universe.

The panel assumes that the Hubble operations will extend until 2010, currently NASA's plan for the facility. It will be the only facility covering the ultraviolet portion of the spectrum, and its imaging power at all wavelengths will remain a necessary adjunct to the suite of facilities planned for the coming decade.

THE SPACE INTERFEROMETRY MISSION

The 1991 survey committee report, *The Decade of Discovery in Astronomy and Astrophysics*,[4] recommended an astrometric interferometry mission as a high priority. In line with that recommendation, a major initiative of the last decade was the development of the Space Interferometry Mission (SIM).[5] While the advanced state of SIM's development and an anticipated launch in 2005 precludes its inclusion among the new missions being prioritized here, its successful completion and execution will require an ongoing major investment by both NASA and the astronomical community throughout this decade. The panel believes that the present high level of commitment to the SIM mission is well merited, and it fully supports the launch of SIM.

The primary scientific objective of the SIM mission is ultrahigh-accuracy astrometry. The spacecraft is a Michelson interferometer with a 10-m baseline, operating in visible light. The performance goals are 4-μarcsec absolute accuracy anywhere on the sky and 1-μarcsec relative precision over a 1-deg field to a limiting visual magnitude of 20. These astrometric goals are approximately 200 times more accurate than the measurements of the Hipparcos mission, and the faint limit is nearly 1000 times more sensitive. It is important to combine high astrometric accuracy with deep sensitivity, for both are required to study the entire Galaxy and local group.

The additional mission goals—to demonstrate interferometric nulling

---

[4]Astronomy and Astrophysics Survey Committee, National Research Council. 1991. *The Decade of Discovery in Astronomy and Astrophysics* (Washington, D.C.: National Academy Press).

[5]Information on SIM can be found online at <http://sim.jpl.nasa.gov>.

over a dynamic range of 10^4 and synthesis imaging in space—are essential tests of the technology for imaging planets around nearby stars. SIM is a technology precursor to the Terrestrial Planet Finder. As the first long-baseline interferometer in space, SIM will be a testbed for precision deployment of distributed optics, optical control systems, laser metrology with relative precision in the tens of picometer range, vibration and thermal control, interferometric delay mechanisms, and synthesis imaging. These capabilities are essential to the TPF mission, and many also support key facets of the NGST.

The scientific capability of SIM is enormous. The measurement of distance is arguably the most fundamental and difficult measurement in astronomy. The SIM goals will provide distance measurements of 1 percent accuracy to distances of several kiloparsecs and of 10 percent accuracy throughout the Galaxy, providing a firm foundation for the understanding of stellar astrophysics. At the same time, luminosity determinations for key classes of stars—for example, Cepheid and RR Lyrae variables—will reduce the calibration uncertainties in the cosmological distance scale. Distances to a selected sample of stars throughout the Galaxy will refine our understanding of galactic structure, and in particular the structure of the Galactic halo, thus tracing the distribution of dark matter.

Searching for planets near stars in the solar neighborhood is the most ambitious of SIM's goals. SIM should generate a preliminary survey of the local planetary population and a more extensive survey of the Jovian-mass planets. By detecting the shifts in stellar positions, the orbital parameters for these planetary systems will be able to determine mass directly when combined with radial velocity techniques and will not leave any ambiguity about the masses of the extrasolar planets.

The majority of exciting astrophysical goals already pose a challenge under the floor requirements currently proposed: 3 µarcsec for narrow-angle astrometry and 10 µarcsec for wide-angle astrometry. Relaxing these floor requirements might eliminate large fractions of the important science. If SIM is unable to meet these requirements within its presently planned budget and schedule, it should be reevaluated by the scientific community to determine if it should remain a high priority as a major mission.

RECOMMENDED NEW INITIATIVES

MAJOR MISSIONS

NEXT GENERATION SPACE TELESCOPE

The parameters of NGST,[6] which is a passively cooled 8-m telescope in an orbit at L_2, are shown in Table 7.1.

Scientific Goals

The primary science problems to be addressed by NGST are the following:

• Detect the light from the first epoch of star formation in the universe and trace the evolution of galaxies from that epoch to the present time.
  • Determine the pattern of production of elements, beginning with the first generation of stars and leading to the current epoch, to understand the history of element creation.
  • Understand how stars and planets are born, from the creation of circumstellar disks, through the time of planet formation, ending in solar-system-like dust clouds and Kuiper Belts.
  • Exploit the enormous discovery potential associated with improving sensitivities in the infrared by several orders of magnitude and image sharpness by a factor of 10.

NGST will reveal the nature of the universe at high redshifts using a combination of multicolor deep surveys and spectroscopy of a representative sample of galaxies. Deep imaging surveys from 1 to ~10 μm will be able to discover galaxies at redshifts as high as 20, if they exist. Larger area surveys will discover enough galaxies to observe the development of structure and construct the history of star formation after the Big Bang. Spectra of these galaxies will determine their masses and internal dynamics to allow searching for giant black holes undergoing assembly.

---

[6]Information on NGST can be found online at <http://ngst.gsfc.nasa.gov> and <http://www.ngst.stsci.edu>.

TABLE 7.1 Parameters of NGST

Parameter	Planned value	Comments
Wavelength range	0.6-10 μm	Minimum: 1-5 μm
		Goal: 0.6-30 μm
Sensitivity	5.3 nJy (1.5 μm)	HST achieves 590 nJy
($\lambda/\Delta\lambda = 5$,	4.9 nJy (3.5 μm)	SIRTF will achieve 3000 nJy
5σ in 10^3 s)	0.2 μJy (8 μm)	SIRTF will achieve 25 μJy
	2.6 μJy (24 μm)	SIRTF will achieve 300 μJy
Angular resolution		
(2-μm, diffraction-	0.05″ (2 μm)	HST provides 0.2″
limited)	0.2″ (8 μm)	SIRTF will provide 1.9″
	0.6″ (24 μm)	SIRTF will provide 5.8″
Spectral resolution	$\lambda/\Delta\lambda = 5$ for imaging	HST provides $\lambda/\Delta\lambda \sim$ 5-50 over
Multiobject spectroscopy	$\lambda/\Delta\lambda = 100$-5000 for	1-2.5 μm
(MOS), integral field	MOS and/or IFS	SIRTF will provide $\lambda/\Delta\lambda \sim$ 600
spectroscopy (IFS)		over 5-40 μm
Temporal resolution	~1 s	
Field of view (FOV)	$> 4' \times 4'$	Telescope delivers a large FOV
		($>10'$); available telemetry rates
		limit the practical FOV
Lifetime	5 years	Goal: 10 years
Cost category	Major	NASA primary funding, ESA and
		CSA participation

NGST will routinely pick up supernovae at high redshift, which can then be used to determine the rate of expansion of the universe and its deceleration at early times and to study the evolution of the supernova rate (Figure 7.1). Spectra of the galaxies and supernovae trace the history of production of the heavy elements in the universe.

NGST's spectral imaging capability in the thermal infrared (5 to 30 μm) will make it possible to study how matter accretes around a star, becomes a disk, creates planets, and eventually disperses after the planet formation phase. Observations of stars at different ages, including older stars with "debris disks" akin to the remnant material in the solar system, will show how the disks evolve throughout their history. The high-resolution images should reveal gaps in the disks created by giant planets that have already assembled in the early history. The excellent sensitivity will suffice to study small objects in the solar system's own Kuiper Belt.

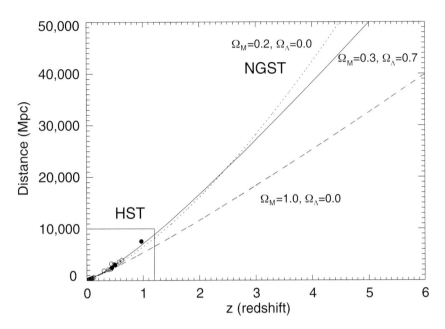

FIGURE 7.1 This diagram plots the distance to supernovae as a function of redshift, used to determine the geometry of the universe. The open circles were measured by ground-based observatories, and the filled circles are measurements with the Hubble Space Telescope. The box in the lower left-hand corner shows that the most distant supernovae that can be observed with HST are at $z = 1.2$. The curved lines show the expected relationship between distance and redshift for three different models of the universe: an open model with no cosmological constant (dotted line), a flat universe with some matter and a cosmological constant making up 70 percent of the energy density (solid line), and a flat universe dominated by matter (dashed line). Only NGST will be able to measure the magnitudes for rest-frame wavelengths and in the region beyond $z = 1.2$, where the models are clearly differentiated from one another. Courtesy of M. Livio, Space Telescope Science Institute (STScI).

By means of mid-infrared observations of the Kuiper Belt objects, from which distributions of sizes, albedos, and collision frequencies can be inferred, it will be possible to watch processes similar to those that created the planets several eons ago and check the theories of assembly.

NGST has superior sensitivity to large ground-based telescopes across its entire wavelength range. Its sensitivities will improve upon those of ground-based facilities by factors of up to 1000, meaning astronomical capability will increase by factors of up to 1 million. Its sensitivities will also exceed those of SIRTF at all operational wavelengths shorter than 30 μm, and it will have an order-of-magnitude better angular resolution.

This angular resolution is essential to resolve the most distant galaxies responsible for the far-infrared background (Figure 7.2). NGST will be the first facility to provide the sensitivity coupled with angular resolution commensurate with the sizes of high-redshift galaxies at wavelengths longer than 2 μm. Of particular importance will be NGST's ability to

FIGURE 7.2 A portion of the Hubble Deep Field. The eight blue circles identify the galaxies whose near-infrared spectra can be observed with an 8-m-class ground-based telescope; the red circles mark the sources for which only NGST will be able to study the spectra at wavelengths longer than about 2.5 μm, allowing physical analysis of these faint galaxies. NGST will also be unique in providing continuous spectral coverage of these faint galaxies to give complete coverage of different redshifts. Courtesy of E. Schreier and H. Stockman, STScI.

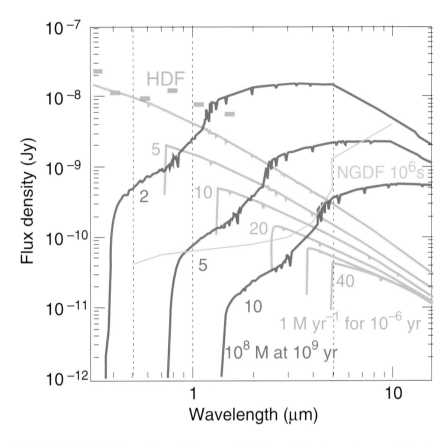

FIGURE 7.3 This plot shows the spectral energy distributions of two types of young galaxies at different redshifts. The blue curves indicate a starburst of 1 solar mass per year for 1 million years at redshifts of 0, 5, 10, 20, 30, and 40. The red curves show a more evolved galaxy with 100 million solar masses after 1 billion years of evolution at redshifts of 2, 5, and 10. The orange dashes show the sensitivity limits of the Hubble Deep Field in its six wavelength bands. The solid yellow curve shows the sensitivity of an NGST Deep Field after an exposure time of 1 million seconds—about 10 days. NGST will be able to detect young starbursts to a redshift of 20 and dwarf galaxies to a redshift of 10. Courtesy of H. Stockman, STScI.

obtain redshifts for the most distant galaxies (Figure 7.3). SIRTF will be unable to measure redshifts in most of these young objects, and the panel expects that the rich rest-frame optical spectrum of galaxies will continue to be the most important signature needed for redshift measurements of newly discovered objects in the distant universe.

Other proposed facilities for studying galaxy evolution and the history of star formation in the universe include FIRST, the Atacama Large Millimeter Array (ALMA), a large far-infrared telescope in space (SAFIR), and a very large (30 to 100 m) telescope on the ground. FIRST is much less sensitive for deep imaging than NGST and lacks the angular resolution to discriminate high-redshift galaxies unambiguously from field galaxies at redshifts near 1. ALMA will be able to observe the Rayleigh-Jeans tails of some of the high-redshift galaxies that NGST will also see, but it will not provide enough data to characterize the galaxies to understand their evolution—for example, it cannot observe the stars in such galaxies and will only get redshifts in some cases. A very large ground-based telescope would be complementary to NGST, since it would be excellent for high-resolution visible and near-infrared spectra of objects discovered in NGST's wide-field images. This space-ground combination would be similar to the currently successful HST-Keck combination.

The panel considered both long- and short-wavelength extensions of NGST's core wavelength range of 1 to 5 μm. The longer-wavelength extension is the more compelling enhancement. The sensitivity and resolution of NGST will be superior to anything proposed for the next decade in the wavelength range 5 to 30 μm. The combination of NGST and a large far-infrared telescope would provide an unmatched combination for studying the high-redshift universe and the star-formation process nearby.

Extension to longer wavelengths is important for both extragalactic research into the origins of galaxies and for studies of star and planet formation in the Galaxy. The rich spectral domain in the thermal infrared—5 to 30 μm—and the transparency of dust make it a compelling window on the physical processes governing cloud collapse, disk evolution, and planet formation.

The diffuse infrared background radiation reaching us from the cosmos is an indicator of the total energy generated by stars and active galactic nuclei since the end of the dark age. Most of this background radiation probably originates in dusty young galaxies and their nuclei, where energy generated by stars and accretion onto massive black holes at optical and x-ray wavelengths is absorbed by dust and reradiated in the infrared. Ultimately a link between this diffuse radiation and the individual sources that generated this radiation will be required if we are to understand how and when these galaxies, their nuclei, and their stellar components originally formed and how they further evolved.

Source counts in the thermal infrared obtained with the Infrared Space Observatory reach galaxies out to $z \sim 1$ but not much farther.

NGST could easily extend this range to much higher redshifts and could resolve the individual objects contributing to the diffuse infrared light. In fact, NGST will be needed to bridge the resolution gap between far infrared and near-infrared/optical telescopes for source identification of cool infrared objects. The large aperture provides adequate spatial resolution to permit clear identification, so that studies at other wavelengths can be carried out as well.

Similarly, star formation in the Galaxy is always associated with opaque clouds of gas and dust, and the study of the youngest phases of new stars takes place mainly at infrared wavelengths long enough to penetrate the dust. The thermal infrared is also the waveband where most of the luminosity of the youngest stars emerges, making thermal infrared observations important to determine the total energy budget of the stars. An extension of NGST capabilities to longer wavelengths is the panel's highest priority for star formation research.

Given the large factors by which NGST improves upon the sensitivity and angular resolution of its precursors, it has an enormous discovery potential. It typically improves observational capability over all precursor missions by a factor of 1000. This discovery potential is one of the main reasons that the panel ranked NGST as the top priority.

Technology Development

NASA recognized that NGST needs substantial development in several areas, and work is now under way to enable a launch by 2008. NGST requires a deployable primary mirror fabricated from lighter materials and structures than have ever been used before. It needs control and image analysis systems to align the primary mirror segments once they are deployed and to keep them aligned in the changing thermal environment to which they will be exposed. Telemetry rates limit the number of detectors and field-of-view coverage for the observatory. Improvements in cryocoolers will be needed to enable operation beyond about 5 mm. Finally, improvements in the dark currents and read noises of near-infrared detectors are required to ensure good performance of NGST's high-resolution spectrometers.

The panel is confident that the NGST project team's technology development will be realized within the time and budget constraints. The project has already fabricated mirror sections with the desired low surface density. Advances in detector technology put the current generation of detectors very close to NGST requirements. Pixel formats of

2048^2 for near infrared and 1024^2 for mid-infrared detectors are available now or will become available in the near future. Extrapolations indicate that the dark currents requirements can be met with another generation of work on materials. The area of cryocoolers may need more attention in the NASA plan if the mid-infrared is to be pursued to the maximum extent possible.

Several of these technologies, once developed, will enable other missions, notably TPF. The deployable sunshade and the deployable primary mirror enable missions at both far-infrared and UV wavelengths. Cryocooler development is of interest to the same two categories of mission.

Cost

NASA presented the following estimates to the panel in FY2000 dollars:

Conceptual design and development	$ 271 million
Construction (assumes an ESA instrument)	575 million
Launch (new mid-size EELV)	92 million
Science and mission operations (10 years)	<u>267 million</u>
Total	$1205 million

The estimates include all U.S. investments by NASA but not Department of Defense investments. The panel notes that ESA and CSA together plan to contribute another $271 million. DOD is funding the bulk of the costs of the advanced mirror development technology in partnership with NASA.

TERRESTRIAL PLANET FINDER

The parameters of TPF—which consists of four 3.5-m telescopes in a free-flying interferometric array, diffraction-limited at 2 μm, and operating at <40 K—are listed in Table 7.2.

Scientific Goals

The primary science problems to be addressed by TPF are the following:

TABLE 7.2 Parameters of TPF

Parameter	Planned Value	Comments
Wavelength range	3-30 μm	General imaging
	7-20 μm	Planet detection
Sensitivity	0.35 μJy	Planet detection (12 μm, 5σ in 10^4 s at $\lambda/\Delta\lambda \sim 3$); better sensitivity for imaging
Angular resolution	7.5×10^{-4} arcsec	3 μm, 1-km baseline
Spectral resolution	$\lambda/\Delta\lambda \sim 3\text{-}20$	Planet detection (imaging)
	$\lambda/\Delta\lambda \sim 3\text{-}300$	Continuum and spectral line imaging
	$\lambda/\Delta\lambda \sim 10^5$	Option for specific lines (e.g., H_2, H, CO)
Temporal resolution	<1 s	Fringe sensing
FOV	0.25″	3 μm
	1.0″	12 μm
Lifetime	>5 years	
Cost category	Major	NASA funded

- Survey of ~150 stars to determine the frequency of planetary systems with planets the size of Earth or larger;
- Low-resolution spectroscopic observations of ~50 planetary systems, looking for broad, strong spectral lines such as CO_2 and H_2O;
- High-resolution spectroscopy of about five planetary systems to search for O_3 or CH_4; and
- Milliarcsecond images of ~1000 astronomical objects at infrared wavelengths providing unprecedented views of protoplanetary disks, galactic nuclei, starburst galaxies, and galaxies at high redshift, as well as many other interesting objects.

The goal of TPF is to observe directly Earth-like planets around other stars and to measure their rudimentary atmospheric properties, looking for disequilibrium chemistry (Figure 7.4). Finding atmospheric species such as oxygen that require continuous production is the best hope at present of discovering life on other planets beyond the solar system in a

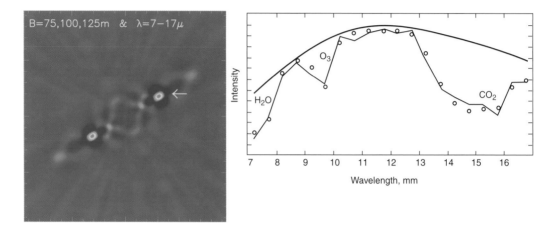

FIGURE 7.4 The left-hand figure shows an image reconstruction of an Earth-like planet around a nearby star. Because of the phase ambiguity in an interferometer, there are two symmetrically placed images. The right-hand figure shows the reconstruction of the spectrum derived from images at many different wavelengths. Note the presence of H_2O, O_3, and CO_2, similar to what is seen in a spectrum of Earth's atmosphere. Courtesy of C. Beichman, Jet Propulsion Laboratory.

passive experiment; active experiments would require any extraterrestrial life form to broadcast its existence either inadvertently or as a targeted attempt to communicate. TPF will be designed to detect Earth-sized planets around any of the several hundred nearest stars (<15 pc). Onboard spectrometers will have the resolution and sensitivity to resolve common molecular species in the atmospheres of most of these planets.

The panel considers direct observations of planets with the possibility of passive experiments to infer life as potentially the most important scientific advance of the next century, let alone the next decade. The philosophical implications are enormous, and it is a goal that will have overwhelming public support. If the technological hurdles can be overcome, the committee believes this facility must be built. It is only a question of time.

TPF will have the ability to make very-high-resolution images with exquisite sensitivity. Four cold, 3.5-m telescopes will be able to image examples of most common astronomical objects, so TPF will be an imaging interferometer with unprecedented resolution and depth. It will have a spatial resolution of 0.1 AU at the distance of the nearest star formation regions and 0.4 pc at a distance of 100 Mpc for extragalactic objects. Examples of observations that TPF might carry out are the following:

- Observations of the disks surrounding young stars at spatial resolutions of ~0.1 AU. Such an observational capability will permit direct study of the morphological, physical, and chemical structure of planet-forming environments. Existing theoretical studies of orbital migration and dynamical scattering as revealed by gaps and other structures in circumstellar disks will be challenged by observational data.
- Observations of the hot dust surrounding a giant black hole in a galaxy nucleus. TPF will provide a direct test of unification theories for active galactic nuclei.
- Detailed studies of star-formation regions in a variety of distant galaxies. TPF will achieve spatial resolution at infrared wavelengths in Virgo Cluster galaxies similar to what we now have for star formation regions in the outer parts of the Milky Way.

There is not room here to list all the potential science targets; however, there will be few areas of astrophysics untouched by the power of an infrared interferometer with the resolution and sensitivity of TPF.

While the primary design drivers for TPF are extrasolar planet studies, the panel noted with satisfaction that TPF will also be able to address a wide variety of other classes of astrophysical problems. The panel recommends that the TPF mission should plan for an allocation of observing time between planet searches and general astrophysics to ensure that astronomical observations are included in the mission goals.

Technology Development

TPF will require more technology development than most missions. The areas where the current state of the art needs substantial improvements include nulling (Figure 7.5), formation flying, large cryooptics, adaptive optics, metrology, control systems and systems engineering, passive cooling, and infrared detectors. NASA has a plan to develop each of these areas in precursor missions that will operate before TPF enters its development phase.

Some of these techniques will be challenging. The panel believes all will be developed as part of a staged plan but is skeptical about the predicted development times. It may well be that TPF technologies will be ready for launch before 2020; NASA did manage to put a man on the

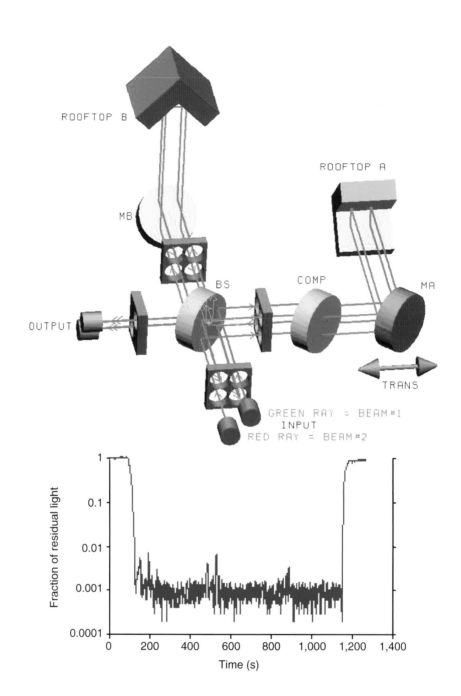

FIGURE 7.5 Lab experiment (shown schematically at the top) from E. Serabyn, Jet Propulsion Laboratory (JPL), on nulling in the visible that achieves 25,000:1 on laser and 5000:1 on 5 percent white-light source at 0.6 μm. The plot below is the amount of residual light versus time for the experiment (in seconds). Since null depth improves as wavelength squared for a given path length stability and optical system accuracy, the 10-μm performance of the white-light system would be $5000 \times (10/0.6)^2 \sim 1.4 \times 10^6$, sufficient to achieve TPF's main goals. Courtesy of C. Beichman, JPL.

Moon in 10 years starting from essentially no base. It seems unlikely, though, that the country would be willing to support TPF at the same level at which it supported the Apollo program.

Table 7.3 lists precursors to the TPF mission that will be used to develop the needed technology. Since the development is incremental, some of the early precursors will not test technology to the precision needed to guarantee a successful TPF. However, at least one mission in each category will test the technology at the needed level (these are denoted in the table with asterisks). These critical missions, including SIM, ST-3, and NGST, must be flown successfully before TPF to demonstrate the technological readiness of the mission.

It would not be surprising if delays of several years occurred because so many new techniques need to be developed. Therefore, it would be

TABLE 7.3 TPF Precursor Missions

Techology	Precursor Mission PTI	Keck I	LBT	SIRTF	SIM	ST-3	NGST
Nulling		√	√		*		
Formation flying						*	
Large cryooptics				√			*
Active optics	√	√	√		*	√	*
Metrology	√	√	√		*	*	
Pointing, stability, and vibration	√	√	√		*	√	*
Passive cooling				√			*
Infrared detectors				√			*
Interferometric system engineering	√	√	√		*	√	

NOTE: *denotes mission critical to success of TPF; √ denotes mission contributing to technology development for TPF.

risky to accord TPF a higher priority for the next decade than NGST, for example, which is farther along its development cycle and will certainly be as important to general astrophysics as viewed today.

Cost

There is no cost cap on the current TPF mission. The panel believes it is exceedingly difficult to estimate the cost of a mission with such ambitious technology goals, and although it did not attempt an independent cost analysis, it believes that cost growth would not be surprising in light of the development work to be done. NASA estimates the life cycle costs at approximately $2.1 billion:

Predevelopment (FY1999 to 2006)	$ 415 million
ST-3, including launch vehicle	180 million
Technology development	110 million
Studies and designs through PDR	125 million
Development cost (FY 2007-2011)	1300 million
Launch vehicle	200 million
Operations (5 years)	200 million
Total	$2115 million

MODERATE MISSIONS

SINGLE-APERTURE FAR-INFRARED OBSERVATORY

The parameters for SAFIR, a passively cooled, 8-m-class, far-infrared telescope in a distant orbit (probably L_2)—it could be an NGST clone with a modified sunshield—are given in Table 7.4.

Scientific Goals

The primary science problems to be addressed by the SAFIR observatory are the following:

• Study the birth and evolution of stars and planetary systems at ages so young that they are invisible even to NGST. The far infrared is sensitive to emission from gas and dust at temperatures between ~20 K and 200 K, which is the temperature range for the majority of the material in protostellar envelopes and protoplanetary disks.

TABLE 7.4 Parameters for SAFIR

Parameter	Planned Value	Comments
Wavelength range	30-300 μm	
Sensitivity[a]	0.3 μJy (30 μm)	SIRTF = 28 μJy at 24 μm
	1.3 μJy (60 μm)	SIRTF = 400 μJy at 70 μm
	18 μJy (100 μm)	FIRST = 2400 μJy at 90 μm
	100 μJy (300 μm)	FIRST = 3500 μJy at 180 μm
Angular resolution	0.8″ (30 μm)	Same as NGST at 30 μm
(30 μm diffraction-limited)	8″ (300 μm)	FIRST will provide ~18″ at 300 μm
Spectral resolution	$\lambda/\Delta\lambda \sim 5$	Imaging
	$\lambda/\Delta\lambda > 10^3$	Atomic and molecular lines
Temporal resolution	1 s	
FOV[b]	$>6' \times 6'$	Assumes 128 × 128 photo-conductors and 32 × 32 bolometer arrays
Lifetime	5 years	
Cost category	Moderate	NGST clone assumed

[a]5 in 10^4 s, confusion-limited by faint galaxies (ISO extrapolated).
[b]To achieve the full field of view, a separate wide-field mode would be present that is not diffraction-limited at all wavelengths owing to the limited number of detectors.

• Determine the bolometric luminosities of the first generations of stars and galaxies in the early universe. Dusty galaxies radiate the bulk of their energy at far-infrared wavelengths.
• Understand the trade-off between star formation and nuclear activity in active galaxies. Far-infrared radiation penetrates circumnuclear and circumstellar dust and provides the tools to distinguish the sources of luminosity in active galaxies.
• Explore the universe in the far infrared, making use of the enormous discovery potential (greater by a factor of ~10^5 than that of SIRTF) to uncover new phenomena.

The far-infrared region lies between the visual wavelengths dominated by starlight and the millimeter wavelengths of the cosmic background radiation. This wavelength region, here taken to be the wavelengths between about 30 and 300 μm, is one of the least explored but potentially most important because of its diagnostic potential. It plays an important role in astronomy because of the prevalence of dust in the space between the stars. The dust is opaque, or nearly so, in the visible spectrum, so it absorbs starlight and reradiates the energy as far-infrared light. In some galaxies and in many regions within our galaxy, the dust is so dense that the far-infrared light dominates the energy output. Equally important, the dust is almost transparent in the far-infrared spectrum, making it possible to penetrate dusty clouds and galaxies with infrared observations and see the objects producing the original radiation: stars and accretion onto giant black holes.

The coming generations of far-infrared telescopes, SIRTF and the European FIRST mission, will be limited by source confusion at the faintest levels (Figure 7.6). The only way to overcome source confusion is to increase the size of the aperture or increase the baseline in an interferometer. The build path proposed here is to construct an 8-m-class telescope that could be used first as an interferometer (by using the

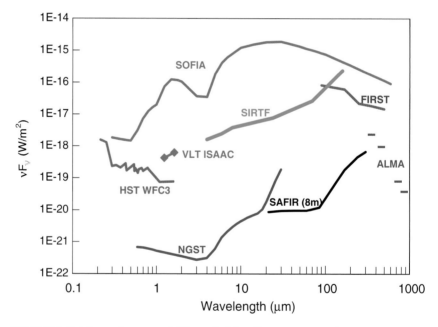

FIGURE 7.6 Relative performance of different facilities. The estimated performance is based on the combination of photon and confusion noise appropriate to the high-Galactic-latitude sky and assumes 5σ detections in 10^4 seconds. The performance is compared with the most sensitive facilities that will be available by 2010 under current plans. ALMA will not have continuous coverage, owing to the opacity of Earth's atmosphere, and is plotted as line segments corresponding to the transmission windows. Courtesy of G. Rieke, University of Arizona, and S. Beckwith, STScI.

outside radii of the mirror) to give a resolution of 1 arcsec at 100 μm. This telescope could, alternatively, be one element of an interferometer to be built after 2010, so that baselines of several hundred meters would become routine. At a resolution of 1 arcsec, it will be possible to measure the far-infrared light from the galaxies in the distant universe discovered by NGST without source confusion and to penetrate the cores of molecular clouds to witness the onset of disk formation when stars are born.

SAFIR will address scientific areas similar to those addressed by ALMA—star formation locally and in very distant galaxies—but the two facilities do so from complementary perspectives. Because ALMA observes wavelengths longward of 300 μm and in only narrow bands, it is best suited to studying the coolest material in the outer regions of collapsing stars and the very highest redshift galaxies. SAFIR will be most sensitive to warmer material at $T \sim 20$ to 100 K. Examples of warmer material include circumstellar disks and Kuiper-Belt-like regions lying 5 to 200 AU from stars. The vast majority of the bolometric luminosity from regions of star formation at redshifts below 2 is in the far infrared. Furthermore, SAFIR provides continuous coverage of the spectrum, especially important to observe spectral features such as the mid-infrared lines that distinguish active galactic nuclei (AGN) from starbursts over a broad range of redshifts. These lines, as well as the fundamental transitions of molecular hydrogen at 28 and 17 μm, are not accessible with ALMA except at redshifts in excess of 12. SAFIR will excel in the $z \sim 1$ to 5 domain, where it will be able to observe the peak in far-infrared emission from dust.

Birth and Evolution of Stars and Planetary Systems   Stars are born in cold interstellar cloud cores that are opaque to all radiation at mid-infrared (~10 μm) and shorter wavelengths. After about 100,000 years, the creation of a young star within the cloud core causes a disruption of the cloud by the action of powerful stellar winds. Surrounding the star is a circumstellar disk, whose subsequent evolution is the source of a planetary system. The initial assembly of the star and disk is hidden from view. To understand how the cloud collapses, how it fragments, and how it builds a disk, we must penetrate the dust and observe the physical conditions directly. To understand when, where, and how frequently these disks give rise to planetary systems, it will be essential to observe these processes at far-infrared wavelengths, where the bulk of the radiation emerges.

A far-infrared 8-m telescope can provide a resolution of 1 arcsec at 100 μm or about 150 AU at the distance to the nearest star-forming regions. Spectral imaging would resolve the density and temperature structure of ~1000 AU collapsing cores. This resolution should uncover fragmentation in the formation of binary systems, since about half the binaries observed have separation greater than 100 AU.

The temperature of the gas in the core is only a few tens of kelvin, so that the primary molecular transitions lie in the far-infrared and submillimeter regions. Emission lines from water and high-order rotational lines of CO, as well as the atomic lines of oxygen, will reveal physical conditions in the same manner as they have for the Orion Nebula (Figure 7.7). Emission from H_2O is the principal means by which clouds lose their energy from the highest temperature and density regions in their interior. Bright water lines between 25 and 180 μm dominate the cooling in the inner cloud, where a broad component is expected from the accretion shock and a narrow one from the disk. The CO lines from 170 to 520 μm are the main coolants for the outer cloud; warmer CO from within the cloud can also be studied because of velocity shifts due to the collapse. This suite of lines therefore provides complete access to the process of star formation.

The First Generations of Stars and Galaxies Even small amounts of dust can shift starlight into the far infrared. This shift occurs for a few percent of the local galaxy population, one of the best examples being M82, where only a few percent of the light escapes in the visual band and the luminosity is dominated by far-infrared radiation. At redshifts between 1 and 3, when star formation appears to have been most vigorous, a substantial fraction of all galaxies may have their light extinguished at short wavelengths only to reemerge in the far infrared.

The measurements of the DIRBE experiment aboard the Cosmic Background Explorer (COBE) indicate that the far-infrared energy density in the distant universe is comparable to that of visible and near-infrared light. This far-infrared background is thought to arise from starburst galaxies at redshifts between 1 and 3, meaning that roughly half the young galaxies radiate primarily in the far infrared. If so, it will be necessary to resolve this background into its components and measure the far-infrared luminosities of the individual sources to understand the history of heavy-element synthesis after the Big Bang. An 8-m telescope with detection limits on the order of 0.1 mJy would probably resolve most of this high-redshift background into individual galaxies, thus showing the

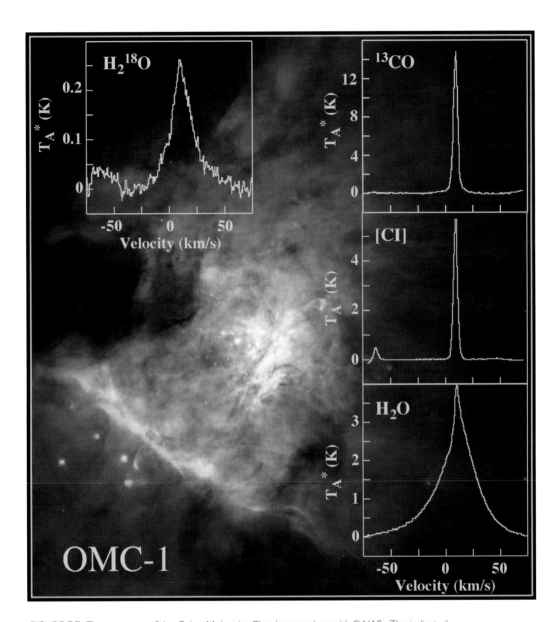

FIGURE 7.7 These spectra of the Orion Molecular Cloud were taken with SWAS. The indicated gases—$H_2^{18}O$, ^{13}CO, H_2O, and atomic carbon, CI—are primary coolants over most of the temperature and density range in which star formation occurs. Cooling by pure molecular hydrogen is believed to play a major role in the earliest phases of galaxy formation in the early universe, before heavier elements like oxygen and carbon had been formed. To observe the formation of these primordial galaxies at high redshifts, far-infrared telescopes with large apertures are essential. Courtesy of G. J. Melnick, Harvard-Smithsonian Center for Astrophysics.

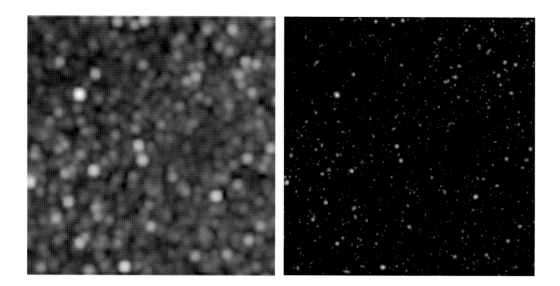

FIGURE 7.8 These simulated 70-μm images span a four-decade range in flux from the peak flux in the field. The brightest galaxy in the field is 39 mJy. The FOV is 413 arcsec for each image. On the left is a SIRTF/MIPS image; on the right is an image from an 8-m far-infrared telescope. At this wavelength, the 8-m gives a 2 arcsec diffraction-limited beam, which provides an angular resolution that is barely suitable for identifying sources at other wavelengths. The SCUBA 12-arcsec beam and the SIRTF 17-arcsec beam will lead to confusion in source identifications and redshifts. Courtesy of E. L. Wright, University of California, Los Angeles.

dominant phases of dust-embedded star formation and nuclear activity throughout the universe (Figure 7.8).

The Hubble Deep Field revealed many galaxies too faint to contribute significantly to the far-infrared background. To complete the study of star formation in the early universe, we must understand these small systems and possible galaxy fragments. The rate of star formation in these small galaxies can best be determined from far-infrared measurements of the energy between 20 and 200 μm. By using an 8-m far-infrared telescope in tandem with NGST and ALMA, the spectral energy distribution from 1 μm to 1 mm can be measured, which will define the peak output wavelength and total energy balance. For the faintest galaxies, the location of the peak can be used to estimate a redshift not possible with NGST and ALMA alone.

Active Galactic Nuclei and Starbursts in the Early Universe   Giant black holes weighing as much as a billion Suns are believed to reside at the center of most galaxies. Astronomers observe galaxy mergers nearby that produce luminous far-infrared radiation through a combination of violent starbursts and energy released as matter falls into these black holes (AGN). Both types of energy source are hidden within cocoons of interstellar dust that are impenetrable at the optical and near-infrared wavelengths. These local events are uncommon, but galaxy mergers are thought to have been very common in the early universe, when the galaxies were assembled from the residue of the Big Bang.

What happened during the much more common mergers that built galaxies in the early universe? The fine structure lines of Ne II (12.8 μm), Ne III (15.6 μm), and Ne V (14.3 μm) are the best tools to distinguish unambiguously whether the luminosity of a dusty galaxy is dominated by a burst of star formation or by an AGN. They are 30 times less susceptible to absorption by dust than lines at visible wavelengths. Observed today, these lines will be redshifted into the 45- to 55-μm range from AGN at the peak of the assembly phase, currently thought to be near redshifts of 3. An 8-m far-infrared telescope would have both the necessary resolution and sensitivity to use this suite of lines to determine the relative roles of star formation and nuclear activity in the early universe.

An 8-m NGST clone will have 10 times better angular resolution than SIRTF and will provide more than 100 times better sensitivity with more than 10 times the number of detectors. As for the speed of an observation—the astronomical capability defined in the 1991 survey committee report, an 8-m far-infrared telescope will be 10^5 times faster than SIRTF and more than 10^3 times faster than FIRST at the longest wavelengths where they overlap. This enormous discovery potential is what most impressed the panel when it was making its recommendation for this mission.

The techniques developed for SIM and, eventually, TPF will later be usable with several of these telescopes to produce an infrared interferometer as a next step to advance the field. In this phased approach, the most expensive technology development will be borne by the Grand Challenge missions: NGST for cooled, deployable mirrors and SIM and TPF for interferometric techniques. For this reason, it can be assumed that the filled-aperture telescope will be a moderate mission (<$500 million).

Technology Development

As noted above, the Grand Challenge missions will develop the most difficult technologies. An NGST clone with a sunshield of twice the efficiency (i.e., doubling the number of layers in the NGST design) could produce a superb far-infrared telescope. A telescope reaching < 10 K passively cooled would probably eliminate the need for a cryocooler for the thermal infrared detectors, although it is uncertain in the absence of a detailed design study if passive cooling will be adequate to reach these temperatures. A cryocooler to take bolometers to subkelvin temperatures is desirable and will have to be developed for other missions as well (e.g., ultraviolet detectors).

Detectors exist with the necessary sensitivity, but they do not have large pixel formats to take advantage of large fields of view. Pixel formats of 32 × 32 are currently state of the art for the far-infrared photoconductors. The largest bolometer arrays are about 10 × 10. The main technology development needed is in increasing these formats. The panel's working assumptions are that bolometers will increase 10-fold, to 32 × 32, and photoconductors will increase 16-fold, to 128 × 128.

Bolometer formats are increasing now because they have important applications in ground-based astronomy. The panel assumed that this technology would develop without additional funding from NASA. To move from the current state of the art in photoconductor arrays, which are 32 × 32 devices, will require a substantial investment on NASA's part. There are no ground-based or commercial applications for these arrays. An investment of $500,000 per year for 5 years would lead to a flight-qualified prototype of a 128 × 128 Ge:Ga photoconductor array. This estimate assumes that the current expertise in building arrays for SIRTF does not get lost and will not have to be relearned. This $2.5 million might be included in the cost of the mission.

Cost

The panel requested a detailed cost estimate from scientists at the Goddard Space Flight Center and the Space Telescope Science Institute based on NGST costs. Its conclusion is that a far-infrared 8-m telescope (30 to 300 μm) based on NGST technology would cost about 50 percent of NGST's cost. It is likely that a pure NGST clone would be inadequate to realize the full potential of a far-infrared telescope, so that some technology development would be needed to cool the telescope to less than 10 K. The technology development cost is unknown and not counted in the budget for projects. The cost estimate below, which has been adjusted to FY2000 dollars, assumes reduced mission operations costs and a 5-year lifetime.

Although this telescope is recommended as a moderate-class mission, its cost is very close to the boundary that separates moderate from major missions, and the panel recognizes that it could just as appropriately be classified as a major mission, as was done in the survey committee report.

Construction (assumes an ESA instrument)	$335 million
Launch (new mid-size EELV)	92 million
Science and mission operations (5 years)	108 million
Total	$535 million

SPACE ULTRAVIOLET OPTICAL TELESCOPE

The parameters for SUVO, an 8-m-class telescope in a distant orbit (probably L_2)—it could also be a spin-off from NGST—are given in Table 7.5.

Scientific Goals

The primary science goals of SUVO are the following:

- Study the evolution of the structure and composition of the intergalactic medium.
- Map out dark matter at cluster and supercluster scales.
- Study feedback effects from star formation on interstellar medium (ISM) and protoplanetary disks.
- Determine processes by which galaxies, clusters, and AGNs are formed.

The SUVO (8-m) mission will focus on high-throughput ultraviolet spectroscopy and wide-field optical and ultraviolet imaging.

Mapping Dark Baryons and Large-Scale Structure   One of the important predictions generated by hydrodynamical simulations of galaxy formation is the existence of a large-scale filamentary network of matter spread throughout intergalactic space (the "cosmic web"). Not only is

TABLE 7.5 Parameters for SUVO

Parameter	Planned Value	Comments
Wavelength range	0.115-1 µm	
Sensitivity	50 pJy	Visual imaging, $m_V(\text{lim}) \sim 35^m$
	0.6 µJy	$\lambda/\Delta\lambda \sim 3 \times 10^4$, S/N = 10 in 4×10^4 s
Angular resolution	0.005″ (0.1 µm)	NGST will give ~0.06″ at 1 µm
	0.03″ (1 µm)	
Spectral resolution	$\lambda/\Delta\lambda \sim 2 \times 10^3$	Very high resolution is a stretch goal
	$\lambda/\Delta\lambda \sim 4 \times 10^4$	
	$\lambda/\Delta\lambda \sim 2 \times 10^5$	
Temporal resolution	33 µs-1 ms	MCP and STJ responses[a]
FOV	10′-15′	
Lifetime	5-10 years	
Cost category	Moderate	NGST spin-off assumed

[a]MCP is microchannel plate and STJ is superconducting tunnel junction.

the universe dominated by unseen (dark) matter, but a large fraction of ordinary (baryonic) matter remains undetected and probably resides in the intergalactic medium (IGM). The infall of this intergalactic gas continues the buildup of modern galaxies and their halos. It may also have triggered "recent" star formation 3 to 10 billion years ago ($z < 2$) and fueled quasar outbursts by supplying gas to the black holes thought to lie at the centers of many galaxies.

At $z < 1$, the gas in the IGM is distributed in roughly equal amounts between warm, photoionized gas at 30,000 to 100,000 K and hot shocked gas at 10^6 to 10^7 K. Such gas is exceedingly difficult to detect optically. By far the best way of detecting the warm IGM during the last 70 to 80 percent of cosmic time ($z < 1.65$) is by measuring the UV resonance absorption lines of hydrogen (Lyman-alpha) and the heavy elements produced by the first stars (e.g., C, O, Si, Mg, and Fe). The ultraviolet band is unique in being able to detect trace amounts of these important chemical elements; UV spectra are several orders of magnitude more sensitive than x rays in detecting heavy elements. In addition, the background quasars used for absorption spectra have far-higher photon fluxes in the UV than in the x ray. Thus, studying the cosmic web of matter and making quantitative measurements of the modern evolution of the chemical products is a problem that can best be done by high-throughput UV spectroscopy with SUVO.

From recent galaxy redshift surveys, astronomers have detected the existence of an organized large-scale structure in the galaxy distribution; this structure takes the form of large filamentary walls and voids. By 2010, these galaxy surveys will outline the distribution of luminous matter in fine detail, but the dark, gaseous universe (the IGM) will remain largely unexplored at $z < 1.65$ (the last 70 to 80 percent of cosmic time). SUVO will measure the distribution of dark baryonic matter in these filaments and voids, tracing the cosmic web using UV spectra of the Lyman-alpha lines toward QSOs and other AGN (Figure 7.9). The SUVO goal is to conduct a survey on subdegree angular scales of baryons in the IGM comparable to that of the MAP explorer and to the structure seen in galaxy surveys. Doing so will allow connecting the high-redshift seeds of galaxies and clusters with the distribution of galaxies and IGM in the modern epoch, at $z < 1$. This project is impossible with the throughput of current UV spectrographs. To achieve a frequency of one QSO every 100 arcmin^2 on the sky requires using QSOs at $m_B = 18$ to 20 as background targets. In the next several years, the GALEX mission is expected to identify 10^5 to 10^6 QSOs with $18 < m_B < 20$, and the Sloan survey will provide redshifts for 10^5 of these targets.

Detecting Unseen Matter in the Modern Universe   SUVO can use large-scale weak gravitational lensing to probe the underlying matter in galaxy clusters and superclusters, extending over 5 to 20 Mpc h^{-1}. A 10 to 15 arcmin field of view (much larger than is achievable with HST or NGST instruments) provides a good match to expected correlation lengths of cosmological structures at $z < 1$. Through its wide-field imaging, SUVO will use weak gravitational lensing to map out the distribution of dark matter in clusters of galaxies and perhaps on the supercluster scale. Light from distant galaxies is bent and focused into small, faint arclets by the gravitational fields of intervening clusters and superclusters. Although some weak lensing studies can be done from the ground, experience with the Hubble Space Telescope has shown that the amount of dark matter on large scales is most accurately determined in space. The availability of a very small, stable point-spread function and the access to the blue-light characteristic of the background galaxies at $z < 1$ gives SUVO wide-field imaging a significant advantage.

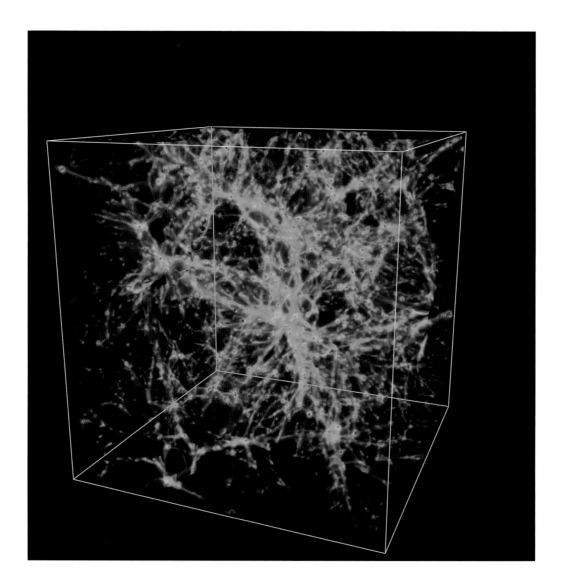

FIGURE 7.9 This simulation shows the expected distribution of intergalactic gas resulting from structure formation in the early universe. Most of the matter is cold and dark but can be detected via its absorption of light from distant quasars. SUVO could map this intergalactic structure that may contain most of the normal matter in the universe through systematic observations of many quasars. Courtesy of R.-Y. Cen and J.P. Ostriker, Princeton University.

Studying the Feedback from Star Formation   An important challenge for UV astronomy in 2010 will be to follow the emergence of the modern universe over the last 10 billion years ($z < 1.65$). As described earlier, the gaseous components of the universe during this epoch can best be studied in the ultraviolet, where one has access to the resonance lines of H (Lyman-alpha) and other key heavy elements. At $z > 1.65$, some of these lines are redshifted into the optical. However, this leaves some 70 to 80 percent of cosmic time inaccessible without UV instruments. The lack of high-throughput UV spectroscopy is a major hindrance to understanding the cosmic chemical evolution of the modern universe at $z < 1.65$.

With SUVO, astronomers will be able to make precision studies of the modern ($z < 1.65$) counterparts of star-forming galaxies in the high-redshift universe studied with NGST. The fact that stars at redshifts less than 1.65 produced over 90 percent of the heavy elements seen today further justifies the need for access to the UV, which will give astronomers the key resonance lines of most elements. For the most recent 70 percent of cosmic time, the UV band provides the best means for them to quantify the rate at which galaxies accrete mass from the IGM, form new heavy elements in stars, and expel mass and radiation as a result of this star formation. UV spectra of both emission and absorption lines provide quantitative measurements of the effects of "feedback" from star formation on the surrounding IGM and on the galaxies themselves. This feedback takes the form of newly synthesized heavy elements (C, O, and so on), of powerful ionizing radiation from massive stars, and of hot gas produced by stellar winds and exploding stars. This feedback of radiation and energy modulates the rate at which galaxies continue to accrete gas and determines how heavy elements are spread throughout the universe.

Studying Galactic Star Formation   With Hubble, spectral imaging in the optical and ultraviolet proved its value in measuring the products of star formation in local regions of the Milky Way. The usefulness of these images is dictated by the physics of line emission and the peak of the radiative cooling function at 10^5 K, arising from strong UV emission lines of C, N, and O. The UV and visible bands also allow astronomers to probe the effects of stellar radiation on exposed molecular clouds and protoplanetary disks. Understanding the effects of Galactic ionizing radiation and stellar outflows is critical to understanding large-scale star formation and the survivability of disks in the Galactic environment.

The narrow-field UV/optical detectors aboard HST provided detailed images of nearby star-forming regions, Orion's protoplanetary disks, planetary nebulae, and shock waves. With its wide-field imagers and filters, SUVO will be able to map, in a single pointing, entire star-forming regions in diagnostic emission lines and broadband colors. Within the Milky Way, but more importantly in external star-forming galaxies, SUVO will be able to study global star formation and the interactions between young star clusters and the ISM. Diffraction-limited imaging will allow astronomers to measure the interactions of stars and the ISM at scale lengths necessary to probe important physical processes (winds, shock waves, thermal conduction, and wind ablation of disks). The dynamic range and wide field of these images will enable astronomers to study the stellar initial mass function, after the molecular cloud is blown away, from the most massive stars down to brown dwarfs. In this respect, UV/optical imaging is complementary to infrared studies of the initial stages of star formation.

Some of these science goals are complementary to those in the IR and x-ray regimes. However, some can only be achieved using the wide-field images and sensitive spectral diagnostics of the UV and optical bands. In particular, images of weak gravitational lensing and large-scale surveys of Lyman-alpha absorption may be the only way to find the dominant components of the invisible universe. SUVO will allow astronomers to map out the large-scale distribution of dark matter and the nearly invisible warm components of the cosmic web of gaseous matter that may dominate the missing baryons.

Technology Development

No high-throughput UV/optical mission will be possible without significant NASA investments in technology, including UV detectors, gratings, mirrors, spectrographs, and imagers. The panel assumes the telescope will use the technology developed for NGST: lightweight, deployable optics. With an 8-m aperture, SUVO will achieve a 10-fold increase in collecting area over HST for imaging and spectroscopy. With improved efficiencies in detectors and gratings and with increases in camera sizes and in spectral multiplex efficiency by means of simultaneous wavelength coverage, SUVO should achieve, overall, a 100-fold increase in power.

The most critical needs for SUVO are the following:

• Develop more sensitive UV detectors with low background noise, high quantum efficiency, large dynamic range, and large formats.

• Develop three-dimensional, energy-resolving detectors such as photon-counting superconducting tunnel junction (STJ) or transition-edge sensor (TES) devices. These cryogenic detectors have the potential to revolutionize UV/optical astrophysics, and the cryogenic technology would be shared with other missions (far-IR and submillimeter).

• Space-qualify large mosaics of low-noise, high-quantum efficiency charge-coupled device (CCD) detectors (at least 16K × 16K) for wide-field imaging cameras in the optical and near UV.

• Develop micromirror arrays for use in multiobject and integral-field spectrographs in the UV and visible.

• Develop large, lightweight precision mirrors for use in the UV/optical. Although SUVO could be done with a 4-m monolith, the extension to an 8-m aperture will probably require segmented deployable optics.

SMALL MISSIONS

ULTRALONG-DURATION BALLOONS

For many small payloads, the top of the stratosphere accrues essentially all the benefits of space, and these payloads could be flown on balloons at a fraction of the cost of satellites or of using a space shuttle. The principal disadvantages of balloons—short-duration flights and lack of control over the flight path—can be alleviated with new technologies that would enable flights of 1-ton payloads at altitudes above 35 km for 100 days or longer with enough control over the flight path to ensure a safe landing at the point of launch. As such, balloons would be more attractive for many small payloads than traditional spacecraft and would give NASA an inexpensive way to carry out novel science experiments as well as to prove new payload technology and train future scientists in instrumentation.

Examples of the kind of science that can be carried out with balloons include searches for planets with a coronagraph on a diffraction-limited telescope a few meters in diameter; the imaging of convective flows and magnetic fields in the Sun's photosphere with a large solar telescope; extragalactic observations with a moderate-size far infrared telescope; and all-sky surveys at hard x-ray wavelengths. The top of the stratosphere is far superior to terrestrial sites and enables a wide range of small-mission science experiments at wavelengths not transmitted to Earth.

The typical balloon science payloads would be similar to those proposed under the Explorer program. The panel recommends that ULDBs be permitted in the competition for Explorer missions. Furthermore, it finds that it would be appropriate to invest money in balloon technology to enable flight path control and very long flights. The investment could come from either the technology program or the Explorer program as part of the cost of a SMEX mission.

LABORATORY ASTROPHYSICS

Much of the data provided by new space missions will have spectral signatures that cannot be interpreted given our current knowledge of atomic and molecular lines. Resonance transitions from small particles are commonly observed that lack counterparts in the limited number of laboratory studies of solid-state lines. It is already the case that space missions reveal spectral lines that are unidentified owing to a lack of laboratory data. This problem spans the spectrum from the x ray to the far infrared and means that the space missions do not realize their full potential for basic research.

Laboratory research can lead to important discoveries. For example, spectra observed in the disks around young stars showed the presence of the mineral forsterite, whose spectrum had been measured in the laboratory. Forsterite is a magnesium-rich silicate that is seen in the spectra of some comets and is also known on Earth. The close match between spectra of the circumstellar disks, thought to be the birthsites of new planetary systems, and those of comets, consisting of the same material in the disk surrounding the proto-Sun, greatly strengthened the belief that the young stars are akin to the early phases of our solar system and that the early solar system consisted of cometlike bodies orbiting the young Sun. The material from these comets is believed to have aggregated in the solar system planets known today.

Similar conditions dominate the optical and ultraviolet spectra of many ionized molecular and atomic species and of radicals that are unstable under normal terrestrial conditions. Such substances, although short-lived, can often be produced in the laboratory and studied in sufficient depth to permit unambiguous identification when astronomically observed. A particular instance of this is the identification of the diffuse interstellar bands. More than 150 of these are now known, though none has to date been unambiguously identified. Preliminary laboratory studies now indicate that some of these bands may be associated with ionized buckyballs, C_{60}^+. Another recent laboratory study indicates that the anion of the linear carbon chain C_7^- may be associated with the narrow bands at 496.4, 561.0, 574.7, 606.5, and 627.0 nm. This study led to the suggestion that other bands may be due to carbon chain anions that are six, eight, and nine atoms long. Suggestions such as these require laboratory follow-up, since there are no theoretical means for calculating the energy levels of such molecular ions or the transition probabilities between these levels.

The cost of laboratory studies tends to be low in comparison with overall mission costs but can make a decisive difference in obtaining the astrophysical insights that these missions were designed to yield. The panel recommends an increased investment in laboratory astrophysics to take advantage of the wealth of new data already being returned by space science missions.

TECHNOLOGY FOR THE FUTURE

ENERGY-SENSITIVE UV/OPTICAL DETECTORS

A revolution is occurring in the technology for ultraviolet detectors. Two technologies, superconducting tunnel junctions (STJ) (Figure 7.10) and transition-edge sensors (TES), make it possible to measure the energy of individual photons as they are detected. Unlike all previous techniques, these detectors simultaneously give the photon rate (brightness) and spectral energy distribution. For many applications, all images will immediately yield modest-resolution spectra of each pixel. The increase in information capacity relative to today's images is huge. These detectors will increase the efficiency of spectral observations by factors on the order of 100.

When fully developed, these detectors will increase the astronomical capabilities of even small telescopes by an equivalent factor. The increased capability will permit many scientific problems to be solved with current telescopes, without the expense of building larger telescopes. This potential is so great that the panel recommends NASA invest aggressively in these technologies, perhaps selecting the most promising after an initial investment, to ensure that the new potential is fully exploited. The panel believes that investments of about $25 million would

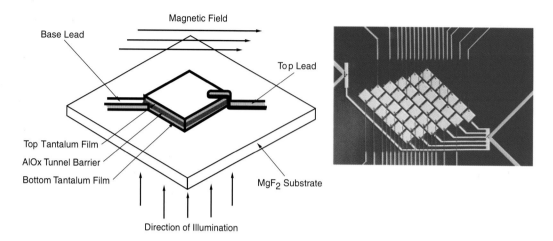

FIGURE 7.10 A schematic diagram of a superconducting tunnel junction detector shows the basic elements for the first of the ultraviolet-energy-resolving detectors (left). An array of 6 × 6 detectors (pixels) has been fabricated and is shown on the right in a microphotograph. Courtesy of ESTEC (<http://www.estec.esa.nl>).

save several times that amount by enabling new science without further increasing the size of telescopes in the next two decades.

Investment in energy-resolving detectors is the top priority of the panel for new technology development.

REFRIGERATORS

Most detectors require cooling to achieve their full sensitivity. The STJ and TES detectors operate at millikelvin temperatures. Far-infrared detectors typically must be cooled to a few kelvin. Far-infrared telescopes require optics below about 10 K to achieve their full potential. For these applications, missions will need refrigerators capable of operating for long periods with minimal power to keep the detectors and optics cold.

NASA should invest in technology to develop space-qualified refrigerators needed for the suite of missions enabled by the new detector technologies. It should be straightforward to develop devices for far-infrared applications. It will be more challenging to make refrigerators capable of millikelvin temperatures, but because the new UV detectors have tremendous potential, it is vitally important to invest in the needed technology. The panel believes an investment comparable to that for detectors, about $25 million, will be needed to develop technologies for next-generation cryocoolers.

SPACECRAFT COMMUNICATIONS

Missions currently under consideration envisage gathering data at rates many orders of magnitude higher than were traditionally considered possible. Missions such as NGST are expected to have focal plane arrays with more than 3×10^8 pixels, and even this number will not capture all the information delivered by the telescope—more than 10 times as many pixels would be needed to critically sample the entire diffraction-limited field of view. At 16 bits per pixel, the characterization of a single frame then would require almost 10^{10} bits of information. Exposure times lasting no more than a minute will be routine to ensure that none of the pixels become saturated, meaning that bandwidths on the order of 1 billion bits per second (GHz) would be needed to transmit the data to Earth.

Current telemetry systems for NASA's science missions are limited to data transmission rates of a few million bits per second (MHz). Moreover, the telemetry bands need to be shared among all the different missions simultaneously operating in space. The rate has not been upgraded significantly in the last few decades and is rapidly becoming the bottleneck that will limit the utility of the more ambitious missions being planned.

Onboard processing and data compression can help but cannot completely alleviate this bottleneck. Most missions encounter unanticipated sources of noise that can be removed only with processing that involves interactive analysis. Interactive processing requires the transfer of all the data to Earth, where sophisticated algorithms can be developed on computers much more powerful than those on board the mission.

Commercial systems are now becoming available that permit transmission at bandwidths of 100 GHz. These normally work at infrared wavelengths and would require an approach to ground stations different from that used by the radio telescopes now in place. However, investment in new ground stations permitting vastly increased telemetry bandwidths is likely to be needed if we are to realize the enormous gains made possible by the advanced detector arrays that will be used with the next generation of space science missions.

A dedicated optical communications satellite at L_2 would cost about $70 million. Operational costs for using a dedicated 3-m telescope as a receiver would need to be added, although it is almost certain that one or more telescopes already exist that could be used for this purpose. The panel urges NASA to make an investment of this magnitude to capitalize on the great potential for science from L_2.

ULTRALIGHTWEIGHT ("GOSSAMER") OPTICS

Filled-aperture telescopes have been the principal sources of information for astronomers for almost 400 years, since Galileo turned the first telescope to the heavens in the early 17th century. Very large telescopes in space have an enormous advantage over those on the ground because they avoid Earth's atmosphere and its concomitant distortion of the light. If it were possible to develop very large optical surfaces for space telescopes—more than 100 m, say—such telescopes would enable astronomical studies of great interest, in particular, the study of Earthlike planets around nearby stars, without having to use special interferometric techniques, where the collection of light is important for high-resolution spectroscopy of the planetary atmospheres. Very large ultraviolet telescopes would be extremely useful for high-throughput spectroscopy of the intergalactic medium, galactic halos, and heavy-element evolution in the ISM and IGM, where the background targets are faint.

The panel supports NASA's initiative to develop gossamer optical surfaces that would be the basis for a new generation of giant telescopes in space. Although how much money would be needed to achieve this breakthrough technology is unknown at present, the panel believes it would be prudent to invest several million dollars per year to explore the means of manufacturing very large, accurate optical surfaces for future space telescopes.

ACRONYMS AND ABBREVIATIONS

AGN—active galactic nuclei
ALMA—Atacama Large Millimeter Array
AU—astronomical unit, a basic unit of distance equal to the separation between Earth and the Sun, about 150 million km

CCD—charge-coupled device, an electronic detector used for low-light-level imaging and astronomical observations; CCDs were developed by NASA for use in the Hubble Space Telescope and the Galileo Probe to Jupiter and are widely used on ground-based telescopes

COBE—Cosmic Background Explorer, a NASA mission launched in 1989 to study the cosmic background radiation from the Big Bang

CSA—Canadian Space Agency

DIRBE—Diffuse Infrared Background Explorer, an instrument aboard COBE

DOD—Department of Defense

EELV—Evolved Expendable Launch Vehicle

ESA—European Space Agency, the European equivalent of NASA

ESTEC—European Space Research and Technology Centre

FIRST—Far-Infrared Space Telescope (ESA)

FOV—field of view

FUSE—Far-Ultraviolet Spectroscopic Explorer

GALEX—Galaxy Evolution Explorer, a space ultraviolet emission and spectroscopic Small Explorer mission

Hipparcos—European Space Agency mission for measuring the distances, motions, and colors of stars

HST—Hubble Space Telescope, a 2.4-m-diameter space telescope designed to study visible, ultraviolet, and infrared radiation; the first of NASA's Great Observatories

IFS—integral field spectrograph

IGM—intergalactic medium

ISM—interstellar medium

ISO—(European) Infrared Space Observatory

L_2 orbit—Lissajous orbit about the L_2 Sun-Earth Lagrange point 1.5 million km from Earth

LBT—Large Binocular Telescope, an American, Italian, and German collaboration

MAP—Microwave Anisotropy Probe

MCP—microchannel plate

MIPS—Multiband Imaging Photometer for SIRTF, a far-infrared instrument capable of imaging photometry and high-resolution imaging

MOS—multiobject spectrograph

NASA—National Aeronautics and Space Administration

NGST—Next Generation Space Telescope, an 8-m infrared space telescope

PTI—Parkes-Tidbinbilla Interferometer (it has a baseline of 275 km)

QE—quantum efficiency

QSOs—quasi-stellar objects; together with active galactic nuclei, they form the group of objects known as active galaxies

SAFIR—Single-Aperture Far-Infrared Observatory, an 8-m space-based telescope

SCUBA—Submillimeter Common-User Bolometer Array, a British-French-Canadian ground-based telescope in Hawaii operating at wavelengths between 350 and 2000 μm

SIM—Space Interferometry Mission

SIRTF—Space Infrared Telescope Facility, NASA's fourth Great Observatory, will study infrared radiation

Sloan survey, also known as the Sloan Digital Sky Survey (SDSS)—a mission to produce high-resolution pictures of one quarter of the sky and to measure the redshift of distant galaxies

SMEX—Small Explorer, a NASA program to fly small, inexpensive satellites on a rapid timescale

ST-3—NASA's Space Technology 3, two spacecraft launched together and put into an orbit around the Sun to demonstrate interferometry

STJ—superconducting tunnel junctions

SUVO—Space Ultraviolet Observatory, a proposed 8-m-class telescope

SWAS—Submillimeter Wave Satellite, one of NASA's Small Explorer missions; it studies interstellar clouds

TDRSS—Telemetry Data Relay Satellite System

TES—transition-edge sensors

TPF—Terrestrial Planet Finder, a free-flying infrared interferometer designed to study terrestrial planets around nearby stars

ULDB—ultralong-duration balloon flights

UV—ultraviolet

UVOIR—ultraviolet, optical, and infrared

VLT—Very Large Telescope, the European Southern Observatory's four 8-m telescopes